# Probiotics and Prebiotics in Animal Health and Food Safety

Diana Di Gioia • Bruno Biavati
Editors

# Probiotics and Prebiotics in Animal Health and Food Safety

Springer

*Editors*
Diana Di Gioia
Department of Agricultural Sciences
University of Bologna
Bologna
Italy

Bruno Biavati
Division of Rural Sciences & Food System
University of Malta
Msida
Malta

ISBN 978-3-030-10134-3 ISBN 978-3-319-71950-4 (eBook)
https://doi.org/10.1007/978-3-319-71950-4



Printed on acid-free paper

This Springer imprint is published by Springer Nature
The registered company is Springer International Publishing AG
The registered company address is: Gewerbestrasse 11, 6330 Cham, Switzerland

# Contents

# Probiotics and Prebiotics: An Overview on Recent Trends

1

Georgia Zoumpopoulou, Maria Kazou, Voula Alexandraki, Angeliki Angelopoulou, Konstantinos Papadimitriou, Bruno Pot, and Effie Tsakalidou

## 1.1 Introduction

Nowadays, the pro- and prebiotic concept is very well known regarding human applications related to preserving or restoring health. However, applications in feed are far less documented. While prebiotics should be considered a more recent concept, the history of probiotics is long and interesting. Thousands of years ago, man discovered already the benefits of "fermented" foods, e.g., in extending shelf life of fresh food products (Ozen and Dinleyici 2015; Gogineni et al. 2013). This observation was extended to the fermentation of feed as well. Ancient evidence from Egyptian drawings and old Carthusian silos shows that more than 1000 years ago, farmers already knew that silage was an excellent way to preserve summer crops for their animals during winter times (Mannetje 2010). It took, however, until the early twentieth century before the Nobel Prize winner Ellie Metchnikoff for the first time defined and studied the role of the fermenting bacteria in health (Metchnikoff 1908). While at that time the "probiotic" concept (live microorganisms that can promote health) was born, it took until 2001 for the concept to be acceptably defined. This

Georgia Zoumpopoulou, Maria Kazou, Voula Alexandraki, and Angeliki Angelopoulou contributed equally to this work.

G. Zoumpopoulou • M. Kazou • V. Alexandraki • A. Angelopoulou •
K. Papadimitriou • E. Tsakalidou (✉)
Laboratory of Dairy Research, Department of Food Science and Human Nutrition,
Agricultural University of Athens, Athens, Greece
e-mail: et@aua.gr

B. Pot
Research Group of Industrial Microbiology and Food Biotechnology (IMDO),
Vrije Universiteit Brussel, Brussels, Belgium



D. Di Gioia, B. Biavati (eds.), *Probiotics and Prebiotics in Animal Health and Food Safety*, https://doi.org/10.1007/978-3-319-71950-4_1

was achieved by an expert panel composed by the Food and Agriculture Organization of the United Nations (FAO) and the World Health Organization (WHO), who defined probiotics as "Live microorganisms that when administered in adequate amounts confer a health benefit on the host" (FAO/WHO 2001). Followed by FAO/WHO (2002) and recently revised by Hill et al. (2014), this definition today is widely accepted by the scientific community and most governmental institutions.

In the area of animal nutrition, however, in the USA in 1989, also the term "direct-fed microbials" (DFM) was introduced by the Food and Drug Administration (FDA). DFM were, similarly as probiotics, defined as "a source of live (viable), naturally occurring microorganisms." Clearly, the health aspect was not maintained in that definition. Consequently, while manufacturers were required to use this term on their ingredient lists, FDA did not allow them to make therapeutic claims. FDA together with the Association of American Feed Control Officials (AAFCO) published a list of microbial species that could be used in DFM products (AAFCO 1999). The interest in DFM was already raised in the 1950s when a positive growth response was noted in animals fed with antibiotics, suggesting that the suppression of pathogens had a positive impact on the animal health and improved its growth response. It was not until much later that it was shown that a healthy intestinal microbiota consists of a certain balance between potentially beneficial commensal and potentially pathogenic bacteria and that supplementation with "beneficial" bacteria might evoke a similar positive growth response, without the need to use antibiotics that left traces in the meat and caused resistance problems (Landers et al. 2012; Tzivara et al. 2013). In general, nowadays, two categories of DFM applications can be distinguished: (1) feed inoculants (e.g., for silage or high-moisture grain), intended to ferment the feed substrate and modify the digestibility and safety of the feed component, and (2) beneficial, viable microorganisms added to feed to improve the health parameters of the consuming animal.

The prebiotic concept is much younger and was first introduced by Gibson and Roberfroid (1995) as "Non-digestible food ingredients that beneficially affect the host by selectively stimulating the growth and/or activity of one or a limited number of bacterial species already resident in the colon, and thus attempt to improve host health." As for probiotics, the original definition has been modified frequently, but so far no consensus has been reached. Bindels et al. (2015) proposed the definition "A non-digestible compound that, through its metabolization by microorganisms in the gut, modulates composition and/or activity of the gut microbiota, thus conferring a beneficial physiological effect on the host." In the latter definition, the "selectivity" is no longer a criterion, and the link with the metabolic degradation of the compound and the results on the ecology of the gut microbiota or the observed physiological effects are more deeply stressed.

Nowadays, in both the pro- and prebiotic definitions, the importance for health (whether man or animal) occupies a central position. Traditional applications of probiotics have been mostly performed through fermented dairy products and were mainly directed to usual digestive processes. However, more recently other foods and food supplements have gained much more attention, increasing the worldwide probiotic market, which exceeded 30 billion dollar in 2015 and expected to account for 52 billion by 2020 (Market Research Report 2016). Moreover, the explosion of the metagenomics approach to study the microbiota niches in man and animal, and their exploration in terms of health and disease, has opened a completely new range of applications, situated in the pharmaceutical field on top of the nutritional field. The applications of pro- and prebiotics in pet animals and their exploitation in farm animals, in aquaculture, or even in plants are equally expanding. The global probiotics in the animal feed market are expected to attain 4.71 billion dollars by 2021 (Market Report 2016). While the potential of these applications is high at the prophylactic as well as therapeutic level, there are currently a number of technological, microbiological, and regulatory bottlenecks, which slow down the developments in the field and which will be explored further in this chapter.

*The technological challenges* for probiotics are mainly related to the requirement that the microorganisms should be viable at the end of the product's shelf life. While spray-dried vegetative cells know a faster decline in numbers over time compared to freeze-dried, the latter is considerably more expensive, which, for agricultural applications, is a considerable bottleneck. The use of spore-forming bacteria has, therefore, been proposed successfully. Spores are better in resisting environmental conditions, such as heat, moisture, and pH changes. Upon digestion by the animals, they will germinate into active vegetative cells. Because of their resistance properties, they are frequently added to pelleted diets, where they survive the thermal treatments that are often necessary in feed compacting and pelleting.

*The strain selection*. Probiotic effects are known to be strain specific (FAO/WHO 2001); therefore, it is important that strains intended for a particular application in a particular animal are carefully selected. Selection criteria will differ depending on the animal species (e.g., farm versus pet animals) and desired application (e.g., growth promotion versus anti-infection).

*The cost issue*. The above research requires a considerable amount of effort, the cost of which has to be borne by the farmer. As financial margins in animal husbandry are shrinking, the cost of probiotics does become an issue. A number of older studies have shown, however, that the supplementation with a mixture of lactobacilli could improve egg production and feed efficiency by 3.03 and 7.41%, respectively, in Leghorn hens, while a large-scale study comprising 101,615 commercial hens showed an egg production increase from 69.5 to 72.2% with a feed reduction from 1.75 to 1.69 kg. For further examples in turkey, pigs, and ruminants, see Ezema (2013). The use of probiotics in pet animals, mainly for health or animal well-being reasons, however, has a different economic reasoning and might be more difficult to calculate in this simple way. In estimating the cost of probiotics in animals, it might also be important to consider arguments related to the reduction of antibiotic usage. In the European Union (EU), in 2006, a ban on the use of antibiotics as growth promoters was introduced (European Parliament and Council 2003). This ban did reduce antibiotic resistance development but also increased the general infection rate in husbandry installations (Bywater et al. 2005; Casewell et al. 2003).

*The short life span* of, e.g., broiler chicken, reaching slaughter weight after 5–7 weeks, leaves little time for probiotics to contribute to the development of a mature immune system, a process, which takes up to 6 weeks. Therefore, rather than being fed with probiotics, hen and broiler chicks are most often vaccinated against infectious pathogens, such as *Salmonella* or a variety of viruses (Breytenbach 1999). Vaccines are mostly administered via drinking water or spraying. However, given the need for fast growing (weight gains of over 50 g per day), selected probiotics could be used for weight gain purpose.

*The regulatory aspect* for probiotics is complex. In human applications, different countries allow different levels of health claims. While in the EU currently no health claims are approved for human applications with probiotics, except for yogurt in relation to lactose intolerance, the use of probiotics in animals is well regulated. Manufacturers of probiotics will need to provide evidence of the identity, safety, and efficacy of the product, which will be assessed by a committee of experts (European Parliament and Council 2003). When approved, probiotic products can be labelled and marketed as "gut flora stabilizers" under the category "zootechnical additives," which is one of the five categories of feed additives defined by this Regulation (EC) No 1831/2003: (1) technological additives, (2) sensory additives, (3) nutritional additives, (4) zootechnical additives, and (5) coccidiostats and histomonostats (European Parliament and Council 2003). In the USA, the Center for Veterinary Medicine (CVM) within the FDA is responsible for the recognition of safety, effectiveness, labelling, and distribution of livestock feeds, pet foods, veterinary drugs, and devices. As mentioned before, the FDA uses the term DFM for probiotics used in animal feed, "products that are purported to contain live (viable) micro-organisms (bacteria and/or yeast)" (FDA 2015). For more detailed information on global regulations, see FAO (2016).

In order to convince the legislator to recognize the benefits of probiotics at a much broader scale and promote their acceptability by the farmer and the consumer, the clarification of the mechanisms underlying the beneficial properties is extremely important. These mechanisms can be more generic in nature, e.g., the production of organic acids, the reduction of toxic amines, or more strain specific, e.g., the production of antimicrobials, such as bacteriocins or hydrogen peroxide (explaining the competitive exclusion of pathogens, most often observed for probiotics), or the production of specific enzymes like amylases, lipases, proteases, and glycosidases, which can assist digestion. Other mechanisms, such as the stimulation of immune responses are strain specific, may be limited to a specific period during the development of the animal, and efficacy is mostly depending on the animal species. Some of these mechanisms are further discussed in this chapter.

## 1.2 Polygastric Animals

Ruminants, the most widely distributed group of mammals on Earth, currently add up to about 150 domestic and wild species, while economic interest lies mainly in the breeding of cattle, sheep, goats, and water buffaloes. Ruminants are able to assimilate nutrients from low-quality plant-based feeds, through their digestive tract, which is uniquely designed and includes, in contrast to other mammals, a four-compartmentalized stomach consisting of the rumen, the reticulum, the omasum, and the abomasum. From the physiological point of view, each chamber performs different processes. The microbial fermentation of the fibers and solid feeds takes place in the rumen, while the liquids are transferred to the reticulum, which serves also to the entrapment of large feed particles, regurgitated subsequently for optimal digestion. In the omasum the liquids are filtered and various nutrients are being absorbed, and, finally, in the abomasum the enzymatic digestion of the feed takes place (Hofmann 1989). While the ruminant gastrointestinal tract (GIT) consists of different niches, the vast microbial diversity is observed in the rumen, where the microbial fermentation of the feed is carried out. The rumen microbiome is composed predominately of bacterial species but also of methanogenic archaea, flagellated and ciliated protozoa, fungi, and bacteriophages (Chaucheyras-Durand and Ossa 2014) with populations at a level of $10^{10}$ (bacteria), $10^8$ (protozoa), $10^7$ (archaea), and $10^3$ (fungal spores) colony-forming units per mL of rumen fluid (Deusch et al. 2015).

The optimized ruminal fermentation is essential in supporting health and productivity in the ruminants, since several physiological parameters of farm animals are highly correlated with the abundance of various bacterial members of the rumen microbiome (Jami et al. 2014). Toward this, the systematic use of antibiotics was gradually adopted as a common practice in animal husbandry, targeting, inter alia, the beneficial manipulation of ruminal metabolism. Nevertheless, their rampant use as growth promoters in animal feed during the last decades gradually raised concerns, not only for the antibiotic residues in animal products and the emergence of drug-resistant microorganisms but also for the well-being of the animals themselves. In recent years, probiotics and DFM are widely used in the livestock production, especially in the EU, where the use of antibiotics in this field has been completely prohibited (Landers et al. 2012; Papatsiros et al. 2013). However, numerous countries, e.g., China, the USA, Australia, etc., still employ antibiotics in livestock production, and an unprecedented increase in usage rate during the next decade is foreseen, mainly in developing countries (Van Boeckel et al. 2015).

The application of probiotics and DFM in ruminant productivity and health includes treatment of digestive disorders and reduction of gut pathogens (Wisener et al. 2015), stabilization of the ruminal pH (Chiquette et al. 2008), enhanced animal performance, increased feed conversion efficiency and fiber digestibility (Zhang et al. 2015b), improved milk yield and composition (Ayad et al. 2013; Maragkoudakis et al. 2010), stimulation of the immune system (Spaniol et al. 2015), treatment of mastitis (Espeche et al. 2012), and methane mitigation (Alazzeh et al. 2012). The potential of some probiotics to bind mutagens either present in feeds or formed due to stress or GIT infections has been recently also reported (Apas et al. 2014). The vast majority of the applications concern cows and the pre-ruminant life of calves, whereas the number of respective studies for lambs, sheep, and goats has increased over the last years. The probiotic preparations are delivered to ruminants mainly orally, directly, or in the feed. However, the oral administration may compromise the probiotic efficacy due to the adverse conditions prevailing in the GIT. For ensuring the stability and viability of probiotics, the microencapsulation technology has come into use, providing protection and controlled deliverance of the probiotic preparation in the GIT (Qi et al. 2011).

An overview of the respective literature reveals the broad applicability of the well-studied lactic acid bacteria (LAB) as probiotics and DFM in ruminants. The use of *Lactobacillus*, *Enterococcus*, *Streptococcus*, and *Bifidobacterium* species has been reported. Besides LAB, several other microorganisms have been studied for their probiotic potential in ruminants. These include lactic acid utilizers, such as *Propionibacterium* and strains of *Megasphaera elsdenii*, as well as other bacteria, such as *Escherichia coli*, *Bacillus*, and fibrolytic *Prevotella* species (Dhama et al. 2008; Puniya et al. 2015; Rafat and Hussain 2013; Seo et al. 2010). In two recent studies, the use of cellulolytic *Ruminococcus* species in buffaloes and reindeers resulted in the beneficial modulation of their rumen microbiome (Kumar and Sirohi 2013; Praesteng et al. 2013). While most bacterial probiotics are highly efficacious in pre-ruminant calves, probiotic yeasts and fungi, such as *Saccharomyces cerevisiae* and *Aspergillus oryzae*, respectively, have shown greater benefits for adult ruminants (Nagaraja 2012). It has been demonstrated that their use positively influences certain bacterial populations and the fermentation patterns in the rumen (Pinloche et al. 2013). Furthermore, non-live products from fermentations of probiotic microorganisms have been efficiently employed in ruminants (Bernard 2015). In a recent study, it has been demonstrated that there was no evident benefit from the supplementation of live LAB when compared to the administration of non-live probiotic extracts (Jenkins and Jenkins 2014). Among the various studies performed, even kefir has been examined as a probiotic supplement in ruminants, but its administration did not affect significantly the physiological parameters of the animals (Atasoglu et al. 2010).

The interest for identifying candidate probiotics for ruminants is gradually focusing on the autochthonous microorganisms from the various niches of the target animal and toward their ensuing use in the digestive tract of the animal, which served as the initial isolation source (Fraga et al. 2014; Nader-Macias et al. 2008). For example, comparison of the probiotic characteristics among isolates from dairy products and animal rumen revealed that the latter were more tolerant to bile salts and exhibited higher inhibition against pathogens (Jose et al. 2015). These findings show that the adaptation of the microorganisms to a specific ecosystem could play a significant role in the selection of probiotic candidates and that the probiotic efficacy of selected isolates might depend to some extent on the original host. Furthermore, the use of rumen inhabitants as probiotics will result in enhancing the existing beneficial gut microbiota, which seems to be a milder method of gut microbiome manipulation than introducing ecosystem-irrelevant microbes (Kumar and Sirohi 2013). Therefore, the niches of the ruminant GIT constitute a rich and diverse reservoir for mining potentially novel probiotics (Tellez et al. 2015). The boost in the development of high-throughput sequencing techniques revealed an abundance of non-culturable bacteria in the rumen ecosystem in comparison to data obtained by conventional microbiology (Kim et al. 2011b). The recent accumulation of metagenomics studies on the rumen microbiome can provide a vast body of information concerning not only the composition and the function of the respective microbiota but also its interaction with the host and its feed (Morgavi et al. 2013).

The concept of using bacteriophages for manipulating certain microbial populations in ruminants has been also studied (Callaway et al. 2008; Sheng et al. 2006). Although phages present high host specificity, their efficient application requires the identification of the bacterial target in the rumen. To prevent bacterial resistance, the use of phage cocktails is recommended. In a recent study, a cocktail of designed bacteriophages was successfully employed in rats as a biocontrol means against the gut pathogen *E. coli*, suggesting further testing for possible use in ruminants (Abdulamir et al. 2014). An effective treatment demands the monitoring of the developing resistance mechanisms, the use of newly isolated phages from the rumen environment, and even the development of new phages in the laboratories. Furthermore, the use of isolated lysins instead of whole bacteriophages could be a promising alternative. However, there are only few data available about the rumen virome. Recent studies on the rumen bacteriophages and their interactions with the rumen bacteria constitute an initial attempt to study the rumen virome in depth, helping to obtain new insights probably exploitable in the manipulation of the rumen microbiome (Berg Miller et al. 2012; Ross et al. 2013). The detailed characterization of the rumen virome would be of great significance, since the endemic ruminal phages could prove to be either a useful tool (Hallewell et al. 2014) or a drawback for the probiotic interventions in the animals (Kropinski et al. 2012). Additionally, further research is needed regarding the potential risk associated with the use of phages in lactating ruminants and the possible contamination of milk and dairy products. If the adverse effect on dairy manufacturing is demonstrated, their application could be limited to meat-producing animals.

The application of probiotics and DFM could also play a decisive role in the mitigation of rumen methanogenesis, since the reduction of the enteric methane emissions could be attained through the enhancement of rumen fermentation efficiency and the augmentation of animals' productivity (Karakurt et al. 2012). The environmental impact of the ruminant-derived methane is of considerable importance for the sustainability of livestock, since it is accountable for 25% of the global methane emissions produced by anthropogenic activities (Buddle et al. 2011). The use of probiotic acetogenic bacteria and yeasts, mainly *S. cerevisiae*, for decreasing rumen's methane emissions has been studied with promising results (Jeyanathan et al. 2014). Another interesting aspect is the use of probiotics for controlling specifically the protozoal population in the rumen, since it has been reported that methanogens found both attached and inside ciliate protozoal cells are responsible for 9–37% of the enteric methane production (Jeyanathan et al. 2014; Martin et al. 2010). The proportional correlation among rumen protozoa and methane emission has been confirmed using a meta-analysis approach (Guyader et al. 2014). Recently, the availability of genome projects on rumen methanogens can provide information about the dominant microorganisms implicated in methane production, e.g., methanogenic archaea (Leahy et al. 2013), leading to a more targeted selection of probiotics and DFM.

The use of recombinant microorganisms with probiotic properties in ruminants has been also documented. The most successful study concerns the genetically modified bacterium *Butyrivibrio fibrisolvens*, in which a dehalogenase for fluoroacetate encoding gene from a *Moraxella* soil species was introduced (Gregg et al. 1994). The modified organism was able to degrade the toxic fluoroacetate present in forage plants. The results were encouraging since the microorganism survived in the rumen of sheep and cattle without the loss of the respective gene (Gregg et al. 1998; Padmanabha et al. 2004). The same species was also used for the creation of a recombinant xylanolytic strain. A plasmid containing a xylanase gene from *Neocallimastix patriciarum* was successfully inserted into *Butyrivibrio fibrisolvens* (Xue et al. 1997), and although the modified microorganism had enhanced capacity for xylan degradation, it failed to persist in the rumen (Krause et al. 2001). The recent information obtained from various sequencing projects and databases reveals the abundance of specialized microorganisms in the rumen. Thus, it would be difficult for genetically engineered superbugs to fully colonize the ruminal microbial ecosystem and exert on the host the benefits for which they have been designed (Krause et al. 2013).

Similar to probiotics, prebiotics, which are nondigestible oligosaccharides, are also effective in altering the composition and activity of the microbiome in the GIT, since they constitute suitable substrates for the enhancement of certain beneficial ruminal microorganisms. However, the ability of ruminants to catabolize most of the common prebiotic compounds creates a limitation in the use of prebiotics as growth promoters in ruminant production. In addition, several nondigestible oligosaccharides found naturally in plant cell wall are included in feeds normally used in ruminant rations (Gaggia et al. 2010), making the implementation of prebiotics in ruminants possibly unnecessary. The administration of prebiotics seems to be beneficial on very young ruminants, since these substrates may contribute to the formation of a desirable intestinal community, which may further improve the performance of older animals (Uyeno et al. 2015). The advance of rumen-protective technologies providing shielding from ruminal digestion, such as encapsulation, may become useful tools for the eventual use of selected prebiotics in ruminant feed, as it has been reported for probiotics (Mustapha et al. 2013).

Despite the wide applicability of probiotics and to a lesser extent of prebiotics in ruminant production and the promising results obtained from various studies, reproducibility issues are raised, since experimental data acquired are often inconsistent (Uyeno et al. 2015). A wide variety of factors, such as the growth environment, the animal species and breed, the age and physiological state of the animal, the diet, the nature of the probiotic preparation used (e.g., type of microorganism, live culture, or lyophilized cells), and even its dose, seem to affect the outcomes of probiotics' utilization in livestock. Obviously, comprehensive research is needed for the reliable and viable use of probiotics and prebiotics in ruminant production.

## 1.3 Monogastric Animals

Monogastrics are classified as animals having one simple or single-chambered stomach with the main agricultural species being pigs, poultry, and horses. The gut microbiota of pigs mainly consists of bacteria, while a small percentage of archaea mostly *Methanomicrobia* and *Thermococci* have been also identified (Isaacson and

Kim 2012; Lamendella et al. 2011). In the poultry GIT, 13 phyla of bacteria were discovered with *Firmicutes*, *Bacteroidetes*, and *Proteobacteria* being the more representative ones with up to 900 and 500 species in chicken and turkey gut, respectively. Of all the species found, only 117 out of 900 and 69 out of 500 are established genera of bacteria with the most predominant genera in both chicken and turkey being *Clostridium*, *Ruminococcus*, *Lactobacillus*, and *Bacteroides*. Besides bacteria, the poultry GIT is also inhabited by methanogenic archaea, fungi, and viruses (Pan and Yu 2014; Yeoman et al. 2012). Furthermore, the horse GIT is inhabited by bacteria as well, but archaea, fungi, and protozoa are also present (Daly et al. 2001).

The composition and activity of intestinal microbiota have a crucial impact on the animal health, growth, and performance as a whole. After the ban of antibiotics as animal growth promoters in the European Union, Korea, and Japan, probiotics gained ground as they present a variety of beneficial effects including, among others, promotion of gut health and homeostasis (Hou et al. 2015). Costs, however, have been a major bottleneck for their routine use.

Over the years, probiotics have been used in a number of different ways in livestock, but in the 1960s, it was demonstrated for the first time that *Lactobacillus* strains were able to improve the growth performance of pigs (Ahasan et al. 2015). The most frequently used probiotics in monogastric animals are yeasts (*Saccharomyces boulardii* and *S. cerevisiae*) and bacteria (*Lactobacillus* spp., *Enterococcus* spp., *Pediococcus* spp., *Bacillus* spp.) targeting the cecum and the colon. The most common benefits of probiotics in monogastric animals are the increase of body weight, the reduction of the risk of diarrhea, the improvement of feed efficiency, and diet digestibility (Ahasan et al. 2015). Furthermore, probiotics have been assigned to play a significant role in providing supportive care to piglets during their initial part of life, while probiotics like *Enterococcus faecium* and *Bacillus subtilis* can reduce the concentration of ammonia in the excreta of poultry (Dhama et al. 2008). There are many microorganisms to be considered as potential probiotics, but only a limited number of microorganisms seem to satisfy the necessary criteria.

In order to identify and detect the GIT microbiota from the animal gut and feces, several techniques have been developed based on biochemical, microbiological, immunological, and molecular biological features. Among them, the expansion of high-throughput sequencing techniques exposed the plethora of non-culturable bacteria enabling the comprehensive characterization of the intestinal microbiota of poultry and other monogastric animals (Danzeisen et al. 2011; Kim et al. 2011a). A full understanding of the intestinal microbiota and the genomic functions of its members, i.e., microbiome, will lead to the development of targeted probiotic strains and novel or improved strategies for effective microbiota modulation (Chambers and Gong 2011; Choi et al. 2015; Pan and Yu 2014; Umu et al. 2015). Next-generation sequencing studies on broilers and pig's gut microbiota shed light on the age-related bacterial diversity revealing the importance of gut modulation to improve animal health (Kim et al. 2011a; Mohd Shaufi et al. 2015). Compared to the other monogastric animals, there is only a limited number of studies characterizing the equine gut microbiota using culture-independent methods (Daly and

Shirazi-Beechey 2003; Hastie et al. 2008; Shepherd et al. 2012; Yamano et al. 2008). However, as these techniques have been recently developed, the results are not always reliable (Sachsenröder et al. 2014).

Although the native gut microbiota is commonly used as a pool for probiotic candidates, the use of genetically modified strains as probiotics in monogastric animals is of ongoing interest (Sieo et al. 2005). A species commonly used for genetic engineering in poultry is *Lactobacillus reuteri*. A lot of research has been conducted using strains of this species expressing heterologous genes in a poultry diet with encouraging results on the growth performance and welfare of animals (Li et al. 2014; Liu et al. 2005, 2007; Yu et al. 2008). Since genetic engineering approaches have positive results in poultry, research is currently focusing on genetically modified strains capable of expressing more than one heterologous genes (Wang et al. 2014). Apart from poultry, genetically engineered probiotics are also used in pigs either therapeutically, e.g., in pancreatic insufficiency, or as feed additives enhancing livestock production (Drouault et al. 2002; Yin et al. 2010).

The idea of using bacteriophages to manage or eliminate zoonotic bacteria in poultry husbandry has been established as a cost-effective approach with significant advantages compared to antibiotics. The chicken gut microbial imbalance frequently caused by broad-spectrum antibiotics is avoided using host-specific bacteriophages. These bacteriophages are naturally self-limiting as they replicate only in the target bacteria and only as long as the bacteria are present (Atterbury et al. 2007). Recently, due to the advantages of bacteriophage biology, a lot of successful research has been made in broiler chickens indicating the ability of host-specific bacteriophages alone or in combination with probiotics to reduce colonization of *Salmonella* and *Campylobacter* (Atterbury et al. 2007; Bardina et al. 2012; Loc Carrillo et al. 2005; Marietto-Gonçalves et al. 2014). It is important that both *Salmonella* and *Campylobacter* phages can be isolated from poultry feces and farm environment resulting in gut microbial stability (Atterbury et al. 2007). Additionally, the use of lytic bacteriophages to prevent or treat colibacillosis in broilers has also been studied (El-Gohary et al. 2014; Lau et al. 2010; Oliveira et al. 2010). It is worth noting that although the successful use of phage therapy in swine dates back to 1920s, it recently regained the attention of the research community (Zhang et al. 2015a). A limited number of studies on pigs indicate that the use of bacteriophages could be a successful strategy against various species of *Salmonella* (Albino et al. 2014; Callaway et al. 2011). In general, a cocktail of phages that use different receptors on the host cell is more effective in reducing pathogens compared to pure phages and also delays the formation of phage resistance (Goodridge 2010). Bacteriophages are used not only therapeutically but also as growth promoters in pigs and poultry (Gebru et al. 2010; Kim et al. 2014a; Wang et al. 2013; Zhao et al. 2012). Yan et al. (2012) suggested that a bacteriophage diet can be used as an antibiotic alternative on growth performance of pigs, and in some cases bacteriophages appeared more effective than probiotics on the performance of growing pigs, as indicated by Kim et al. (2014b). In addition, the further understanding of the biology underlying phage therapy, safe practice, quality control, and accumulation of knowledge and experience remains future challenges (Chambers and Gong 2011).

It should be noted, however, that the oral use of probiotics or bacteriophages can be effective only if they manage to survive during the passage through the digestive system. Therefore, a successful delivery system is of utmost importance. A number of studies have been performed on poultry or swine simulated GIT conditions showing that a microencapsulation technique can protect the bacteriophages or probiotics against gastric environment (Ma et al. 2008; Musikasang et al. 2009; Ross et al. 2008). In a similar study, a microencapsulated phage cocktail administered to swine feed remained effective after the passage through the GIT and successfully reduced *Salmonella* colonization (Saez et al. 2011). The same results were observed in poultry with a cocktail of liposome-encapsulated bacteriophages (Colom et al. 2015).

Although the concept of functional foods has been introduced a long time ago, scientific evidence for the use of prebiotics in animal feed exists from the late 1990s for poultry and pigs (Hajati and Rezaei 2010). The majority of research in prebiotics has been performed in poultry, as this is the most studied monogastric animal. Prebiotics were found to increase the stool volume of chicken by regulating intestinal microbiota through selective stimulation of beneficial bacteria and inhibiting undesirable bacteria, such as *Salmonella* (Park et al. 2013; Totton et al. 2012). The most common prebiotics used in monogastric animals are inulin, fructo-oligosaccharides (FOS), mannan-oligosaccharides (MOS), and galacto-oligosaccharides (GOS), with GOS being the less investigated in the poultry industry compared to FOS (Park et al. 2013). It is difficult, however, to draw conclusions for the prebiotic effects in animals from the published studies due to the wide variety of these studies regarding subjects, age, diet, outcome parameters, substances tested, dose, and duration of the experiments (Allaart et al. 2013; Samanta et al. 2013). Such inconsistent results have been mainly recorded after the use of MOS to reduce the intestinal numbers of *Clostridium perfringens* in poultry and after the use of inulin as prebiotic to improve growth performance of layers and broilers, indicating that the effects are both dose- and diet-dependent (Allaart et al. 2013; Biggs et al. 2007; Ortiz et al. 2009; Yusrizal and Chen 2003). The application of prebiotics in animal feed is a relatively recent effort, and although the results are promising, many issues must be solved, such as the establishment of the efficacy of prebiotics in routine diets of livestock. The advanced techniques like next-generation sequencing could be very useful to substantiate any prebiotic effect on animal microbiota, while at the same time future research of prebiotics in livestock should be focused on immunological aspects, changes at the gut epithelium, and at livestock product quality (Samanta et al. 2013).

## 1.4 Aquaculture

Aquaculture is the farming of aquatic organisms, and it involves the cultivation of freshwater and saltwater populations under controlled conditions. Compared to commercial fishing, this activity allows a selective increase in the production of species used for human consumption, industry, or sport fishing. Due to overfishing of wild populations, aquaculture has become an economic activity of great importance around the world (FAO 2012). The possibility to use feed supplements to improve animal health, welfare, and productivity also addressed the manipulation of the GIT microbial ecosystem in fish (Chaucheyras-Durand and Durand 2010). To our knowledge, the first empirical application of probiotics in aquaculture (Kozasa 1986) is relatively recent and built on their benefits exerted on humans and poultry. Additionally, the first study on prebiotics in aquaculture was reported in 1995 (Hanley et al. 1995). However, a growing number of scientific papers currently deal with probiotics and prebiotics in aquaculture, passing from their empirical use to a science-based approach.

When looking for probiotics intended for an aquatic usage, it is important to consider certain influencing factors that are fundamentally different from probiotics used in, e.g., mammals. Aquatic animals have a much closer relationship with their external environment (Kesarcodi-Watson et al. 2008). This intensive interaction between the environment and the farmed aquatic animals inspired some to change the definition of probiotics when applied to aquaculture. One of these "alternative" definitions proposed is "a live microbial adjunct which has a beneficial effect on the host by modifying the host-associated or ambient microbial community, by ensuring improved use of the feed or enhancing its nutritional value, by enhancing the host response towards disease, or by improving the quality of its ambient environment" (Verschuere et al. 2000). Apart from the FAO/WHO requirement of probiotics to be a live culture, this and other definitions are a (too) lengthy way of describing a probiotic, e.g., "a microorganism or components thereof that is/are beneficial to the health of the host" (Irianto and Austin 2002). The most recent definition identifies aquaculture-probiotics as "live or dead, or even a component of the microorganism that acts under different modes of action in conferring beneficial effects to the host or to its environment" (Lazado and Caipang 2014), again in contradiction with the generally accepted definition of probiotics (FAO/WHO 2002). In fact, nowadays, there is no definition of aquaculture-probiotics that is accepted by the majority of the aquaculture community. Hence, it is important, if really required, to develop a definition of aquaculture-probiotics that is eliminating ambiguity on the term and makes clear the difference with the FAO/WHO definition for "classical" probiotics, used in human and other animals.

The development of probiotics applicable to commercial use in aquaculture is a multistep and multidisciplinary process requiring both empirical and fundamental research, full-scale field trials, and an economic assessment of its use. Defined procedural strategies have been proposed on the selection and evaluation of probiotic candidates for farmed aquatic animals (Lazado et al. 2015).

A good pool of candidate probiotics is of major importance in the selection process, and for aquaculture it is vital to examine isolates that are both autochthonous and allochthonous to the aquatic environment (Gatesoupe 2008). Whereas humans and terrestrial farm animals tend to have an intestinal microbiota dominated by Gram-positive obligate or facultative anaerobes, that of aquatic animals consists mainly of Gram-negative aerobes as well as obligate and facultative anaerobic bacteria (Vine et al. 2006). Bacteria, such as *Vibrio*, *Pseudomonas*, and *Acinetobacter* constitute the predominant indigenous microbiota of a variety of marine fish species and crustaceans

(Pandiyan et al. 2013), while, in contrast to saltwater species, the indigenous microbiota of freshwater animals is dominated by members of the genera *Aeromonas* and *Plesiomonas*, representatives of the family *Enterobacteriaceae*, and obligate anaerobic bacteria of the genera *Bacteroides*, *Fusobacterium*, and *Eubacterium* (Sakata 1990). LAB are generally subdominant in aquatic organisms and represented essentially by the genus *Carnobacterium* (Balcázar et al. 2006). Interestingly, despite the indigenous Gram-negative species, probiotics used in aquaculture belong mainly to the Gram-positive genera *Bacillus*, *Enterococcus*, *Lactobacillus*, and *Carnobacterium* as well as to yeast species when used as biological control or immunostimulatory agents. In contrast, probiotics used as antimicrobials in aquaculture belong essentially to the aforementioned Gram-negative genera (De et al. 2014). Nevertheless, recently, the Gram-positive bacteria belonging to *Streptomyces* genus demonstrated promising results as probiotics in aquaculture regarding not only the production of antagonistic and antimicrobial compounds against pathogens but also the improved growth of the aquatic organisms (Tan et al. 2016).

In the past, the information available on the intestinal microbiota of aquatic species was based on the use of conventional culture-dependent methods. Nowadays, molecular-based approaches are used successfully for the analysis of bacterial communities (Martínez Cruz et al. 2012): (1) 16S rDNA clone libraries (Han et al. 2010; Iehata et al. 2015); (2) fingerprinting methods, such as denaturing gradient gel electrophoresis (DGGE) (McIntosh et al. 2008; Sun et al. 2012) and temporal temperature gradient electrophoresis (TTGE) (Navarrete et al. 2010); and (3) fluorescent in situ hybridization (FISH) (Payne et al. 2007). Also, in a limited number of recent studies, next-generation sequencing (NGS) 16S-amplicon metagenomics-based approaches have been used and reveal a far greater level of diversity in the gut microbiota of animals than previous studies that lacked an NGS approach (Zhou et al. 2014).

The use of gnotobiotic systems (animals cultured in axenic conditions or with a known reconstituted microbiota) can be an excellent tool to extend the understanding of mechanisms involved in host-microbe interactions of cultured animals (Dimitroglou et al. 2011). This approach in parallel to the use of mutant strains, e.g., non-motile *Pseudomonas* mutants (Rawls et al. 2007) or yeast mutants (Soltanian et al. 2007), led to the clarification of genes involved in specific probiotic mechanisms in fishes and crustaceans in the past, respectively. Further mechanistic understanding might also result from the use of tissue- or cell- specific mutants expressing green fluorescent protein (GFP) or GFP variants as a powerful method for in situ monitoring of the presence and behavior of microbes that are intentionally introduced into the host organisms (Mulero et al. 2007). According to Tinh et al. (2008), GFP translational fusions of genes of interest in probiotics might provide additional data on gene functioning when introduced into translucent larvae.

The ability of probiotics to affect the ontogenetic development of animals by interfering with their gonad differentiation and maturation or progression to puberty and aging gains interest for future studies (Avella et al. 2010). Indeed, microarray analysis was used in the past to evaluate alterations on the expression of genes involved in immune response, protein folding, cytoskeletal/structural proteins, and other vital cellular processes such as lipid metabolism, cell proliferation, and apoptosis in aquatic organisms (Rodriguez-Lanetty et al. 2006; Skugor et al. 2008).

The genomic information that is generated from the sequences of known probiotic bacteria provides clear understanding on the inherent probiotic properties (Ventura et al. 2012). In aquaculture, the concept of probiogenomics is not yet widely recognized or applied, although recently the relevance of this perspective in aquaculture has been raised (Lazado and Caipang 2014).

Features correlated to certain modes of probiotic action in the aquatic environment are under investigation (Kesarcodi-Watson et al. 2008). Enhancement of colonization resistance and/or direct inhibitory activity against pathogens are considered important factors when probiotics are used for the prevention of bacterial diseases (Balcázar et al. 2006). Potential probiotics can also be correlated to the growth promotion of cultivated fishes by producing a variety of extracellular enzymes (i.e., proteases, lipases, carbohydrases, phosphatases, esterases, and peptidases) that facilitate the efficient absorption of nutrients (Bairagi et al. 2002; Giri et al. 2013). For instance, the use of plant protein sources in the diets (Gatlin et al. 2007) led to the investigation of the metabolic capabilities of probiotics, such as the degradation of anti-nutritional factors, a feature interrelated with the improvement of the nutritional value of the feed of aquatic animals in the past (Refstie et al. 2005). Immunomodulation by probiotics has also gained great attention, and the assessment of phagocytic, respiratory burst, lysozyme, serum peroxidase, and complement activities and modulation of cytokine production have been referred to as potential strategies for finding novel probiotic strains for aquaculture (Akhter et al. 2015; Magnadottir 2010; Nayak 2010). Furthermore, the theory that probiotic bacteria, through biofilm formation, enhance the survival rate and growth performance in aquatic organisms (Pandey et al. 2014) has been also established (Boutin et al. 2013).

Taking into consideration that probiotics for aquaculture are marketed in two forms, dry and liquid (Sahu et al. 2008), an appropriate route of delivery of the probiotic to the host should be proposed. So far, literature refers to several ways in which probiotics can be provided to the host or added to its aquatic environment, such as addition via live food, bathing, and addition to culture water and to artificial diet (Balcázar et al. 2006). Bioencapsulation of probiotics has also been demonstrated to be a more effective way to introduce probiotics in the animal gut; in the case of some allochthonous bacteria, this may be the only efficient route (Pintado et al. 2014).

The current literature is heavily focused on the bacterial microbiota, and considerably less information is available on indigenous yeast, bacteriophages, archaea, microalgae, and protozoans in aquaculture. Although it is debatable whether or not bacteriophages constitute bona fide probiotics, their influence on indigenous and/or probiotic bacteria must be taken into account for future studies, especially after the "kill the winner" hypothesis about their important role in shaping the mammalian gut microbiota (Mills et al. 2013). Moreover, bacteriophage therapy has been suggested in the past as an alternative for the prevention and treatment of microbial diseases in aquaculture (Nakai and Park 2002). Even if many recent studies indicate their promising application (Oliveira et al. 2012), caution must be taken for their use in the future (Madhusudana Rao and Lalitha 2015).

Despite the potential benefits of prebiotics to health and performance as noted in various terrestrial species, less information is available about the effect of prebiotics in aquatic organisms. The most common prebiotics used in aquatic species are inulin, FOS, short-chain FOS (scFOS), mannan-oligosaccharides (MOS), trans-galacto-oligosaccharides (TOS), GOS, xylo-oligosaccharides (XOS), arabinoxylo-oligosaccharides (AXOS), iso-malto-oligosaccharides (IMO), and various commercial products containing multiple prebiotic combinations. Prebiotic applications in aquaculture improve animal growth performance and survival, feed conversion and digestibility, GIT enzyme activities and GIT morphology, as well as the suppression of potentially pathogenic bacteria due to the presence of beneficial gut bacteria (Ganguly et al. 2013; Ringø et al. 2014). The role of prebiotics as immunostimulants in aquaculture is also well studied with promising results (Akhter et al. 2015; Song et al. 2014).

One major issue that needs to be addressed is whether the prebiotic supplementation effect can vary in regard to age- and size-related responses, appropriate doses, and timing of administration. The life stage of the animal was highlighted in a study where inulin was used (Hoseinifar et al. 2010). Furthermore, the surrounding environment, i.e., water temperature and salinity and oxygen availability, might have greater influences than the diet on animal health or potentially confound interpretations of the prebiotic findings (Daniels and Hoseinifar 2014). Further research is needed in order to differentiate the health-promoting effects from potentially deleterious responses toward prebiotics (Ringø et al. 2010) as observed for oligosaccharides from soybeans, causing diarrhea in Atlantic salmon (Refstie et al. 2005).

Another recent concept with regard to the manipulation of gut microbiota of animals in aquaculture are synbiotics. The use of synbiotics is an important approach in order to explore in what way prebiotic administration may seed and maintain probiotic strains as the dominant species in the fish GIT (Rurangwa et al. 2009). Despite recent progress in the field of synbiotic administration in aquaculture, there is limited information available on different aspects of synbiotics on fish species (Llewellyn et al. 2014). To our knowledge, few studies so far have investigated the effect of synbiotics only in fish species since the first one in 2009 (Cerezuela et al. 2011). In those studies, probiotics belonging to the genera *Enterococcus*, *Bacillus*, and *Pediococcus*, as well as FOS and MOS prebiotics, were used. The studied fish species were rainbow trout (Mehrabi et al. 2012; Rodriguez-Estrada et al. 2009), Japanese flounder (Ye et al. 2011), yellow croaker (Ai et al. 2011), cobia (Geng et al. 2011), sea bream (Cerezuela et al. 2013), Nile tilapia (Aly et al. 2008), and Atlantic salmon (Abid et al. 2013), indicating better growth, feed efficiency ratio, improved immune responses, and disease resistance of aquatic animals after synbiotic supplementation.

## 1.5 Pets

Companion animals have high numbers of microorganisms in the GIT, which in fact exceed in quantity those living in the human gut. Nonetheless, both cats and dogs have distinct bacterial species that also vary among different dog and cat breeds, gut niches, and geographical areas. Microbial diversity and concentration increase along the length of the GIT. The prevalent bacterial phyla in the colon and feces of both dogs and cats are represented by *Firmicutes*, *Bacteroidetes*, *Proteobacteria*, and *Fusobacteria* as well as *Eubacterium* in cats. The microbial differences between dogs and cats are manifested in the microbial groups and on the species levels (Grzeskowiak et al. 2015). Molecular fingerprinting has revealed that every individual pet has a unique and stable microbial ecosystem (Suchodolski 2011). A recent metagenomics approach estimated that, besides bacteria, the feline GIT microbiota comprises 0.02% fungi, 0.09% archaea, and 0.09% viruses with 99% of them being bacteriophages. The most commonly observed archaeal phyla belonged to *Crenarchaeota* and *Euryarchaeota*, with the most abundant families being *Desulfurococcaceae* (54.8% of sequences), *Methanobacteriaceae* (40.6%), *Methanosarcinaceae* (5%), and *Halobacteriaceae* (2.7%) (Tun et al. 2012). According to Handl et al. (2011), among fungi, *Aspergillus* and *Saccharomyces* are the most abundant genera in the feline GIT microbiota. As for other animals, any disturbances within the gut microbiota of the pets may lead to the development of a multitude of diseases and disorders, such as diarrhea, allergies, obesity, and stress symptoms (Lee and Hase 2014).

Possible benefits of the probiotic use in pets include modulation of the immune system, assistance in stress maintenance, protection from infections caused by enteropathogens, increased growth and development, control of allergic disorders, and recently obesity (Grzeskowiak et al. 2015).

So far, the common mode of administration of probiotics to pets is oral by adding them to the pets' feed (Arslan et al. 2012; Biagi et al. 2013; Bybee et al. 2011; Hutchins et al. 2013; Strompfova et al. 2012). Regarding the genera, which are used as probiotics in companion animals, these include mainly *Bacillus* spp. (Biourge et al. 1998; Gonzalez-Ortiz et al. 2013), *Lactobacillus* spp. (Gómez-Gallego et al. 2016; Kumar et al. 2017; Marsella et al. 2012; Ohshima-Terada et al. 2015; Strompfova et al. 2012) *Bifidobacterium* spp. (Biagi et al. 2013), and *Enterococcus faecium* (Bybee et al. 2011; Gonzalez-Ortiz et al. 2013), and only recently scientists started using as probiotics *Weissella confusa* (Manninen et al. 2006) and *Streptococcus thermophilus* (Arslan et al. 2012).

In order to enhance survival of probiotics during passage through the GIT of pets, encapsulation of bacteria has been used so that a larger number of viable bacteria can reach the intestine. Starch, alginate, carrageenan, and chitosan are included among the hydrocolloids used to encapsulate or to obtain films and coatings (González-Forte et al. 2014; Ma et al. 2015).

The use of prebiotics in companion animal nutrition was reviewed comprehensively by Swanson et al. (2002). Studies evaluating prebiotics have utilized several outcome variables to assess efficacy in canine and feline diets, including (1) food intake, (2) fecal output, (3) stool consistency, (4) macronutrient digestibility (ideal and total tract apparent digestibility), (5) fermentative end products, (6) immune indices, and (7) intestinal microbial populations (Fahey and Vester 2009). From the limited number of reports in the field, it appears that prebiotic supplementation has several beneficial effects on the GIT of dogs and cats, such as positive shifts in microbial populations, decreases in fecal protein catabolites, and changes in immune status. However, more research is required to determine optimal doses, life stages most likely to benefit, and disease states likely to be avoided or treated with prebiotic supplementation. In the future, experiments must also investigate prebiotic supplementation on animals at different life stages and disease states.

Among prebiotics, FOS are the most studied in dogs and cats. They have been used to alleviate small intestinal bacterial overgrowth (Willard et al. 1994), to promote reduction of clostridia, to increase bifidobacteria and lactobacilli populations (Sparkes et al. 1998; Swanson et al. 2002; Twomey et al. 2003), and to reduce the concentrations of protein catabolites produced in the colon (Swanson et al. 2002). MOS may beneficially change the enteric microbiota, since its addition to the diet can reduce *Clostridium perfringens* counts in dog feces (Strickling et al. 2000). In vitro studies suggest that MOS are moderately fermentable by canine and feline microbiota (Vickers et al. 2001) being thus a source of energy to lactate-producing bacteria. This explains the reduced fecal pH and fecal ammonia excretion verified in dogs, improving indices of colonic health (Zentek et al. 2002). Furthermore, Swanson and Fahey (2006) reported the immunomodulatory effect of MOS in dogs, in particular on the concentrations of IgA, IgG, and plasma lymphocytes. The use of MOS and FOS in diseased or immunocompromised animals (Apanavicius et al. 2007; Gouveia et al. 2006) has revealed a protective effect for immunocompromised dogs.

Molecular techniques have been also employed to evaluate the effect of prebiotics on the GIT microbial consortia in cats and dogs (Middelbos et al. 2007; Vanhoutte et al. 2005). Vanhoutte et al. (2005) observed potential alterations in fecal microbiota of seven adult healthy dogs related to the administration of oligofructose and inulin. Middelbos et al. (2007) appraised spray-dried yeast cell wall (YCW) supplementation of diets to adult dogs where it was noted that YCW altered digest flow through the intestinal tract, decreased quadratically total white blood cell and eosinophil counts, and was responsible for the reduction of fecal microbial populations. However, further work is necessary to confirm the above results and also elucidate the effect of prebiotics in other diseases, such as inflammatory bowel disease, small intestinal bacterial overgrowth, etc.

Regarding the combination of probiotics and prebiotics, Swanson et al. (2002) were the first to study the effect of synbiotics, namely, administration of FOS and/or *Lactobacillus acidophilus*, on the gut microbial populations, end products, and nutrient digestibility in healthy adult dogs. It was shown that FOS enhanced indices of gut health by positively reshaping gut microbial ecology and fecal protein catabolites, whereas *Lactobacillus acidophilus* was more effective when fed in combination with FOS rather than fed alone. Later on, Ogue-Bon et al. (2010) showed that GOS supplementation in dogs can sustain the growth of *Bifidobacterium bifidum*, when used as a synbiotic combination, while Biagi et al. (2013) reported that the combination of GOS with a strain of *Bifidobacterium pseudocatenulatum* had some positive effects on the intestinal microbiota in cats.

## 1.6 Bees

As a pollinator, the honeybee, *Apis mellifera*, is a key species for agricultural production and contributes to the human food supply (Aizen et al. 2008; Klein et al. 2007). Recent losses of *A. mellifera* and bumble bees (genus *Bombus*), and the potential association of these declines with various infectious agents, call for a better understanding of the bees' microbiota (Evans and Schwarz 2011; Genersch 2010). Honeybees pool resources, divide labor, and communicate in highly structured social colonies. Sterile female worker bees dominate within colonies, in which they initially clean cells, rear brood, and store food; then they leave the hive and search for pollen and nectar (Seeley 1985). Independent studies of bacterial community profiles based on 16S rRNA sequences demonstrate that workers of *A. mellifera* and some *Bombus* species consistently harbor an offbeat gut microbiota not shared with solitary bees (Cox-Foster et al. 2007; Koch and Schmid-Hempel 2011; Martinson et al. 2011). This microbiota consists of eight distinct species or phylotypes, i.e., closely related strains with ≥97% sequence identity in 16S rRNA sequences, hereafter referred to as species. These include three Gram-positive species, namely, two closely related *Firmicutes* within *Lactobacillus* and one within *Bifidobacterium*, and five Gram-negative species, namely, one β-proteobacterium with the Candidatus name "*Snodgrassella alvi*," two closely related γ-proteobacteria, one with the Candidatus name "*Gilliamella apicola*," and two α-proteobacteria (Martinson et al. 2012).

The application of probiotics in bees is achieved through feeding, with *Lactobacillus* and *Bifidobacterium* being the major genera used until now. Machova et al. (1997) were the first who added probiotics, without specifying the microorganisms used though, into sugar syrup in order to feed honeybees (*Apis mellifera*) and noticed that this ameliorated bee survival. The next attempt was not until 7 years later, and it was demonstrated that probiotics including *Bifidobacterium infantis*, *B. longum*, *B. breve*, *Lactobacillus rhamnosus*, *L. acidophilus*, *L. reuteri*, *L. casei*, and *L. plantarum* enhance immune responses in bees by stimulating the production of antimicrobial peptides against *Paenibacillus* and *Ascosphaera apis* infections (Evans and Lopez 2004). Kaznowski et al. (2005) used *Lactobacillus* spp., *Pediococcus acidilactici*, *B. bifidum*, and *E. faecium* as supplements to pollen substitute in feeding honeybees. It was shown that in order to accomplish increase in dry mass and crude fat level, it was sufficient to supply probiotics only in the beginning of the feeding period, directly after bee emergence. These results have been confirmed by Kazimierczak-Baryczko and Szymaś (2006), who used the same species and showed that the addition of probiotics in pollen substitute prolonged bee life span and stimulated the growth of the faucial gland and fat body. Moreover, administration of *Lactobacillus* spp., *Bifidobacterium* spp., *Saccharomyces boulardii*, and *Streptococcus thermophilus* through sugar syrup resulted in better colony development, a longer life span, and enhanced development of wax production (Patruica et al. 2011a, b, 2012, 2013; Patruica and Mot 2012). It seems, however, that in order to be efficient, probiotics have to be tailored for bees (Johnson et al. 2014).

In recent years, molecular methods offer great potential for the phylogenetic identification of probiotic microorganisms in bees (Mattila et al. 2012; Olofsson and Vasquez 2008; Tajabadi et al. 2013). Olofsson and Vasquez (2008) detected and identified novel LAB, mainly lactobacilli, as well as bifidobacteria in the honey stomach of honeybees by employing 16S rRNA sequencing. Using the same method, Tajabadi et al. (2013) detected *Lactobacillus* spp. in *Apis dorsata* honeycomb, which could be explored as a source of new bacteria with probiotic potential in honeybees. Moreover, deeper comprehension of the complex host-microbial interactions might also result from the use of tissue- or cell- specific mutants expressing GFP or GFP variants. In this direction, Hyrsl et al. (2015) have successfully used a mutant of *Photorabdum luminscens* that expressed GFP in order to track the nematobacterial infection in bees.

## Conclusions

The overview of the existing literature regarding the studies performed so far with probiotics and prebiotics in monogastric and polygastric animals, aquaculture, pets, and bees highlights (a) the variety of microorganisms, comprising both bacteria, which are mainly lactic acid bacteria and yeasts, and to a lesser degree the array of oligosaccharides, mainly inulin, FOS, MOS, and GOS, employed; (b) the administration method, e.g., feed, water, and aquatic environment (for fish); (c) the origin of strains, although mainly allochthonous; (d) the target body function, e.g., balance of GIT microbiota, reduction of diarrhea risk, improvement of feed efficiency, and diet digestibility to increase body weight, growth and development, immune system, assistance in stress maintenance, protection from infections, control of allergic disorders, obesity, etc.; and (e) the assays used for the elucidation of these beneficial actions.

Despite the stimulating data accumulated so far, further studies are needed using more standardized protocols. These protocols should consider, among others, the age and size of the animal, the rearing conditions, the dose and composition of the active compound used, the route and matrix of delivery, etc. This will allow a more reliable comparison of results, thus facilitating the consistency and conclusions that can be drawn about the beneficial impact of both probiotics and prebiotics. Moreover, the application of advanced techniques, e.g., metagenomics, to shed full light to the indigenous microbiota of the animal under consideration, comprising a complex mixture of indigenous bacteria, yeasts, bacteriophages, archaea, microalgae, and protozoans, will drive the use of autochthonous and/or animal tailored probiotic strains. Indeed, a full understanding of the intestinal microbiota and the genomic functions of its members will lead to the development of targeted probiotic strains and novel or improved strategies for effective microbiota modulation. It should be also stressed that the detailed characterization of the animal virome would be of great significance, since the endemic phages could prove to be either a useful tool or a drawback for the probiotic interventions in animals. Additionally, the use of well-targeted recombinant probiotics is expected to receive further attention in the near future.

Furthermore, it should be stressed that the elucidation and interpretation of the mechanisms underlying the beneficial properties are extremely important. Learning from the human applications of probiotics and prebiotics, these mechanisms can be either of generic nature or strain specific, while some mechanisms may be limited to a specific life stage during the development of the animal, and their efficacy critically depends on the animal species. For instance, the use of gnotobiotic systems can be an excellent tool to extend the understanding of mechanisms involved in host-microbe interactions in animals and to study cause and consequence of specific interventions. Thus, this field needs further investigation, which will help to understand the interactions between probiotics and prebiotics as well as the interactions of the host with either probiotics and prebiotics separately or in combination.

Unravelling and overcoming problems existing at the science level are needed to facilitate the applications and subsequently smoothen the regulatory actions. The legislator needs to clearly recognize the benefits of probiotics and prebiotics at a much broader scale in order to promote their acceptability by the farmer and the consumer.

## References


AAFCO (1999) Official publication: Association of American Feed Control Officials, Inc. C.P. Frank, Georgia Department of Agriculture, Plant Food, Feed and Grain Dividion, Capitol Square, Atlanta

Abdulamir AS, Jassim SA, Abu Bakar F (2014) Novel approach of using a cocktail of designed bacteriophages against gut pathogenic *E. coli* for bacterial load biocontrol. Ann Clin Microbiol Antimicrob 13:39. https://doi.org/10.1186/s12941-014-0039-z

Abid A, Davies SJ, Waines P, Emery M, Castex M, Gioacchini G, Carnevali O, Bickerdike R, Romero J, Merrifield DL (2013) Dietary synbiotic application modulates Atlantic salmon (*Salmo salar*) intestinal microbial communities and intestinal immunity. Fish Shellfish Immunol 35(6):1948–1956. https://doi.org/10.1016/j.fsi.2013.09.039

Ahasan A, Agazzi A, Invernizzi G, Bontempo V, Savoini G (2015) The beneficial role of probiotics in monogastric animal nutrition and health. J Dairy Vet Anim Res 2(4):00041. 10.15406/jdvar.2015.02.00041

Ai Q, Xu H, Mai K, Xu W, Wang J, Zhang W (2011) Effects of dietary supplementation of *Bacillus subtilis* and fructooligosaccharide on growth performance, survival, non-specific immune response and disease resistance of juvenile large yellow croaker, *Larimichthys crocea*. Aquaculture 317(1–4):155–161. https://doi.org/10.1016/j.aquaculture.2011.04.036

Aizen MA, Garibaldi LA, Cunningham SA, Klein AM (2008) Long-term global trends in crop yield and production reveal no current pollination shortage but increasing pollinator dependency. Curr Biol 18(20):1572–1575. https://doi.org/10.1016/j.cub.2008.08.066

Akhter N, Wu B, Memon AM, Mohsin M (2015) Probiotics and prebiotics associated with aquaculture: a review. Fish Shellfish Immunol 45(2):733–741. https://doi.org/10.1016/j.fsi.2015.05.038

Alazzeh AY, Sultana H, Beauchemin KA, Wang Y, Holo H, Harstad OM, McAllister TA (2012) Using strains of Propionibacteria to mitigate methane emissions *in vitro*. Acta Agric Scand Sect A Anim Sci 62(4):263–272. https://doi.org/10.1080/09064702.2013.773056

Albino LA, Rostagno MH, Hungaro HM, Mendonca RC (2014) Isolation, characterization, and application of bacteriophages for *Salmonella* spp. biocontrol in pigs. Foodborne Pathog Dis 11(8):602–609. https://doi.org/10.1089/fpd.2013.1600

Allaart JG, van Asten AJAM, Gröne A (2013) Predisposing factors and prevention of *Clostridium perfringens*-associated enteritis. Comp Immunol Microbiol Infect Dis 36(5):449–464. https://doi.org/10.1016/j.cimid.2013.05.001

Aly SM, Ahmed AGY, Ghareebb AAA, Mohameda MF (2008) Studies on *Bacillus subtilis* and *Lactobacillus acidophilus*, as potential probiotics, on the immune response and resistance of Tilapia nilotica (*Oreochromis niloticus*) to challenge infections. Fish Shellfish Immunol 25(1–2):128–136. https://doi.org/10.1016/j.fsi.2008.03.013

Apanavicius CJ, Powell KL, Vester BM, Karr-Lilienthal LK, Pope LL, Fastinger ND, Wallig MA, Tappenden KA, Swanson KS (2007) Fructan supplementation and infection affect food intake, fever, and epithelial sloughing from *Salmonella* challenge in weanling puppies. J Nutr 137(8):1923–1930

Apas AL, Gonzalez SN, Arena ME (2014) Potential of goat probiotic to bind mutagens. Anaerobe 28:8–12. https://doi.org/10.1016/j.anaerobe.2014.04.004

Arslan H, Aksu D, Terzi G, Nisbet C (2012) Therapeutic effects of probiotic bacteria in parvoviral enteritis in dogs. Rev Med Vet-Toulouse 2(163):55–59

Atasoglu C, Akbag HI, Tolu C, Das G, Savas T, Yurtman IY (2010) Effects of kefir as a probiotic source on the performance of goat kids. S Afr J Anim Sci 40(4):363–370

Atterbury RJ, Van Bergen MA, Ortiz F, Lovell MA, Harris JA, De Boer A, Wagenaar JA, Allen VM, Barrow PA (2007) Bacteriophage therapy to reduce *Salmonella* colonization of broiler chickens. Appl Environ Microbiol 73(14):4543–4549. https://doi.org/10.1128/AEM.00049-07

Avella MA, Olivotto I, Silvi S, Place AR, Carnevali O (2010) Effect of dietary probiotics on clownfish: a molecular approach to define how lactic acid bacteria modulate development in a marine fish. Am J Physiol Regul Integr Comp Physiol 298(2):R359–R371

Ayad MA, Benallou B, Saim MS, Smadi MA, Meziane T (2013) Impact of feeding yeast culture on milk yield, milk components, and blood components in Algerian dairy herds. J Vet Sci Technol 4(2):1–5

Bairagi A, Ghosh KS, Sen SK, Ray AK (2002) Enzyme producing bacterial flora isolated from fish digestive tracts. Aquac Int 10(2):109–121. https://doi.org/10.1023/A:1021355406412

Balcázar JL, de Blas I, Ruiz-Zarzuela I, Cunningham D, Vendrell D, Múzquiz JL (2006) The role of probiotics in aquaculture. Vet Microbiol 114(3–4):173–186. https://doi.org/10.1016/j.vetmic.2006.01.009

Bardina C, Spricigo DA, Cortes P, Llagostera M (2012) Significance of the bacteriophage treatment schedule in reducing *Salmonella* colonization of poultry. Appl Environ Microbiol 78(18):6600–6607. https://doi.org/10.1128/aem.01257-12

Berg Miller ME, Yeoman CJ, Chia N, Tringe SG, Angly FE, Edwards RA, Flint HJ, Lamed R, Bayer EA, White BA (2012) Phage-bacteria relationships and CRISPR elements revealed by a metagenomic survey of the rumen microbiome. Environ Microbiol 14(1):207–227. https://doi.org/10.1111/j.1462-2920.2011.02593.x

Bernard JK (2015) Milk yield and composition of lactating dairy cows fed diets supplemented with a probiotic extract. Prof Anim Sci 31(4):354–358. 10.15232/pas.2015-01404

Biagi G, Cipollini I, Bonaldo A, Grandi M, Pompei A, Stefanelli C, Zaghini G (2013) Effect of feeding a selected combination of galacto-oligosaccharides and a strain of *Bifidobacterium pseudocatenulatum* on the intestinal microbiota of cats. Am J Vet Res 74(1):90–95. https://doi.org/10.2460/ajvr.74.1.90

Biggs P, Parsons CM, Fahey GC (2007) The effects of several oligosaccharides on growth performance, nutrient digestibilities, and cecal microbial populations in young chicks. Poult Sci 86(11):2327–2336. https://doi.org/10.3382/ps.2007-00427

Bindels LB, Delzenne NM, Cani PD, Walter J (2015) Towards a more comprehensive concept for prebiotics. Nat Rev Gastroenterol Hepatol 12(5):303–310. https://doi.org/10.1038/nrgastro.2015.47

Biourge V, Vallet C, Levesque A, Sergheraert R, Chevalier S, Roberton JL (1998) The use of probiotics in the diet of dogs. J Nutr 128(12 Suppl):2730s–2732s

Boutin S, Audet C, Derome N (2013) Probiotic treatment by indigenous bacteria decreases mortality without disturbing the natural microbiota of *Salvelinus fontinalis*. Can J Microbiol 59(10):662–670. https://doi.org/10.1139/cjm-2013-0443

Breytenbach JH (1999) Guidelines for effective vaccination of broilers. Int Poult Prod 13(7):7–9. http://www.positiveaction.info/pdfs/articles/pp13.7p7.pdf

Buddle BM, Denis M, Attwood GT, Altermann E, Janssen PH, Ronimus RS, Pinares-Patino CS, Muetzel S, Neil Wedlock D (2011) Strategies to reduce methane emissions from farmed ruminants grazing on pasture. Vet J 188(1):11–17. https://doi.org/10.1016/j.tvjl.2010.02.019

Bybee SN, Scorza AV, Lappin MR (2011) Effect of the probiotic *Enterococcus faecium* SF68 on presence of diarrhea in cats and dogs housed in an animal shelter. J Vet Intern Med 25(4):856–860. https://doi.org/10.1111/j.1939-1676.2011.0738.x

Bywater R, McConville M, Phillips I, Shiyock T (2005) The susceptibility to growth-promoting antibiotics of *Enterococcus faecium* isolates from pigs and chickens in Europe. J Antimicrob Chemother 56(3):538–543. https://doi.org/10.1093/jac/dki273

Callaway TR, Edrington TS, Brabban AD, Anderson RC, Rossman ML, Engler MJ, Carr MA, Genovese KJ, Keen JE, Looper ML, Kutter EM, Nisbet DJ (2008) Bacteriophage isolated from feedlot cattle can reduce *Escherichia coli* O157:H7 populations in ruminant gastrointestinal tracts. Foodborne Pathog Dis 5(2):183–191. https://doi.org/10.1089/fpd.2007.0057

Callaway TR, Edrington TS, Brabban A, Kutter B, Karriker L, Stahl C, Wagstrom E, Anderson R, Poole TL, Genovese K, Krueger N, Harvey R, Nisbet DJ (2011) Evaluation of phage treatment as a strategy to reduce *Salmonella* populations in growing swine. Foodborne Pathog Dis 8(2):261–266. https://doi.org/10.1089/fpd.2010.0671

Casewell M, Friis C, Marco E, McMullin P, Phillips I (2003) The European ban on growth-promoting antibiotics and emerging consequences for human and animal health. J Antimicrob Chemother 52(2):159–161. https://doi.org/10.1093/jac/dkg313

Cerezuela R, Meseguer J, Esteban MÁ (2011) Current knowledge in synbiotic use for fish aquaculture: a review. J Aquac Res Dev S1:008. https://doi.org/10.4172/2155-9546.S1-008

Cerezuela R, Fumanal M, Tapia-Paniagua ST, Meseguer J, Moriñigo MÁ, Esteban MÁ (2013) Changes in intestinal morphology and microbiota caused by dietary administration of inulin and *Bacillus subtilis* in gilthead sea bream (*Sparus aurata* L.) specimens. Fish Shellfish Immunol 34(5):1063–1070. https://doi.org/10.1016/j.fsi.2013.01.015

Chambers JR, Gong J (2011) The intestinal microbiota and its modulation for *Salmonella* control in chickens. Food Res Int 44(10):3149–3159. https://doi.org/10.1016/j.foodres.2011.08.017

Chaucheyras-Durand F, Durand H (2010) Probiotics in animal nutrition and health. Benef Microbes 1(1):3–9. https://doi.org/10.3920/bm2008.1002

Chaucheyras-Durand F, Ossa F (2014) Review: the rumen microbiome composition, abundance, diversity, and new investigative tools. Prof Anim Sci 30:1–12. 10.15232/S1080-7446(15)30076-0

Chiquette J, Allison MJ, Rasmussen MA (2008) *Prevotella bryantii* 25A used as a probiotic in early-lactation dairy cows: effect on ruminal fermentation characteristics, milk production, and milk composition. J Dairy Sci 91(9):3536–3543. https://doi.org/10.3168/jds.2007-0849

Choi KY, Lee TK, Sul WJ (2015) Metagenomic analysis of chicken gut microbiota for improving metabolism and health of chickens—a review. Asian Australas J Anim Sci 28(9):1217–1225. https://doi.org/10.5713/ajas.15.0026

Colom J, Cano-Sarabia M, Otero J, Cortes P, Maspoch D, Llagostera M (2015) Liposome-encapsulated bacteriophages for enhanced oral phage therapy against *Salmonella* spp. Appl Environ Microbiol 81(14):4841–4849. https://doi.org/10.1128/aem.00812-15

Cox-Foster DL, Conlan S, Holmes EC, Palacios G, Evans JD, Moran NA, Quan P-L, Briese T, Hornig M, Geiser DM, Martinson V, vanEngelsdorp D, Kalkstein AL, Drysdale A, Hui J, Zhai J, Cui L, Hutchison SK, Simons JF, Egholm M, Pettis JS, Lipkin WI (2007) A metagenomic survey of microbes in honey bee colony collapse disorder. Science 318(5848):283–287. https://doi.org/10.1126/science.1146498

Daly K, Shirazi-Beechey SP (2003) Design and evaluation of group-specific oligonucleotide probes for quantitative analysis of intestinal ecosystems: their application to assessment of equine colonic microflora. FEMS Microbiol Ecol 44(2):243–252. https://doi.org/10.1016/S0168-6496(03)00032-1

Daly K, Stewart CS, Flint HJ, Shirazi-Beechey SP (2001) Bacterial diversity within the equine large intestine as revealed by molecular analysis of cloned 16S rRNA genes. FEMS Microbiol Ecol 38(2–3):141–151. https://doi.org/10.1016/S0168-6496(01)00178-7

Daniels C, Hoseinifar SH (2014) Prebiotic applications in shellfish. In: Merrifield D, Ringø E (eds) Aquaculture nutrition: gut health, probiotics and prebiotics. Wiley, Chichester, pp 401–418. https://doi.org/10.1002/9781118897263.ch15

Danzeisen JL, Kim HB, Isaacson RE, Tu ZJ, Johnson TJ (2011) Modulations of the chicken cecal microbiome and metagenome in response to anticoccidial and growth promoter treatment. PLoS One 6(11):e27949. https://doi.org/10.1371/journal.pone.0027949

De BC, Meena DK, Behera BK, Das P, Das Mohapatra PK, Sharma AP (2014) Probiotics in fish and shellfish culture: immunomodulatory and ecophysiological responses. Fish Physiol Biochem 40(3):921–971. https://doi.org/10.1007/s10695-013-9897-0

Deusch S, Tilocca B, Camarinha-Silva A, Seifert J (2015) News in livestock research—use of omics-technologies to study the microbiota in the gastrointestinal tract of farm animals. Comput Struct Biotechnol J 13:55–63. https://doi.org/10.1016/j.csbj.2014.12.005

Dhama K, Mahendran M, Tomar S, Chauhan RS (2008) Beneficial effects of probiotics and prebiotics in livestock and poultry: the current perspectives. Intas Polivet 9(1):1–12

Dimitroglou A, Merrifield DL, Carnevali O, Picchietti S, Avella M, Daniels C, Güroy D, Davies SJ (2011) Microbial manipulations to improve fish health and production—a Mediterranean perspective. Fish Shellfish Immunol 30(1):1–16. https://doi.org/10.1016/j.fsi.2010.08.009

Drouault S, Juste C, Marteau P, Renault P, Corthier G (2002) Oral treatment with *Lactococcus lactis* expressing *Staphylococcus hyicus* lipase enhances lipid digestion in pigs with induced pancreatic insufficiency. Appl Environ Microbiol 68(6):3166–3168. https://doi.org/10.1128/AEM.68.6.3166-3168.2002

El-Gohary FA, Huff WE, Huff GR, Rath NC, Zhou ZY, Donoghue AM (2014) Environmental augmentation with bacteriophage prevents colibacillosis in broiler chickens. Poult Sci 93(11):2788–2792. https://doi.org/10.3382/ps.2014-04282

Espeche MC, Pellegrino M, Frola I, Larriestra A, Bogni C, Nader-Macias ME (2012) Lactic acid bacteria from raw milk as potentially beneficial strains to prevent bovine mastitis. Anaerobe 18(1):103–109. https://doi.org/10.1016/j.anaerobe.2012.01.002

European Parliament and Council (2003) Regulation (EC) No 1831/2003 of the European parliament and of the council of 22 September 2003 on additives for use in animal nutrition. Off. J Eur Union L268:29–43

Evans JD, Lopez DL (2004) Bacterial probiotics induce an immune response in the honey bee (Hymenoptera: Apidae). J Econ Entomol 97(3):752–756

Evans JD, Schwarz RS (2011) Bees brought to their knees: microbes affecting honey bee health. Trends Microbiol 19(12):614–620. https://doi.org/10.1016/j.tim.2011.09.003

Ezema C (2013) Probiotics in animal production: a review. J Vet Med Anim Health 5(11):308–316. https://doi.org/10.5897/JVMAH2013.0201

Fahey GC, Vester B (2009) Prebiotics and probiotics in companion animal nutrition. In: Cho SS, Finocchiaro ET (eds) Handbook of prebiotics and probiotics ingredients. CRC Press, Boca Raton, pp 355–381. https://doi.org/10.1201/9781420062151.ch17

FAO (2012) The state of world fisheries and aquaculture. Food and Agriculture Organization of the United Nations, Fisheries and Aquaculture Department, Rome. http://www.fao.org/3/a-i2727e.pdf

FAO (2016) Probiotics in animal nutrition. Production, impact and regulation. FAO Animal Production and Health Paper No. 179 Rome. http://www.fao.org/3/a-i5933e.pdf

FAO/WHO (2001) Probiotics in food. Health and nutritional properties and guidelines for evaluation. Report of a joint FAO/WHO expert consultation on evaluation of health and nutritional properties of probiotics in food including powder milk with live lactic acid bacteria; Cordoba, Argentina. http://www.who.int/foodsafety/publications/fs_-management/en/probiotics.pdf

FAO/WHO (2002) Guidelines for the evaluation of probiotics in food. Report of a joint FAO/WHO working group on drafting guidelines for the evaluation of probiotics in food; London. http://www.who.int/foodsafety/fs_management/en/probiotic_guidelines.pdf

FDA (2015) Compliance Policy Guides Sec. 689.100 Direct-Fed Microbial Products. https://www.fda.gov/iceci/compliancemanuals/compliancepolicyguidancemanual/ucm074707.htm

Fraga M, Perelmuter K, Valencia M, Martínez M, Abin-Carriquiry A, Cajarville C, Zunino P (2014) Evaluation of native potential probiotic bacteria using an *in vitro* ruminal fermentation system. Ann Microbiol 64(3):1149–1156. https://doi.org/10.1007/s13213-013-0753-3

Gaggia F, Mattarelli P, Biavati B (2010) Probiotics and prebiotics in animal feeding for safe food production. Int J Food Microbiol 141(Suppl 1):S15–S28. https://doi.org/10.1016/j.ijfoodmicro.2010.02.031

Ganguly S, Dora KC, Sarkar S, Chowdhury S (2013) Supplementation of prebiotics in fish feed: a review. Rev Fish Biol Fish 23(2):195–199. https://doi.org/10.1007/s11160-012-9291-5

Gatesoupe FJ (2008) Updating the importance of lactic acid bacteria in fish farming: natural occurrence and probiotic treatments. J Mol Microbiol Biotechnol 14(1–3):107–114. https://doi.org/10.1159/000106089

Gatlin DM, Barrows FT, Brown P, Dabrowski K, Gaylord TG, Hardy RW, Herman E, Hu G, Krogdahl Å, Nelson R, Overturf K, Rust M, Sealey W, Skonberg D, Souza EJ, Stone D, Wilson R, Wurtele E (2007) Expanding the utilization of sustainable plant products in aquafeeds: a review. Aquac Res 38(6):551–579. https://doi.org/10.1111/j.1365-2109.2007.01704.x

Gebru E, Lee JS, Son JC, Yang SY, Shin SA, Kim B, Kim MK, Park SC (2010) Effect of probiotic-, bacteriophage-, or organic acid-supplemented feeds or fermented soybean meal on the growth performance, acute-phase response, and bacterial shedding of grower pigs challenged with *Salmonella enterica* serotype Typhimurium. J Anim Sci 88(12):3880–3886. https://doi.org/10.2527/jas.2010-2939

Genersch E (2010) Honey bee pathology: current threats to honey bees and beekeeping. Appl Microbiol Biotechnol 87(1):87–97. https://doi.org/10.1007/s00253-010-2573-8

Geng X, Dong X-H, Tan B-P, Yang Q-H, Chi S-Y, Liu H-Y, Liu X-Q (2011) Effects of dietary chitosan and *Bacillus subtilis* on the growth performance, non-specific immunity and disease resistance of cobia, *Rachycentron canadum*. Fish Shellfish Immunol 31(3):400–406. https://doi.org/10.1016/j.fsi.2011.06.006

Gibson GR, Roberfroid MB (1995) Dietary modulation of the human colonic microbiota: introducing the concept of prebiotics. J Nutr 125(6):1401–1412

Giri SS, Sukumaran V, Oviya M (2013) Potential probiotic *Lactobacillus plantarum* VSG3 improves the growth, immunity, and disease resistance of tropical freshwater fish, *Labeo rohita*. Fish Shellfish Immunol 34(2):660–666. https://doi.org/10.1016/j.fsi.2012.12.008

Gogineni VK, Morrow LE, Gregory PJ, Malesker MA (2013) Probiotics: history and evolution. J Anc Dis Prev Rem 1:1–7. https://doi.org/10.4172/2329-8731.1000107

Gómez-Gallego C, Junnila J, Männikkö S, Hämeenoja P, Valtonen E, Salminen S, Beasley S (2016) A canine-specific probiotic product in treating acute or intermittent diarrhea in dogs: a double-blind placebo-controlled efficacy study. Vet Microbiol 197:122–128. https://doi.org/10.1016/j.vetmic.2016.11.015

González-Forte L, Bruno E, Martino M (2014) Application of coating on dog biscuits for extended survival of probiotic bacteria. Anim Feed Sci Technol 195:76–84. https://doi.org/10.1016/j.anifeedsci.2014.05.015

Gonzalez-Ortiz G, Castillejos L, Mallo JJ, Angels Calvo-Torras M, Dolores Baucells M (2013) Effects of dietary supplementation of *Bacillus amyloliquefaciens* CECT 5940 and *Enterococcus faecium* CECT 4515 in adult healthy dogs. Arch Anim Nutr 67(5):406–415. https://doi.org/10.1080/1745039x.2013.830517

Goodridge LD (2010) Application of bacteriophages to control pathogens in food animal production. In: Sabour PM, Griffiths M (eds) Bacteriophages in the control of food-and waterborne pathogens. ASM Press, Washington, DC, pp 61–77. https://doi.org/10.1128/9781555816629.ch4

Gouveia EM, Silva IS, Onselem VJ, Correa RA, Silva CJ (2006) Use of mannanoligosacharides as an adjuvant treatment for gastrointestinal diseases and their effects on *E. coli* inactivated in dogs. Acta Cir Bras 21(Suppl 4):23–26. https://doi.org/10.1590/S0102-86502006001000006

Gregg K, Cooper CL, Schafer DJ, Sharpe H, Beard CE, Allen G, Xu J (1994) Detoxification of the plant toxin fluoroacetate by a genetically modified rumen bacterium. Biotechnology 12(13):1361–1365

Gregg K, Hamdorf B, Henderson K, Kopecny J, Wong C (1998) Genetically modified ruminal bacteria protect sheep from fluoroacetate poisoning. Appl Environ Microbiol 64(9): 3496–3498

Grzeskowiak L, Endo A, Beasley S, Salminen S (2015) Microbiota and probiotics in canine and feline welfare. Anaerobe 34:14–23. https://doi.org/10.1016/j.anaerobe.2015.04.002

Guyader J, Eugene M, Noziere P, Morgavi DP, Doreau M, Martin C (2014) Influence of rumen protozoa on methane emission in ruminants: a meta-analysis approach. Animal 8(11):1816–1825. https://doi.org/10.1017/s1751731114001852

Hajati H, Rezaei M (2010) The application of prebiotics in poultry production. Int J Poult Sci 9(3):298–304. https://doi.org/10.3923/ijps.2010.298.304

Hallewell J, Niu YD, Munns K, McAllister TA, Johnson RP, Ackermann HW, Thomas JE, Stanford K (2014) Differing populations of endemic bacteriophages in cattle shedding high and low numbers of *Escherichia coli* O157:H7 bacteria in feces. Appl Environ Microbiol 80(13):3819–3825. https://doi.org/10.1128/aem.00708-14

Han S, Liu Y, Zhou Z, He S, Cao Y, Shi P, Yao B, Ringø E (2010) Analysis of bacterial diversity in the intestine of grass carp (*Ctenopharyngodon idellus*) based on 16S rDNA gene sequences. Aquac Res 42(1):47–56. https://doi.org/10.1111/j.1365-2109.2010.02543.x

Handl S, Dowd SE, Garcia-Mazcorro JF, Steiner JM, Suchodolski JS (2011) Massive parallel 16S rRNA gene pyrosequencing reveals highly diverse fecal bacterial and fungal communities in healthy dogs and cats. FEMS Microbiol Ecol 76(2):301–310. https://doi.org/10.1111/j.1574-6941.2011.01058.x

Hanley F, Brown H, Carbery J (1995) First observations on the effects of mannan oligosaccharide added to hatchery diets for warmwater hybrid red tilapia. In: Proceedings of the 11th annual symposium on nutritional biotechnology in the feed and food industries, Lexington

Hastie PM, Mitchell K, Murray JA (2008) Semi-quantitative analysis of *Ruminococcus flavefaciens*, *Fibrobacter succinogenes* and *Streptococcus bovis* in the equine large intestine using real-time polymerase chain reaction. Br J Nutr 100(3):561–568. https://doi.org/10.1017/S0007114508968227

Hill C, Guarner F, Reid G, Gibson GR, Merenstein DJ, Pot B, Morelli L, Canani RB, Flint HJ, Salminen S, Calder PC, Sanders ME (2014) Expert consensus document. The International Scientific Association for Probiotics and Prebiotics consensus statement on the scope and appropriate use of the term probiotic. Nat Rev Gastroenterol Hepatol 11(8):506–514. https://doi.org/10.1038/nrgastro.2014.66

Hofmann RR (1989) Evolutionary steps of ecophysiological adaptation and diversification of ruminants: a comparative view of their digestive system. Oecologia 78(4):443–457. https://doi.org/10.1007/BF00378733

Hoseinifar SH, Zare P, Merrifield DL (2010) The effects of inulin on growth factors and survival of the Indian white shrimp larvae and postlarvae (*Fenneropenaeus indicus*). Aquac Res 41(9):e348–e352. https://doi.org/10.1111/j.1365-2109.2010.02485.x

Hou C, Zeng X, Yang F, Liu H, Qiao S (2015) Study and use of the probiotic *Lactobacillus reuteri* in pigs: a review. J Anim Sci Biotechnol 6(1):14. https://doi.org/10.1186/s40104-015-0014-3

Hutchins RG, Bailey CS, Jacob ME, Harris TL, Wood MW, Saker KE, Vaden SL (2013) The effect of an oral probiotic containing *Lactobacillus*, *Bifidobacterium*, and *Bacillus* species on the vaginal microbiota of spayed female dogs. J Vet Intern Med 27(6):1368–1371. https://doi.org/10.1111/jvim.12174

Hyrsl P, Dobes P, Vojtek L, Berka J, Hurychova J, Havlik J, Kamler M, Hroncova Z, Killer J, Tyl J, Titera D (2015) Plant alcaloid and probiotics increase resistance of honeybees to nematobacterial infection. In: 15th meeting of the IOBCWPRS working group "Microbial and Nematode Control of Invertebrathe Pests", University of Massaryk, Czech Republic

Iehata S, Valenzuela F, Riquelme C (2015) Analysis of bacterial community and bacterial nutritional enzyme activity associated with the digestive tract of wild Chilean octopus (*Octopus mimus* Gould, 1852). Aquac Res 46(4):861–873. https://doi.org/10.1111/are.12240

Irianto A, Austin B (2002) Probiotics in aquaculture. J Fish Dis 25(11):633–642. https://doi.org/10.1046/j.1365-2761.2002.00422.x

Isaacson R, Kim HB (2012) The intestinal microbiome of the pig. Anim Health Res Rev 13(1):100–109. https://doi.org/10.1017/S1466252312000084

Jami E, White BA, Mizrahi I (2014) Potential role of the bovine rumen microbiome in modulating milk composition and feed efficiency. PLoS One 9(1):e85423. https://doi.org/10.1371/journal.pone.0085423

Jenkins TA, Jenkins V (2014) Ruminant response to non-live probiotic microorganism extracts. Proc N Z Soc Anim Prod 74:148–153

Jeyanathan J, Martin C, Morgavi DP (2014) The use of direct-fed microbials for mitigation of ruminant methane emissions: a review. Animal 8(2):250–261. https://doi.org/10.1017/s1751731113002085

Johnson BR, Synk W, Jasper WC, Müssen E (2014) Effects of high fructose corn syrup and probiotics on growth rates of newly founded honey bee colonies. J Apic Res 53(1):165–170. https://doi.org/10.3896/IBRA.1.53.1.18

Jose N, Bunt C, Hussain M (2015) Comparison of microbiological and probiotic characteristics of Lactobacilli isolates from dairy food products and animal rumen contents. Microorganisms 3(2):198. https://doi.org/10.3390/microorganisms3020198

Karakurt I, Aydin G, Aydiner K (2012) Sources and mitigation of methane emissions by sectors: a critical review. Renew Energy 39(1):40–48. https://doi.org/10.1016/j.renene.2011.09.006

Kazimierczak-Baryczko M, Szymaś B (2006) Improvement of the composition of pollen substitute for honey bee (*Apis mellifera* L.), through implementation of probiotic preparations. J Apic Sci 50(1):15–23

Kaznowski A, Szymas B, Jazdzinska E, Kazimierczak M, Paetz H, Mokracka J (2005) The effects of probiotic supplementation on the content of intestinal microflora and chemical composition of worker honey bees (*Apis mellifera*). J Apic Res 44(1):10–14. https://doi.org/10.1080/00218839.2005.11101139

Kesarcodi-Watson A, Kaspar H, Lategan MJ, Gibson L (2008) Probiotics in aquaculture: the need, principles and mechanisms of action and screening processes. Aquaculture 274(1):1–14. https://doi.org/10.1016/j.aquaculture.2007.11.019

Kim HB, Borewicz K, White BA, Singer RS, Sreevatsan S, Tu ZJ, Isaacson RE (2011a) Longitudinal investigation of the age-related bacterial diversity in the feces of commercial pigs. Vet Microbiol 153(1–2):124–133. https://doi.org/10.1016/j.vetmic.2011.05.021

Kim M, Morrison M, Yu Z (2011b) Status of the phylogenetic diversity census of ruminal microbiomes. FEMS Microbiol Ecol 76(1):49–63. https://doi.org/10.1111/j.1574-6941.2010.01029.x

Kim JH, Kim JW, Lee BB, Lee GI, Lee JH, Kim GB, Kil DY (2014a) Effect of dietary supplementation of bacteriophage on growth performance and cecal bacterial populations in broiler chickens raised in different housing systems. Livest Sci 170:137–141. https://doi.org/10.1016/j.livsci.2014.09.005

Kim KH, Ingale SL, Kim JS, Lee SH, Lee JH, Kwon IK, Chae BJ (2014b) Bacteriophage and probiotics both enhance the performance of growing pigs but bacteriophage are more effective. Anim Feed Sci Technol 196:88–95. https://doi.org/10.1016/j.anifeedsci.2014.06.012

Klein A-M, Vaissière BE, Cane JH, Steffan-Dewenter I, Cunningham SA, Kremen C, Tscharntke T (2007) Importance of pollinators in changing landscapes for world crops. Proc Biol Sci 274(1608):303–313

Koch H, Schmid-Hempel P (2011) Bacterial communities in central European bumblebees: low diversity and high specificity. Microb Ecol 62(1):121–133. https://doi.org/10.1007/s00248-011-9854-3

Kozasa M (1986) Toyocerin (*Bacillus toyoi*) as growth promotor for animal feeding. Microbiol Aliment Nutr 4:121–135

Krause DO, Bunch RJ, Dalrymple BD, Gobius KS, Smith WJ, Xue GP, McSweeney CS (2001) Expression of a modified *Neocallimastix patriciarum* xylanase in *Butyrivibrio fibrisolvens* digests more fibre but cannot effectively compete with highly fibrolytic bacteria in the rumen. J Appl Microbiol 90(3):388–396. https://doi.org/10.1046/j.1365-2672.2001.01257.x

Krause DO, Nagaraja TG, Wright AD, Callaway TR (2013) Board-invited review: rumen microbiology: leading the way in microbial ecology. J Anim Sci 91(1):331–341. https://doi.org/10.2527/jas.2012-5567

Kropinski AM, Lingohr EJ, Moyles DM, Ojha S, Mazzocco A, She YM, Bach SJ, Rozema EA, Stanford K, McAllister TA, Johnson RP (2012) Endemic bacteriophages: a cautionary tale for evaluation of bacteriophage therapy and other interventions for infection control in animals. Virol J 9:207. https://doi.org/10.1186/1743-422x-9-207

Kumar B, Sirohi SK (2013) Effect of isolate of ruminal fibrolytic bacterial culture supplementation on fibrolytic bacterial population and survivability of inoculated bacterial strain in lactating Murrah buffaloes. Vet World 6(1):14–17. https://doi.org/10.5455/vetworld.2013.14-17

Kumar S, Pattanaik AK, Sharma S, Jadhav SE, Dutta N, Kumar A (2017) Probiotic potential of a Lactobacillus bacterium of canine faecal-origin and its impact on select gut health indices and immune response of dogs. Probiotics Antimicrob Proteins, 9(3):262–277. https://doi.org/10.1007/s12602-017-9256-z

Lamendella R, Domingo JW, Ghosh S, Martinson J, Oerther DB (2011) Comparative fecal metagenomics unveils unique functional capacity of the swine gut. BMC Microbiol 11:103. https://doi.org/10.1186/1471-2180-11-103

Landers TF, Cohen B, Wittum TE, Larson EL (2012) A review of antibiotic use in food animals: perspective, policy, and potential. Public Health Rep 127(1):4–22. https://doi.org/10.1177/003335491212700103

Lau GL, Sieo CC, Tan WS, Hair-Bejo M, Jalila A, Ho YW (2010) Efficacy of a bacteriophage isolated from chickens as a therapeutic agent for colibacillosis in broiler chickens. Poult Sci 89(12):2589–2596. https://doi.org/10.3382/ps.2010-00904

Lazado CC, Caipang CMA (2014) Mucosal immunity and probiotics in fish. Fish Shellfish Immunol 39(1):78–89. https://doi.org/10.1016/j.fsi.2014.04.015

Lazado CC, Caipang CMA, Estante EG (2015) Prospects of host-associated microorganisms in fish and penaeids as probiotics with immunomodulatory functions. Fish Shellfish Immunol 45(1):2–12. https://doi.org/10.1016/j.fsi.2015.02.023

Leahy SC, Kelly WJ, Ronimus RS, Wedlock N, Altermann E, Attwood GT (2013) Genome sequencing of rumen bacteria and archaea and its application to methane mitigation strategies. Animal 7:235–243. https://doi.org/10.1017/s1751731113000700

Lee WJ, Hase K (2014) Gut microbiota-generated metabolites in animal health and disease. Nat Chem Biol 10(6):416–424. https://doi.org/10.1038/nchembio.1535

Li LB, Xu QQ, Yang CJ, Yang X, Lv L, Yin CH, Liu XL, Yan H (2014) Effects of probiotics on the growth performance and intestinal microflora of broiler chickens. Pak J Pharm Sci 27(3):713–717

Liu JR, Yu B, Liu FH, Cheng KJ, Zhao X (2005) Expression of rumen microbial fibrolytic enzyme genes in probiotic *Lactobacillus reuteri*. Appl Environ Microbiol 71(11):6769–6775. https://doi.org/10.1128/AEM.71.11.6769-6775.2005

Liu JR, Lai SF, Yu B (2007) Evaluation of an intestinal *Lactobacillus reuteri* strain expressing rumen fungal xylanase as a probiotic for broiler chickens fed on a wheat-based diet. Br Poult Sci 48(4):507–514. https://doi.org/10.1080/00071660701485034

Llewellyn MS, Boutin S, Hoseinifar SH, Derome N (2014) Teleost microbiomes: the state of the art in their characterization, manipulation and importance in aquaculture and fisheries. Front Microbiol 5:207. https://doi.org/10.3389/fmicb.2014.00207

Loc Carrillo C, Atterbury RJ, El-Shibiny A, Connerton PL, Dillon E, Scott A, Connerton IF (2005) Bacteriophage therapy to reduce *Campylobacter jejuni* colonization of broiler chickens. Appl Environ Microbiol 71(11):6554–6563. https://doi.org/10.1128/aem.71.11.6554-6563.2005

Ma Y, Pacan JC, Wang Q, Xu Y, Huang X, Korenevsky A, Sabour PM (2008) Microencapsulation of bacteriophage Felix O1 into chitosan-alginate microspheres for oral delivery. Appl Environ Microbiol 74(15):4799–4805. https://doi.org/10.1128/AEM.00246-08

Ma Y, Xing Y, Xu Q, Wang T, Cai Y, Cao D, Che Z (2015) Preparation and application characteristics of microencapsulated *Lactobacillus acidophilus* as probiotics for dogs. Pak J Pharm Sci 28(Suppl 1):341–347

Machova M, Rada V, Huk J, Smekal F (1997) Entwicklung der bienenprobiotik. Apiacta 32:99–111
Madhusudana Rao B, Lalitha KV (2015) Bacteriophages for aquaculture: are they beneficial or inimical. Aquaculture 437:146–154. https://doi.org/10.1016/j.aquaculture.2014.11.039
Magnadottir B (2010) Immunological control of fish diseases. Mar Biotechnol 12(4):361–379. https://doi.org/10.1007/s10126-010-9279-x
Mannetje L (2010) Silage for animal feed. In: Encyclopedia of life support systems [Online], vol 8. Eolss Publishers, Oxford, pp 123–135. http://www.eolss.net/sample-chapters/c17/E6-58-07-05.pdf
Manninen TJK, Rinkinen ML, Beasley SS, Saris PEJ (2006) Alteration of the canine small-intestinal lactic acid bacterium microbiota by feeding of potential probiotics. Appl Environ Microbiol 72(10):6539–6543. https://doi.org/10.1128/AEM.02977-05
Maragkoudakis PA, Mountzouris KC, Rosu C, Zoumpopoulou G, Papadimitriou K, Dalaka E, Hadjipetrou A, Theofanous G, Strozzi GP, Carlini N, Zervas G, Tsakalidou E (2010) Feed supplementation of *Lactobacillus plantarum* PCA 236 modulates gut microbiota and milk fatty acid composition in dairy goats-a preliminary study. Int J Food Microbiol 141(Suppl 1):S109–S116. https://doi.org/10.1016/j.ijfoodmicro.2010.03.007
Marietto-Gonçalves G, Curotto SMR, Baptista A, Donato T, Takahira R, Sequeira J, Andreatti Filho R (2014) Effects of lactobacillus probiotic, P22 bacteriophage and *Salmonella typhimurium* on the heterophilic burst activity of broiler chickens. Rev Bras Cienc Avic 16:257–264. https://doi.org/10.1590/1516-635x1603257-264
Market Report (2016) Probiotics in animal feed market by bacteria (Lactobacilli, *Streptococcus* Thermophiles, and Bifidobacteria), livestock (cattle, poultry, swine, and aquaculture), form (dry and liquid), function (yield, immunity, and productivity), and by region—global forecast to 2021 MarketsandMarkets Research Private Ltd., Pune. Report code: AGI 3616. http://www.marketsandmarkets.com/Market-Reports/probiotics-animal-feed-market-85832335.html
Market Research Report (2016) Probiotics market analysis by application (probiotic foods & beverages (dairy products, non-dairy products, cereals, baked food, fermented meat products, dry food), probiotic dietary supplements (food supplements, nutritional supplements, specialty nutrients, infant formula), animal feed probiotics), by end-use (human probiotics, animal probiotics) and segment forecast to 2024. Grand View Research Inc., San Francisco. Report ID: 978-1-68038-093-4. http://www.grandviewresearch.com/industry-analysis/probiotics-market
Marsella R, Santoro D, Ahrens K (2012) Early exposure to probiotics in a canine model of atopic dermatitis has long-term clinical and immunological effects. Vet Immunol Immunopathol 146(2):185–189. https://doi.org/10.1016/j.vetimm.2012.02.013
Martin C, Morgavi DP, Doreau M (2010) Methane mitigation in ruminants: from microbe to the farm scale. Animal 4(3):351–365. https://doi.org/10.1017/s1751731109990620
Martínez Cruz P, Ibáñez AL, Monroy Hermosillo OA, Ramírez Saad HC (2012) Use of probiotics in aquaculture. ISRN Microbiol 2012:916845. https://doi.org/10.5402/2012/916845
Martinson VG, Danforth BN, Minckley RL, Rueppell O, Tingek S, Moran NA (2011) A simple and distinctive microbiota associated with honey bees and bumble bees. Mol Ecol 20(3):619–628. https://doi.org/10.1111/j.1365-294X.2010.04959.x
Martinson VG, Moy J, Moran NA (2012) Establishment of characteristic gut bacteria during development of the honeybee worker. Appl Environ Microbiol 78(8):2830–2840. https://doi.org/10.1128/aem.07810-11
Mattila HR, Rios D, Walker-Sperling VE, Roeselers G, Newton IL (2012) Characterization of the active microbiotas associated with honey bees reveals healthier and broader communities when colonies are genetically diverse. PLoS One 7(3):e32962. https://doi.org/10.1371/journal.pone.0032962
McIntosh D, Ji B, Forward BS, Puvanendran V, Boyce D, Ritchie R (2008) Culture-independent characterization of the bacterial populations associated with cod (*Gadus morhua* L.) and live feed at an experimental hatchery facility using denaturing gradient gel electrophoresis. Aquaculture 275(1–4):42–50. https://doi.org/10.1016/j.aquaculture.2007.12.021
Mehrabi Z, Firouzbakhsh F, Jafarpour A (2012) Effects of dietary supplementation of synbiotic on growth performance, serum biochemical parameters and carcass composition in rainbow

trout (*Oncorhynchus mykiss*) fingerlings. J Anim Physiol Anim Nutr 96(3):474–481. https://doi.org/10.1111/j.1439-0396.2011.01167.x

Metchnikoff E (1908) The prolongation of life; optimistic studies. English translation by Mitchell Chalmers P (ed), G. P. Putnam's Sons New York & London, The Knickerbocker Press

Middelbos IS, Godoy MR, Fastinger ND, Fahey GC Jr (2007) A dose-response evaluation of spray-dried yeast cell wall supplementation of diets fed to adult dogs: effects on nutrient digestibility, immune indices, and fecal microbial populations. J Anim Sci 85(11):3022–3032. https://doi.org/10.2527/jas.2007-0079

Mills S, Shanahan F, Stanton C, Hill C, Coffey A, Ross RP (2013) Movers and shakers: influence of bacteriophages in shaping the mammalian gut microbiota. Gut Microbes 4(1):4–16. https://doi.org/10.4161/gmic.22371

Mohd Shaufi MA, Sieo CC, Chong CW, Gan HM, Ho YW (2015) Deciphering chicken gut microbial dynamics based on high-throughput 16S rRNA metagenomics analyses. Gut Pathog 7:4. https://doi.org/10.1186/s13099-015-0051-7

Morgavi DP, Kelly WJ, Janssen PH, Attwood GT (2013) Rumen microbial (meta)genomics and its application to ruminant production. Animal 7(Suppl 1):184–201. https://doi.org/10.1017/S1751731112000419

Mulero I, García-Ayala A, Meseguer J, Mulero V (2007) Maternal transfer of immunity and ontogeny of autologous immunocompetence of fish: a minireview. Aquaculture 268(1–4):244–250. https://doi.org/10.1016/j.aquaculture.2007.04.046

Musikasang H, Tani A, H-kittikun A, Maneerat S (2009) Probiotic potential of lactic acid bacteria isolated from chicken gastrointestinal digestive tract. World J Microbiol Biotechnol 25(8):1337–1345. https://doi.org/10.1007/s11274-009-0020-8

Mustapha A, Kerley M, Ahn J (2013) Microencapsulated probiotics for reducing fecal shedding of pathogenic microbes in animals. US Patent 20,130,115,328 A1, 9 May 2013

Nader-Macias ME, Otero MC, Espeche MC, Maldonado NC (2008) Advances in the design of probiotic products for the prevention of major diseases in dairy cattle. J Ind Microbiol Biotechnol 35(11):1387–1395. https://doi.org/10.1007/s10295-008-0438-2

Nagaraja T (2012) A microbiologist's view on improving nutrient utilization in ruminants. In: 23rd annual ruminant nutrition symposium, Department of Animal Sciences, University of Florida, Gainesville

Nakai T, Park SC (2002) Bacteriophage therapy of infectious diseases in aquaculture. Res Microbiol 153(1):13–18. https://doi.org/10.1016/S0923-2508(01)01280-3

Navarrete P, Toledo I, Mardones P, Opazo R, Espejo R, Romero J (2010) Effect of *Thymus vulgaris* essential oil on intestinal bacterial microbiota of rainbow trout, *Oncorhynchus mykiss* (Walbaum) and bacterial isolates. Aquac Res 41(10):e667–e678. https://doi.org/10.1111/j.1365-2109.2010.02590.x

Nayak SK (2010) Probiotics and immunity: a fish perspective. Fish Shellfish Immunol 29(1):2–14. https://doi.org/10.1016/j.fsi.2010.02.017

Ogue-Bon E, Khoo C, McCartney AL, Gibson GR, Rastall RA (2010) *In vitro* effects of synbiotic fermentation on the canine faecal microbiota. FEMS Microbiol Ecol 73(3):587–600. https://doi.org/10.1111/j.1574-6941.2010.00915.x

Ohshima-Terada Y, Higuchi Y, Kumagai T, Hagihara A, Nagata M (2015) Complementary effect of oral administration of *Lactobacillus paracasei* K71 on canine atopic dermatitis. Vet Dermatol 26(5):350–353, e374–355. https://doi.org/10.1111/vde.12224

Oliveira A, Sereno R, Azeredo J (2010) *In vivo* efficiency evaluation of a phage cocktail in controlling severe colibacillosis in confined conditions and experimental poultry houses. Vet Microbiol 146(3–4):303–308. https://doi.org/10.1016/j.vetmic.2010.05.015

Oliveira J, Castilho F, Cunha A, Pereira MJ (2012) Bacteriophage therapy as a bacterial control strategy in aquaculture. Aquac Int 20(5):879–910. https://doi.org/10.1007/s10499-012-9515-7

Olofsson TC, Vasquez A (2008) Detection and identification of a novel lactic acid bacterial flora within the honey stomach of the honeybee *Apis mellifera*. Curr Microbiol 57(4):356–363. https://doi.org/10.1007/s00284-008-9202-0

Ortiz LT, Rodriguez ML, Alzueta C, Rebole A, Trevino J (2009) Effect of inulin on growth performance, intestinal tract sizes, mineral retention and tibial bone mineralisation in broiler chickens. Br Poult Sci 50(3):325–332. https://doi.org/10.1080/00071660902806962

Ozen M, Dinleyici EC (2015) The history of probiotics: the untold story. Benef Microbes 6(2):159–165. https://doi.org/10.3920/BM2014.0103

Padmanabha J, Gregg K, McSweeney CS, Prideaux C, Ford M (2004) Protection of cattle from fluoroacetate poisoning by genetically modified ruminal bacteria. Anim Prod Aust 25:293. https://doi.org/10.1071/SA0401138

Pan D, Yu Z (2014) Intestinal microbiome of poultry and its interaction with host and diet. Gut Microbes 5(1):108–119. https://doi.org/10.4161/gmic.26945

Pandey PK, Bharti V, Kumar K (2014) Biofilm in aquaculture production. Afr J Microbiol Res 8(13):1434–1443. https://doi.org/10.5897/AJMR2013.6445

Pandiyan P, Balaraman D, Thirunavukkarasu R, George EGJ, Subaramaniyan K, Manikkam S, Sadayappan B (2013) Probiotics in aquaculture. Drug Invent Today 5(1):55–59. https://doi.org/10.1016/j.dit.2013.03.003

Papatsiros VG, Katsoulos PD, Koutoulis KC, Karatzia M, Dedousi A, Christodoulopoulos G (2013) Alternatives to antibiotics for farm animals. CAB Rev 8:1–15. https://doi.org/10.1079/PAVSNNR20138032

Park SH, Hanning I, Perrota A, Bench BJ, Alm E, Ricke SC (2013) Modifying the gastrointestinal ecology in alternatively raised poultry and the potential for molecular and metabolomic assessment. Poult Sci 92(2):546–561. https://doi.org/10.3382/ps.2012-02734

Patruica S, Mot D (2012) The effect of using prebiotic and probiotic products on intestinal microflora of the honeybee (*Apis mellifera carpatica*). Bull Entomol Res 102(6):619–623. https://doi.org/10.1017/s0007485312000144

Patruica S, Bogdan AT, Bura M, Banatean-Dunea I, Popovici D (2011a) Research on the influence of probiotic products on bee colonies development in spring. Bull UASVM Anim Sci Biotechnol 68:451–456

Patruica S, Bogdan AT, Bura M, Popovici D (2011b) Evaluating the complementary effect of some prebiotic and probiotic products on the development of bee colonies during spring. Bull UASVM Anim Sci Biotechnol 68:457–458

Patruica S, Dumitrescu G, Stancu A, Bura M, Banatean-Dunea I (2012) The effect of prebiotic and probiotic feed supplementation on the wax glands of worker bees (*Apis Mellifera*). Sci Anim Sci Biotechnol 45(2):268–271

Patruica S, Dumitrescu G, Popescu R, Filimon NM (2013) The effect of prebiotic and probiotic products used in feed to stimulate the bee colony (*Apis mellifera*) on intestines of working bees. J Food Agric Environ 11(3–4):2461–2464

Payne MS, Hall MR, Sly L, Bourne DG (2007) Microbial diversity within early-stage cultured *Panulirus ornatus* phyllosomas. Appl Environ Microbiol 73(6):1940–1951. https://doi.org/10.1128/AEM.02520-06

Pinloche E, McEwan N, Marden JP, Bayourthe C, Auclair E, Newbold CJ (2013) The effects of a probiotic yeast on the bacterial diversity and population structure in the rumen of cattle. PLoS One 8(7):e67824. https://doi.org/10.1371/journal.pone.0067824

Pintado J, Planas M, Makridis P (2014) Live feeds: microbial assemblages, probiotics and prebiotics. In: Merrifield D, Ringø E (eds) Aquaculture nutrition: gut health, probiotics and prebiotics. Wiley, Chichester, pp 419–442. https://doi.org/10.1002/9781118897263.ch16

Praesteng KE, Pope PB, Cann IK, Mackie RI, Mathiesen SD, Folkow LP, Eijsink VG, Sundset MA (2013) Probiotic dosing of *Ruminococcus flavefaciens* affects rumen microbiome structure and function in reindeer. Microb Ecol 66(4):840–849. https://doi.org/10.1007/s00248-013-0279-z

Puniya AK, Salem AZM, Kumar S, Dagar SS, Griffith GW, Puniya M, Ravella SR, Kumar N, Dhewa T, Kumar R (2015) Role of live microbial feed supplements with reference to anaerobic fungi in ruminant productivity: a review. J Integr Agric 14(3):550–560. https://doi.org/10.1016/S2095-3119(14)60837-6

Qi X, Jin Y, Liu H, Wang A, Zhao X (2011) Microencapsulation of *Lactobacillus brevis* and preliminary evaluation of their therapeutic effect on the diarrhea of neonatal calf. J Anim Vet Adv 10(2):151–156. https://doi.org/10.3923/javaa.2011.151.156

Rafat A, Hussain MA (2013) Environmental ethics in animal farming: agricultural probiotics. Am J Agric Sci 1:1–6

Rawls JF, Mahowald MA, Goodman AL, Trent CM, Gordon JI (2007) *In vivo* imaging and genetic analysis link bacterial motility and symbiosis in the zebrafish gut. Proc Natl Acad Sci U S A 104(18):7622–7627. https://doi.org/10.1073/pnas.0702386104

Refstie S, Sahlström S, Bråthen E, Baeverfjord G, Krogedal P (2005) Lactic acid fermentation eliminates indigestible carbohydrates and antinutritional factors in soybean meal for Atlantic salmon (*Salmo salar*). Aquaculture 246(1–4):331–345. https://doi.org/10.1016/j.aquaculture.2005.01.001

Ringø E, Dimitroglou A, Hoseinifar SH, Davies SJ (2014) Prebiotics in finfish: an update. In: Merrifield D, Ringø E (eds) Aquaculture nutrition: gut health, probiotics and prebiotics. Wiley, Chichester, pp 360–400. https://doi.org/10.1002/9781118897263.ch14

RingØ E, Olsen RE, Gifstad TØ, Dalmo RA, Amlund H, Hemre GI, Bakke AM (2010) Prebiotics in aquaculture: a review. Aquac Nutr 16(2):117–136. https://doi.org/10.1111/j.1365-2095.2009.00731.x

Rodriguez-Estrada U, Satoh S, Haga Y, Fushimi H, Sweetman J (2009) Effects of single and combined supplementation of Enterococcus faecalis, mannanoligosaccharide and polyhydrobutyric acid on growth performance and immune response of rainbow trout *Oncorhynchus mykiss*. Aquac Sci 57(4):609–617

Rodriguez-Lanetty M, Phillips WS, Weis VM (2006) Transcriptome analysis of a cnidarian—dinoflagellate mutualism reveals complex modulation of host gene expression. BMC Genomics 7:23–23. https://doi.org/10.1186/1471-2164-7-23

Ross GR, Gusils C, Gonzalez SN (2008) Microencapsulation of probiotic strains for swine feeding. Biol Pharm Bull 31(11):2121–2125

Ross EM, Petrovski S, Moate PJ, Hayes BJ (2013) Metagenomics of rumen bacteriophage from thirteen lactating dairy cattle. BMC Microbiol 13:242. https://doi.org/10.1186/1471-2180-13-242

Rurangwa E, Laranja JL, Van Houdt R, Delaedt Y, Geraylou Z, Van de Wiele T, Van Loo J, Van Craeyveld V, Courtin CM, Delcour JA, Ollevier F (2009) Selected nondigestible carbohydrates and prebiotics support the growth of probiotic fish bacteria mono-cultures *in vitro*. J Appl Microbiol 106(3):932–940. https://doi.org/10.1111/j.1365-2672.2008.04034.x

Sachsenröder J, Twardziok SO, Scheuch M, Johne R (2014) The general composition of the faecal virome of pigs depends on age, but not on feeding with a probiotic bacterium. PLoS One 9(2):e88888. https://doi.org/10.1371/journal.pone.0088888

Saez AC, Zhang J, Rostagno MH, Ebner PD (2011) Direct feeding of microencapsulated bacteriophages to reduce *Salmonella* colonization in pigs. Foodborne Pathog Dis 8(12):1269–1274. https://doi.org/10.1089/fpd.2011.0905

Sahu MK, Swarnakumar NS, Sivakumar K, Thangaradjou T, Kannan L (2008) Probiotics in aquaculture: importance and future perspectives. Indian J Microbiol 48(3):299–308. https://doi.org/10.1007/s12088-008-0024-3

Sakata T (1990) Microflora in the digestive tract of fish and shellfish. In: Lesel R (ed) Microbiology in Poecilotherms. Elsevier, Amsterdam, pp 171–176

Samanta AK, Jayapal N, Senani S, Kolte AP, Sridhar M (2013) Prebiotic inulin: useful dietary adjuncts to manipulate the livestock gut microflora. Braz J Microbiol 44(1):1–14. https://doi.org/10.1590/S1517-83822013005000023

Seeley TD (1985) Honeybee ecology: a study of adaptation in social life. Princeton University Press, Princeton

Seo JK, Kim S-W, Kim MH, Upadhaya SD, Kam DK, Ha JK (2010) Direct-fed microbials for ruminant animals. Asian Australas J Anim Sci 23(12):1657–1667. https://doi.org/10.5713/ajas.2010.r.08

Sheng H, Knecht HJ, Kudva IT, Hovde CJ (2006) Application of bacteriophages to control intestinal *Escherichia coli* O157:H7 levels in ruminants. Appl Environ Microbiol 72(8):5359–5366. https://doi.org/10.1128/aem.00099-06

Shepherd ML, Swecker WS Jr, Jensen RV, Ponder MA (2012) Characterization of the fecal bacteria communities of forage-fed horses by pyrosequencing of 16S rRNA V4 gene amplicons. FEMS Microbiol Lett 326(1):62–68. https://doi.org/10.1111/j.1574-6968.2011.02434.x

Sieo CC, Abdullah N, Tan WS, Ho YW (2005) Effects of beta-glucanase-producing *Lactobacillus* strains on growth, dry matter and crude protein digestibilities and apparent metabolisable energy in broiler chickens. Br Poult Sci 46(3):333–339. https://doi.org/10.1080/00071660500127316

Skugor S, Glover KA, Nilsen F, Krasnov A (2008) Local and systemic gene expression responses of Atlantic salmon (*Salmo salar* L.) to infection with the salmon louse (*Lepeophtheirus salmonis*). BMC Genomics 9:498–498. https://doi.org/10.1186/1471-2164-9-498

Soltanian S, Thai TQ, Dhont J, Sorgeloos P, Bossier P (2007) The protective effect against *Vibrio campbellii* in *Artemia* nauplii by pure β-glucan and isogenic yeast cells differing in β-glucan and chitin content operated with a source-dependent time lag. Fish Shellfish Immunol 23(5):1003–1014. https://doi.org/10.1016/j.fsi.2007.04.002

Song SK, Beck BR, Kim D, Park J, Kim J, Kim HD, Ringø E (2014) Prebiotics as immunostimulants in aquaculture: a review. Fish Shellfish Immunol 40(1):40–48. https://doi.org/10.1016/j.fsi.2014.06.016

Spaniol J, Oltramari C, Locatelli M, Volpato A, Campigotto G, Stefani L, Da Silva A (2015) Influence of probiotic on somatic cell count in milk and immune system of dairy cows. Comp Clin Pathol 24(3):677–681. https://doi.org/10.1007/s00580-014-1966-y

Sparkes AH, Papasouliotis K, Sunvold G, Werrett G, Gruffydd-Jones EA, Egan K, Gruffydd-Jones TJ, Reinhart G (1998) Effect of dietary supplementation with fructo-oligosaccharides on fecal flora of healthy cats. Am J Vet Res 59(4):436–440

Strickling JA, Harmon DL, Dawson KA, Gross KL (2000) Evaluation of oligosaccharide addition to dog diets: influences on nutrient digestion and microbial populations. Anim Feed Sci Technol 86(3):205–219. https://doi.org/10.1016/S0377-8401(00)00175-9

Strompfova V, Laukova A, Gancarcikova S (2012) Effectivity of freeze-dried form of *Lactobacillus fermentum* AD1-CCM7421 in dogs. Folia Microbiol (Praha) 57(4):347–350. https://doi.org/10.1007/s12223-012-0139-0

Suchodolski JS (2011) Intestinal microbiota of dogs and cats: a bigger world than we thought. Vet Clin North Am Small Anim Pract 41(2):261–272. https://doi.org/10.1016/j.cvsm.2010.12.006

Sun Y-Z, Yang H-L, Ma R-L, Huang K-P, Ye J-D (2012) Culture-independent characterization of the autochthonous gut microbiota of grouper *Epinephelus coioides* following the administration of probiotic *Enterococcus faecium*. Aquac Int 20(4):791–801. https://doi.org/10.1007/s10499-012-9503-y

Swanson KS, Fahey GC (2006) Prebiotic impacts on companion animals. In: Prebiotics: development & application. Wiley, pp 213–236. https://doi.org/10.1002/9780470023150.ch10

Swanson KS, Grieshop CM, Flickinger EA, Bauer LL, Chow J, Wolf BW, Garleb KA, Fahey GC Jr (2002) Fructooligosaccharides and *Lactobacillus acidophilus* modify gut microbial populations, total tract nutrient digestibilities and fecal protein catabolite concentrations in healthy adult dogs. J Nutr 132(12):3721–3731

Tajabadi N, Mardan M, Manap MYA, Mustafa S (2013) Molecular identification of *Lactobacillus* spp. isolated from the honey comb of the honey bee (*Apis dorsata*) by 16S rRNA gene sequencing. J Apic Res 52(5):235–241. https://doi.org/10.3896/IBRA.1.52.5.10

Tan LTH, Chan KG, Lee LH, Goh BH (2016) *Streptomyces* bacteria as potential probiotics in aquaculture. Front Microbiol 7:79. https://doi.org/10.3389/fmicb.2016.00079

Tellez G, Laukova A, Latorre JD, Hernandez-Velasco X, Hargis BM, Callaway T (2015) Food-producing animals and their health in relation to human health. Microb Ecol Health Dis 26:25876. https://doi.org/10.3402/mehd.v26.25876

Tinh NTN, Dierckens K, Sorgeloos P, Bossier P (2008) A review of the functionality of probiotics in the larviculture food chain. Mar Biotechnol 10(1):1–12. https://doi.org/10.1007/s10126-007-9054-9

Totton SC, Farrar AM, Wilkins W, Bucher O, Waddell LA, Wilhelm BJ, McEwen SA, Rajic A (2012) The effectiveness of selected feed and water additives for reducing *Salmonella* spp. of public health importance in broiler chickens: a systematic review, meta-analysis, and meta-regression approach. Prev Vet Med 106(3–4):197–213. https://doi.org/10.1016/j.prevetmed.2012.07.007

Tun HM, Brar MS, Khin N, Jun L, Hui RK, Dowd SE, Leung FC (2012) Gene-centric metagenomics analysis of feline intestinal microbiome using 454 junior pyrosequencing. J Microbiol Methods 88(3):369–376. https://doi.org/10.1016/j.mimet.2012.01.001

Twomey LN, Pluske JR, Rowe JB, Choct M, Brown W, Pethick DW (2003) The effects of added fructooligosaccharide (Raftilose®P95) and inulinase on faecal quality and digestibility in dogs. Anim Feed Sci Technol 108(1–4):83–93. https://doi.org/10.1016/S0377-8401(03)00162-7

Tzivara A, Papatsiros VG, Katsoudas V, Dritsa L, Katsoulos PD, Evaggelidou M, Aravidou E, Christodoulopoulos G (2013) Residues depletion study and withdrawal period determination of sulphadiazine and trimethoprim premix in pigs. Am J Anim Vet Sci 8(1):37–44. https://doi.org/10.3844/ajavsp.2013.37.44

Umu OC, Frank JA, Fangel JU, Oostindjer M, da Silva CS, Bolhuis EJ, Bosch G, Willats WG, Pope PB, Diep DB (2015) Resistant starch diet induces change in the swine microbiome and a predominance of beneficial bacterial populations. Microbiome 3:16. https://doi.org/10.1186/s40168-015-0078-5

Uyeno Y, Shigemori S, Shimosato T (2015) Effect of probiotics/prebiotics on cattle health and productivity. Microbes Environ 30(2):126–132. https://doi.org/10.1264/jsme2.ME14176

Van Boeckel TP, Brower C, Gilbert M, Grenfell BT, Levin SA, Robinson TP, Teillant A, Laxminarayan R (2015) Global trends in antimicrobial use in food animals. Proc Natl Acad Sci U S A 112(18):5649–5654. https://doi.org/10.1073/pnas.1503141112

Vanhoutte T, Huys G, De Brandt E, Fahey GC Jr, Swings J (2005) Molecular monitoring and characterization of the faecal microbiota of healthy dogs during fructan supplementation. FEMS Microbiol Lett 249(1):65–71. https://doi.org/10.1016/j.femsle.2005.06.003

Ventura M, Turroni F, van Sinderen D (2012) Probiogenomics as a tool to obtain genetic insights into adaptation of probiotic bacteria to the human gut. Bioeng Bugs 3(2):73–79. https://doi.org/10.4161/bbug.18540

Verschuere L, Rombaut G, Sorgeloos P, Verstraete W (2000) Probiotic bacteria as biological control agents in aquaculture. Microbiol Mol Biol Rev 64(4):655–671

Vickers RJ, Sunvold GD, Kelley RL, Reinhart GA (2001) Comparison of fermentation of selected fructooligosaccharides and other fiber substrates by canine colonic microflora. Am J Vet Res 62(4):609–615

Vine NG, Leukes WD, Kaiser H (2006) Probiotics in marine larviculture. FEMS Microbiol Rev 30(3):404–427

Wang JP, Yan L, Lee JH, Kim IH (2013) Evaluation of bacteriophage supplementation on growth performance, blood characteristics, relative organ weight, breast muscle characteristics and excreta microbial shedding in broilers. Asian Australas J Anim Sci 26(4):573–578. https://doi.org/10.5713/ajas.2012.12544

Wang L, Yang Y, Cai B, Cao P, Yang M, Chen Y (2014) Coexpression and secretion of endoglucanase and phytase genes in *Lactobacillus reuteri*. Int J Mol Sci 15(7):12842–12860. https://doi.org/10.3390/ijms150712842

Willard MD, Simpson RB, Delles EK, Cohen ND, Fossum TW, Kolp D, Reinhart G (1994) Effects of dietary supplementation of fructo-oligosaccharides on small intestinal bacterial overgrowth in dogs. Am J Vet Res 55(5):654–659

Wisener LV, Sargeant JM, O'Connor AM, Faires MC, Glass-Kaastra SK (2015) The use of direct-fed microbials to reduce shedding of *Escherichia coli* O157 in beef cattle: a systematic review and meta-analysis. Zoonoses Public Health 62(2):75–89. https://doi.org/10.1111/zph.12112

Xue GP, Johnson JS, Bransgrove KL, Gregg K, Beard CE, Dalrymple BP, Gobius KS, Aylward JH (1997) Improvement of expression and secretion of a fungal xylanase in the rumen bacterium *Butyrivibrio fibrisolvens* OB156 by manipulation of promoter and signal sequences. J Biotechnol 54(2):139–148

Yamano H, Koike S, Kobayashi Y, Hata H (2008) Phylogenetic analysis of hindgut microbiota in Hokkaido native horses compared to light horses. Anim Sci 79(2):234–242. https://doi.org/10.1111/j.1740-0929.2008.00522.x

Yan L, Hong SM, Kim IH (2012) Effect of bacteriophage supplementation on the growth performance, nutrient digestibility, blood characteristics, and fecal microbial shedding in growing pigs. Asian Australas J Anim Sci 25(10):1451–1456. https://doi.org/10.5713/ajas.2012.12253

Ye JD, Wang K, Li FD, Sun YZ (2011) Single or combined effects of fructo- and mannan oligosaccharide supplements and *Bacillus clausii* on the growth, feed utilization, body composition, digestive enzyme activity, innate immune response and lipid metabolism of the Japanese flounder *Paralichthys olivaceus*. Aquac Nutr 17(4):e902–e911. https://doi.org/10.1111/j.1365-2095.2011.00863.x

Yeoman CJ, Chia N, Jeraldo P, Sipos M, Goldenfeld ND, White BA (2012) The microbiome of the chicken gastrointestinal tract. Anim Health Res Rev 13(1):89–99. https://doi.org/10.1017/S1466252312000138

Yin QQ, Chang J, Zuo RY, Chen LY, Chen QX, Wei XY, Guan QF, Sun JW, Zheng QH, Yang X, Ren GZ (2010) Effect of the transformed lactobacillus with phytase gene on pig production performance, nutrient digestibility, gut microbes and serum biochemical indexes. Asian Australas J Anim Sci 23(2):246–252. https://doi.org/10.5713/ajas.2010.90372

Yu B, Liu JR, Hsiao FS, Chiou PWS (2008) Evaluation of *Lactobacillus reuteri* Pg4 strain expressing heterologous β-glucanase as a probiotic in poultry diets based on barley. Anim Feed Sci Technol 141(1–2):82–91. https://doi.org/10.1016/j.anifeedsci.2007.04.010

Yusrizal, Chen TC (2003) Effect of adding chicory fructans in feed on broiler growth performance, serum cholesterol and intestinal length. Int J Poult Sci 2(3):214–219. https://doi.org/10.3923/ijps.2003.214.219

Zentek J, Marquart B, Pietrzak T (2002) Intestinal effects of mannanoligosaccharides, transgalactooligosaccharides, lactose and lactulose in dogs. J Nutr 132(6 Suppl 2):1682s–1684s

Zhang J, Li Z, Cao Z, Wang L, Li X, Li S, Xu Y (2015a) Bacteriophages as antimicrobial agents against major pathogens in swine: a review. J Anim Sci Biotechnol 6(1):39. https://doi.org/10.1186/s40104-015-0039-7

Zhang R, Zhou M, Tu Y, Zhang NF, Deng KD, Ma T, Diao QY (2015b) Effect of oral administration of probiotics on growth performance, apparent nutrient digestibility and stress-related indicators in Holstein calves. J Anim Physiol Anim Nutr 100(1):33–38. https://doi.org/10.1111/jpn.12338

Zhao PY, Baek HY, Kim IH (2012) Effects of bacteriophage supplementation on egg performance, egg quality, excreta microflora, and moisture content in laying hens. Asian Australas J Anim Sci 25(7):1015–1020. https://doi.org/10.5713/ajas.2012.12026

Zhou Z, Yao B, Romero J, Waines P, Ringø E, Emery M, Liles MR, Merrifield DL (2014) Methodological approaches used to assess fish gastrointestinal communities. In: Merrifield D, Ringø E (eds) Aquaculture nutrition: gut health, probiotics and prebiotics. Wiley, Chichester, pp 101–127. https://doi.org/10.1002/9781118897263.ch5

# Role of the Gut Microbiota in Health and Disease

# 2

Hyunjoon Park, Soyoung Yeo, Karina Arellano, Hyeon Rae Kim, and Wilhelm Holzapfel

## 2.1 Introduction

### 2.1.1 Defining the Gut [Gastrointestinal Tract (GIT)] (from Invertebrates to Vertebrates)

Commonly, the 'gut' refers to the digestive (alimentary) canal between the pylorus and the anus. Other terms such as the 'stomach' and 'intestinal tract' are also frequently used in a more general sense. When referring to different kinds of animals, a more specific definition seems necessary. A clear distinction needs to be drawn between vertebrates and non-vertebrates, also with reference to the relative complexity and size of the alimentary tract. For vertebrates, the alimentary canal is generally being referred to as the gastrointestinal tract (GIT), implying a clear distinction between the stomach (with a low pH and generally a low microbial population) and both the small and large intestines (with an increasing pH and microbial numbers). The GI tract harbours the highest numbers of immune cells in our body. These sites are colonised by large numbers of autochthonous commensal microbiota, a healthy population that protects their ecological niche by different mechanisms such as reinforcing barrier immunity (Belkan and Hand 2014).

Complex microbial communities, also collectively referred to as 'microbiota', are associated with animals and humans. They colonise diverse body sites, in particular the various epithelial tissues ranging from the outer layer of the skin to the lining of 'open' cavities of the digestive and respiratory systems. By this close and long-term

H. Park • K. Arellano • H.R. Kim • W. Holzapfel (✉)
Advanced Green Energy and Environment Institute (AGEE), Handong Global University, Pohang, Gyeongbuk, South Korea
e-mail: wilhelm@woodapple.net

S. Yeo
Research Institute of Eco-friendly Livestock Science, Institute of Green-Bio Science and Technology, Seoul National University, Pyeongchang, Gangwon-do, South Korea



D. Di Gioia, B. Biavati (eds.), *Probiotics and Prebiotics in Animal Health and Food Safety*, https://doi.org/10.1007/978-3-319-71950-4_2

association, these microbial communities decisively influence the physiology and immune functions of their host organism (Brown and Clarke 2016). The highest numbers of microorganisms are associated with the GIT, where numbers and diversity are increasing from the proximal to the distal part. The large intestine (colon) harbours the largest concentration where the microbial biomass comprises roughly a concentration of up to 50% (or around $5 \times 10^{11}$/g) of the total contents. As omnivores, humans probably host the largest and most complex microbial population in their GIT of all creatures. The obvious major research focus on the human microbiome has resulted in a vast amount of rapidly increasing information on the gut microbiota, part of which also benefits the understanding of the gut microbiota of the animal (Turnbaugh et al. 2007). The bacterial phyla *Bacteroidetes* and the *Firmicutes* are considered to be the major bacterial representatives within the gut microbiota, with *Actinobacteria*, *Proteobacteria*, *Verrucomicrobia*, *Fusobacteria* and *Cyanobacteria* present in lower numbers (Eckburg et al. 2005). Novel approaches such as those based on culturomics revealed numerous new (hitherto undetected) species and genera, accumulating to a total number of known species at more than 1500 at present (Lagier et al. 2016).

The wide range of foods/feeds and feeding patterns in nature is related to functional diversification of digestive systems of the diverse hosts and their adaptation to their environment. According to Penry and Jumars (1987), most guts resemble one of three kinds of ideal chemical reactors, or combinations of them, comprising batch reactors (e.g. the gut of the hydra and caecum of a rabbit), plug-flow reactors (PFRs) (e.g. the tubular intestine of many invertebrates and all vertebrates) and continuous-flow stirred tank reactors (CSTRs) (e.g. the rumen of a cow or the hindgut of a termite).

### 2.1.2 Host Specificity

It is still not clear in which way multiple factors, defined by environmental conditions and host genetic make-up, combine to determine the shape of the complex gut ecosystem and its microbiota. Benson et al. (2010) found a 'core measurable microbiota' (CMM) of 64 conserved taxonomic groups that vary quantitatively across most individuals in a mouse population. Representing a complex polygenic trait, the population is therefore shaped by a range of environmental and host genetic factors. Even when particular effects derived related to litter and cohort may partly explain this variation, the contribution of the individual host genotype was found significant and measurable (Benson et al. 2010). This basis appears essential for explaining host specificity when studying pathogens, but still only rare information is available for mutualistic symbionts, such as for the commensal bacterium *Sodalis glossinidius* to colonise the tsetse fly gut (Maltz et al. 2012) and also was shown in the squid–vibrio model (McFall-Ngai 2013). Intraspecies (strain-level) diversification in the vertebrate GIT has been highlighted by the so-called *Lactobacillus reuteri* host-specific paradigm for explaining specifics of host–microbial symbiosis (Walter et al. 2011). Host phylogeny and bacterial community composition are joint factors in many animals by which strain-level host specificity in gut bacteria is defined, ranging, e.g. from 'ants to apes' (Sanders et al. 2014) and to bees (Kwong et al. 2014).

## 2.2 Gut Microbiome Homeostasis vs. Dysbiosis

The microbiota of a healthy individual comprise the microbial taxa typically associated with the host. An established healthy microbial population represents the commensal microbes with supporting physiological functions, ranging from 'neutral' to beneficial to essential. The 'microbiome' comprises the catalogue of these microbes and their genes. The physiological functions, although probably best studied in humans, may also apply in principle to most animals and include contribution to food digestion and maintenance of the immune system, playing a key role in the energy household and practically all physiological functions of the host (Turnbaugh et al. 2009; Qin et al. 2010; Clarke et al. 2014).

Initiated during and immediately after birth, microbial colonisation will progress until reaching a balanced or stable state at adulthood. The microbiome of a healthy adult is characterised by its stability even when challenged by diverse factors and stimuli such as a changing lifestyle and diet, stress, physical activity, travel, seasonal changes, hormonal cycles, and even some disorders. This underlines the important role of the 'mature' microbiome in gut homeostasis and thereby in maintaining a state of well-being. Disturbing of the balance in the microbial population may lead to dysbiotic conditions and thereby lead to negative effects on the host's health. A clear qualitative and quantitative definition of a healthy microbial ecosystem has not been established, yet some major microbial groups may serve as 'indicators' of detrimental changes in the gut microbial population, also by their suggested role in health and/or disease (Clarke et al. 2014; D'Argenio and Salvatore 2015; Ohland and Jobin 2015; Wang and Roy 2016).

### 2.2.1 Gut Microbiome Homeostasis

The mammalian GIT is a complex and dynamic system inhabited by diverse and numerous microbial communities. All-over, the gut microbiota is composed principally of bacteria but also include protozoa, archaea, eukaryotes, fungi and viruses (Gordon 2012). The mutually beneficial connection with the host is based on symbiotic principles by which the microbes benefit from the intestinal environment and its regulation by body physiological processes, temperature and moisture regulation and a steady supply of available nutrients. Gut microbiota are beneficial and even essential to host health in diverse ways, such as the digestion and absorption of indigestible nutrients, the synthesis of essential vitamins, the detoxification of xenobiotic compounds, the protection against pathogenic microorganisms and contribution to the development and maturation of the immune system (Walter et al. 2011).

According to Wang and Roy (2016), gut homeostasis is 'the state of resilience and resistance to external and endogenous disturbances'. A stable state of gut homeostasis is guaranteed by a healthy commensal microbiota. Stability is supported and maintained by the integration of diverse mechanisms by which pathogens are eliminated and, at the same time, the indigenous microbiome 'tolerated'. Based on a symbiotic relationship, host–microbial and bacteria–bacteria communication is essential to preserve intestinal tissue homeostasis and a healthy state (Buchon et al. 2013). Host species maintain intestinal tissue homeostasis by keeping the diversity between microbial groups through competition for nutrients and the expression of antimicrobial components such as bacteriocins, microcins and colicins that control the growth of some pathogenic strains without causing damage to host cells (Ohland and Jobin 2015). On the other hand, the contact with microorganisms in the gastrointestinal tract influences the development and maturation of the immune system. In this way, the immune system recognises and tolerates non-harmful microbes and responds to pathogens and opportunistic organisms. In order to maintain homeostatic interactions between the host and the gut microbiota and prevent irregular inflammation, tolerance of the normal gut microbiota is vital (Mann et al. 2013).

#### Immune Activation

Intestinal homeostasis is strongly dependent of a delicate balance between immune activation and regulation and is essential in the prevention of intestinal inflammation. The commensal bacteria within the intestinal lumen play a key role in providing non-inflammatory protection of the mucosal membrane through immunomodulation. Appropriate acquired response of the host depends on its ability to discriminate between pathogenic and commensal bacteria, while at the same time, the proper inflammatory responses are initiated and regulated. Commensal bacteria play a modulating role in the immune response by regulating the amounts of mediators secreted by intestinal immune system cells and T helper and regulatory cell stimulation. Colonisation resistance and the resulting protection against pathogen invasion are thus strongly based on the proper (chicken) response of the gut-associated immune system as shown by Brisbin et al. (2007) for the chicken. A range of stressors related to the environment, nutrition and infection may affect the health and growth of animals, also by altering immune systems associated with the gastrointestinal tract. Increase in gut permeability, oxidative stress, and inflammatory responses in the gut, as well as infections by pathogenic bacteria (e.g., Enterotoxigenic *E. coli*, ETEC) and viruses (e.g., porcine epidemic diarrhoea virus) may damage the tight junction proteins and thereby reduce the condition of the intestinal epithelial barrier. On the other hand, stabilisation of the gut microbiota, e.g. by supplementation of beneficial (probiotic) *Lactobacillus* strains, may reduce the production of pro-inflammatory cytokines and chemokines in mucosal system in pigs and improve the integrity of the gut barrier (Lee et al. 2016). Lactobacilli from swine intestine and faeces showed immunomodulatory activity in porcine intestinal epithelial (PIE) cells; among the isolated strains, *L. plantarum* MPL16 modulated the production of inflammatory cytokines in PIE cells (Villena et al. 2017).

### 2.2.2 Dysbiosis or Disequilibrium

Since the intestine is a dynamic niche, the host diet, antibiotics, lifestyle and hygiene may affect gut microbiome composition (Sommer and Bäckhed 2013).

The disruption of gut homeostasis by the change or alteration in the diversity, structure or function in the gut microbiota is referred as microbial dysbiosis (Wang and Roy 2016), also implying a state of disequilibrium. The lack of balance among the microbiota may affect the less abundant beneficial species, leading to pathological states to the host (Montalban et al. 2015). Microbial disequilibrium has been associated with susceptibility to diseases such as obesity, diabetes, autoimmune diseases, neurological disorders, allergies and inflammatory and infectious diseases in humans (Wang and Roy 2016).

Numerous environmental stress factors may influence the condition and well-being of farm animals, in particular neonatal and weaned animals. Physiological stress conditions include feeding practices, farm management and dietary requirements and may lead to the invasion of pathogenic bacteria and thereby destabilise the commensal gut microbial population (Yang et al. 2015; Yeo et al. 2016). Dysbiosis of the gut microbiota may correlate with several diseases and inflammatory conditions and with resulting growth retardation in young animals (Chaucheyras-Durand and Durand 2010; Yeo et al. 2016).

Diet is one of the main factors that cause variation in the composition of the gut microbiota and their relative gene content in vertebrates and invertebrates (Montalban et al. 2015). Diet composition benefits specific microorganism groups by providing them with nutritional conditions that promote their growth and give them nutritional advantages over other species.

Antibiotic therapy for the control of infection diseases causes disequilibrium in the intestinal microbiota community since not only pathogens are destroyed but it may also kill or reduce commensal and beneficial microbes. The losses in commensal microbial richness reduce the metabolism and absorption of complex nutrients and also lower the production of essential vitamins, thus triggering a pathological state in the host (McFarland 2014).

Host genetics deficiency interrupts the host–microbial communication and the tolerance to the gut microbiota, leading to dysbiosis and pathologic conditions. The overproduction of pro-inflammatory mediators or mutations in regulatory immune proteins affects the gut microbiota composition and can induce chronic inflammation and metabolic dysfunction (Sommer and Bäckhed 2013).

## 2.3 Function and Role of the Gut Microbiome

### 2.3.1 Metabolic Activity

Based on their enormous metabolic capacity, the gut microbiota are recognised as the 'neglected endocrine organ' (Clarke et al. 2014) or 'our forgotten organ' (D'Argenio and Salvatore 2015). This realisation has prompted special focus on the human microbiome (and also to some extent on that of most animals of economic importance), as a site of complex and far-reaching microbial interactions (Rajilic-Stojanovic and De Vos 2014). With an estimated >$10^{14}$ microbial cells (around ten times as many as in the rest of our bodies), and with >1000 bacterial

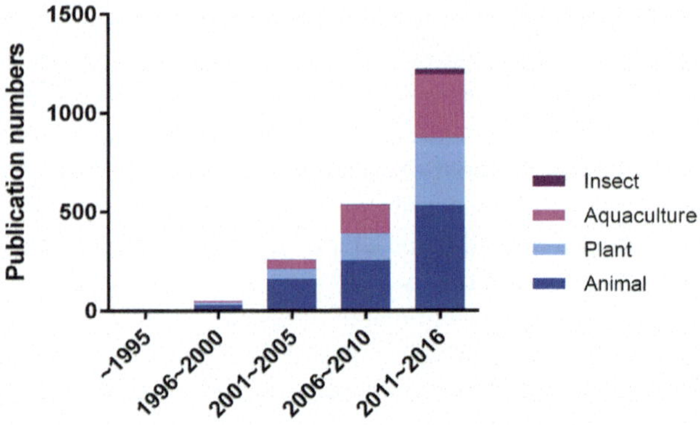


**Fig. 2.1** The number of scientific publications recorded in the PubMed database regarding probiotics for non-human hosts

species, the human microbiome accounts for around 150 times more genes (encoding a multifold more versatile metabolome) than that found in the total human genome. This principle also applies to most animal hosts, where, essentially, the (self) body cells also contribute only a comparatively minor part to homeostasis as compared to the microbiota. The number of research studies using non-human hosts comprises about 70% of that of human probiotic studies, thereby giving some indication of the relative research output dealing with the animal gut microbiota. Still, the number of studies is increasing in diverse target hosts such as livestock, plants, insects and fish (Fig. 2.1). Detecting and identifying potentially beneficial gut microbes may overlap with the 'prospecting' process for potential probiotics. Some studies have not used the term 'probiotics', but the applied strains corresponded to the definition of probiotics with regard to their beneficial influence on a host. Potential benefits and/or applications with regard to several hosts are summarised in Table 2.1. Functional categories such as control of disease and immunomodulation require individual requisite criteria for claims of beneficial effects, safety for the host and user and impact on the environment. The application and prospects of probiotics for animal gut health will be treated in other chapters of this book. However, in the broader context of gut microbiota and health, individual host specificity and the range of potentially beneficial microbes may serve as indication of the broad diversity of animal-related probiotics research.

The 'inability' to culture the majority of the gut microbiota has been considered a major hurdle towards comprehensively studying and understanding physiological interactions of gut microbiota (Cerf-Bensussan and Gaboriau-Routhiau 2010; Sekirov et al. 2010). However, recent research reports have shown that a major part of the (human) gut microbiota can now be cultivated by, e.g. using culture-enriched molecular profiling (a combination of culture and 16S rRNA gene sequencing). These approaches have surprisingly revealed a greater bacterial diversity than culture-independent sequencing techniques (Lau et al. 2016). This approach has also been successfully applied for targeting the recovery of a particular bacterial group, thereby enabling the isolation of specific gut bacteria and opening the way for diversity and mechanistic studies on the interactions between microbiota the host (Lau et al. 2016).

**Table 2.1** Research reports on microbes conferring specific beneficial effects on the host

| Functionality | Target hosts | Microorganisms | Referred 'probiotics' | Representative references |
|---|---|---|---|---|
| *Pig* | | | | |
| Nutrition digestion | Piglet, weaned pig | *Lactobacillus plantarum*, *Lactobacillus acidophilus*, *Lactobacillus casei*, *Lactobacillus fermentum*, *Streptococcus faecium* | Y | Collington et al. (1990), Yu et al. (2008) |
| Growth improvement | (Newborn) Piglet, sow | *Bifidobacterium pseudolongum*, *Lactobacillus acidophilus*, *Bacillus mesentericus*, *Clostridium butyricum*, *Enterococcus faecalis* | Y | Abe et al. (1995), Hayakawa et al. (2016) |
| Control of disease | Sow, weaned and suckling piglet | *Enterococcus faecium*, *Escherichia coli*, cocktail of bacteriophages | Y | Pollmann et al. (2005), Bhandari et al. (2010), Lee et al. (2016) |
| Immunomodulation | Adult and weaned pig | *Escherichia coli*, *Saccharomyces cerevisiae boulardii* (yeast) | Y | Duncker et al. (2006), Collier et al. (2011) |
| Cholesterol assimilation | Mini pig | *Lactobacillus johnsonii*, *Lactobacillus reuteri* | Y | du Toit et al. (1998) |
| *Ruminant* | | | | |
| Nutrition digestion | Calves (Holstein–Friesians, Friesian–Jersey) | *Bacillus amyloliquefaciens* | Y | Lee et al. (2016) |
| Growth improvement | Calves (Holstein–Friesian, Holstein bull, *Bos taurus*, *Bubalus bubalis*) | *Lactobacillus acidophilus*, *Propionibacterium jensenii*, *Lactobacillus casei*, *Lactobacillus salivarius*, *Pediococcus acidilactici* | Y | Cruywagen et al. (1996), Adams et al. (2008), Frizzo et al. (2010), Malik and Bandla (2010) |

(continued)

**Table 2.1** (continued)

| Functionality | Target hosts | Microorganisms | Referred 'probiotics' | Representative references |
|---|---|---|---|---|
| Yield improvement | Sheep, lamb, cow | *Bacillus licheniformis*, *Bacillus subtilis*, *Prevotella bryantii* | Y | Kritas et al. (2006), Chiquette et al. (2008) |
| Control of disease | Feedlot cattle, beef steer, Lithuanian Black-and-White calve | *Propionibacterium* spp., *Enterococcus faecium*, *Lactobacillus acidophilus* | Y | Ghorbani et al. (2002), Elam et al. (2003), Jatkauskas and Vrotniakiene (2010) |
| Immunomodulation | Sheep | *Bacillus cereus* var. Toyoi, *Saccharomyces boulardii* (yeast) | Y | Roos et al. (2010) |
| *Chicken* | | | | |
| Growth improvement | Broiler | *Lactobacillus* spp., *Bifidobacterium* spp., *Enterococcus* spp., *Pediococcus* spp., *Lactobacillus acidophilus*, *Bacillus subtilis*, *Clostridium butyricum* | Y | Mountzouris et al. (2007), Zhang and Kim (2014), Wang et al. (2016) |
| Control of disease | White leghorn chicken, fertile eggs, broiler | *Lactobacillus acidophilus*, *Bifidobacterium bifidum*, *Enterococcus faecalis*, *Lactobacillus reuteri*, phages | Y | Toro et al. (2005), Akbari et al. (2008), Borie et al. (2008, 2009), Mappley (2013) |
| Immunomodulation | Broiler, young chickens | *Lactobacillus crispatus*, *Lactobacillus acidophilus*, *Lactobacillus reuteri*, *Lactobacillus salivarius* | Y | Taheri et al. (2010), Brisbin et al. (2011), Asgari et al. (2016) |
| Intestinal health | Broiler | *Bacillus subtilis* | Y | Sen et al. (2012) |
| *Fish* | | | | |
| Growth improvement | Nile tilapia (*Oreochromis niloticus*), rohu (*Labeo rohita*) | *Bacillus subtilis*, *Lactobacillus acidophilus*, *Lactococcus lactis*, *Lactobacillus plantarum*, *Saccharomyces cerevisiae* (yeast) | Y | Aly et al. (2008), Mohapatra et al. (2012), Giri et al. (2013), Ran et al. (2016) |

**Table 2.1** (continued)

| Functionality | Target hosts | Microorganisms | Referred 'probiotics' | Representative references |
|---|---|---|---|---|
| Control of disease | Turbot (*Scophthalmus maximus* L.) larvae, ayu (*Plecoglossus altivelis*), rainbow trout (*Oncorhynchus mykiss* Walbaum), Japanese flounder (*Paralichthys olivaceus*), carp (*Cyprinus carpio*) | *Vibrio pelagius*, *Myoviridae* and *Podoviridae* (phages), *Pseudomonas* spp., *Bacillus subtilis*, *Lactobacillus acidophilus*, *Clostridium butyricum*, *Enterococcus faecium*, *Saccharomyces cerevisiae* (yeast) | Y/N phages were not | Ringø and Vadstein (1998), Park et al. (2000), Spanggaard et al. (2001), Taoka et al. (2006), Gopalakannan and Arul (2011) |
| Immunomodulation | Rainbow trout (*Oncorhynchus mykiss*), hybrid tilapia, European sea bass (*Dicentrarchus labrax*) larvae, olive flounder (*Paralichthys olivaceus*), rohu (*Labeo rohita*) | *Lactobacillus rhamnosus*, *Lactococcus garvieae*, *Lactococcus lactis*, *Leuconostoc mesenteroides*, *Lactobacillus sakei*, *Streptococcus iniae*, *Bacillus subtilis*, *Enterococcus faecium*, *Lactobacillus casei*, *Saccharomyces cerevisiae* (yeast) | Y/N yeast was not | Nikoskelainen et al. (2003), Brunt and Austin (2005), Balcázar et al. (2007a), Nayak et al. (2007), Panigrahi et al. (2007), He et al. (2011), Kim et al. (2013), Lamari et al. (2016) |
| Microflora modulation | Brown trout (*Salmo trutta*) | *Lactococcus lactis* ssp. *lactis*, *Lactobacillus sakei*, *Leuconostoc mesenteroides* | Y | Balcázar et al. (2007b) |
| Survival improvement | Barred knifejaw (*Oplegnathus fasciatus*) | *Lactobacillus sakei* | Y | Harikrishnan et al. (2011) |

Browne et al. (2016) combined special microbial cultivation techniques (including broad-range agar media) with metagenomics sequencing to show that most known species of the gut microbiota can be grown and preserved in vitro. The results revealed that many gut bacteria, hitherto considered to be 'unculturable', belong to novel groups of which about 60% appear to be endospore producers, this being a probable survival strategy for conditions outside of the host (Browne et al. 2016). Microbial culturomics has recently been introduced as a new approach of using multiple culture techniques combined with matrix-assisted laser desorption/ionisation–time of flight (MALDI–TOF) and 16S rRNA for the identification of growing colonies (Lagier et al. 2015). Using 'best culture' conditions, Lagier et al. (2016) could increase the output of sample analysis, while limitations of former studies could be overcome by applying new protocols such as fresh-sample inoculation and the detection of micro-colonies. In this way 531 species have been added to those previously reported for the human gut (Lagier et al. 2016).

### 2.3.2 Metabolites with Hormonal Functions

Multiple products of microbial metabolism are considered to be of hormonal nature; they are delivered via the bloodstream and may exert an influence on the function of distal organs and systems. By carbohydrate metabolism, short-chain fatty acids (SCFAs) (e.g. butyrate and propionate) are produced, thus providing important nutrients (effective in low concentrations) and supporting regulation of organs ranging from the enteric nervous system to the brain. Biogenic amines, acting as neurotransmitters, are converted by gut microbial amino acid decarboxylases and include serotonin, dopamine, tryptamine and GABA ($\gamma$-amino-butyric acid). Serotonin, for example, is formed by conversion of a precursor such as tryptophan, while the balance within the gut microbiota may exert a modulatory influence on its plasma concentration. Other metabolites with a potential hormonal function include cortisol (a HPA hormone), involved in stress response, anti-inflammatory activities and anabolic and catabolic effects at several body sites, ghrelin, playing a role in host metabolism, and leptin, an appetite regulator (Clarke et al. 2014; Sudo 2014; Williams et al. 2014; Smith 2015). Moreover, the gut microbiota–brain axis has been proposed to play a role in bidirectional signalling and neural homeostasis (Bauer et al. 2016).

Information is increasing on the bidirectional communication between the gut microbiota and the brain and enables a deeper understanding of the important role of the gut–microbiota–brain axis in various disorders ranging from depression, obesity, autism and other neuropsychiatric conditions. Pathways of communication include the vagus nerve, the immune system, neuroendocrine pathways and bacteria-derived metabolites. Yet, at present it remains difficult to determine whether the change in microbiota is the cause or consequence of conditions such as coeliac disease, gut–brain axis and behaviour (Fasano 2017; Sandhu et al. 2017). Modulations of the gut microbial composition/balance may have a substantial impact on the central physiology (Dinan and Cryan 2017). By new approaches such as transplantation of the gut microbiota, a behavioural or physiological phenotype may be transferred.

Microbial communities, collectively called the microbiota, either directly or indirectly have an impact on the various tissues and organ systems of the host. Probably the most far-reaching of all microbial effects are those influencing both the innate and adaptive immune systems (Brown and Clarke 2016).

### 2.3.3 Regulation of Host Defences

In contrast to earlier views of associating microbes with infections and diseases, the vital role of microbiota as beneficial 'regulators' of the host's physiology and immune response is now generally recognised (Ha et al. 2014). In this respect the gut microbiota play a key role in host defence response against infections, with an important regulating contribution by pattern recognition receptors (Brown and Clarke 2016).

**Signalling in the Gut**

As a signalling 'centre', the intestinal microbiome functions as a system that integrates environmental factors (e.g. diet, xenobiotics) with genetic and immune signals. Signalling processes have a decisive impact on the host's metabolism, immunity and responses to infection. Strategically located at the host–microbiome interface, cells of the innate immune system can sense both microorganisms and their metabolites. Resulting signals may then serve to induce host physiological responses and regulate the ecological balance within the microbial community. Well-functioning crosstalk in the intestine between the innate myeloid and lymphoid cells and the commensal gut microbiota is essential for gut homeostasis. Disturbance/disruption of this fine balance may open the way to inflammatory conditions and result in multifactorial disorders (Mortha et al. 2014; Thaiss et al. 2016).

In cohabitation with their multicellular (human and animal) hosts, the complex multispecies microbial communities interact in various ways and, among others, by communication based on quorum sensing. This secretory signalling system regulates the expression of certain target genes with auto-inducers in a cell density-dependent manner and is also operative among the gut microbiota (Miller and Bassler 2001; Yeo et al. 2015). Dysbiosis of the gut microbiota and thus disturbance of the 'normal' gut physiology may contribute to the development of MS (multiple sclerosis), also by disrupting the sensitive relationship between the central nervous system, the immune system and the gut. It appears that by regulating the endocannabinoid system, the interaction between gut microbiota and the immune system may be positively influenced (Adamczyk-Sowa et al. 2017).

Thanks to the recent rapid technical developments such as high-throughput sequencing, members of complex microbial communities and their dynamic shifts are increasingly being studied in depth (Morgan and Huttenhower 2012). Thereby, relationships between dysfunctions of the human microbiota and inflammatory conditions such as metabolic syndrome, bowel disease and antibiotic-resistant infections can now be more accurately linked. As early detection biomarker, the microbiome is an obvious target for therapeutic intervention. Its indirect effects are, however, more far-reaching than formerly expected. Even tissues not in direct contact with microbes form part of host circuits that are impacted by microbial energy degradation products and affected by signals from microbial metabolites. Analysing the signalling process in host-associated communities may provide a strong basis for understanding their diverse functions. This may widen prospects for developing therapies directed towards the combating of disease and sustaining of health by stabilising microbial communities and their metabolic functions (Fischbach and Segre 2016).

### Factors Influencing the Gut Microbiota

Diversity and metabolic capacity between diverse and/or closely related microbial species could differ due to genetic, environmental and nutritional factors (Lozupone et al. 2012).

Comparing bacterial diversity in faecal samples of 71 vertebrate species (with a body mass range of 5.6 log and including mammals, birds and reptiles), Godon et al. (2016) have found some correlation between the increase in gut volume and microbial diversity. Karasov and Douglas (2013) have shown taxon richness of the gut microbiota (identified by, e.g. 16S rRNA gene sequencing) to be an order of magnitude greater in vertebrates than invertebrates, with the diet as a major influencing factor of interspecific variation in microbial composition.

Dietary factors may impact the composition of the microbiota in diverse ways, depending on the basic type of diet and kind of the digestive system (e.g. carnivores and omnivores vs. herbivores). Studying the impact of diet in 33 different mammalian species and 18 humans, Muegge et al. (2011) found similar adaptation to the diet in the different lineages with a sharing of the major functional genes in the gut microbiome. Yet, the interaction between nutrients and microbiota influences the stability of the microbial population and thus the health condition of the host. Both a high-fat and a high-fat–high-sugar diets have been reported to cause a shift in the population towards an increase in the *Firmicutes* (in particular some *Clostridium* groups, while high carbohydrates—in addition—increased numbers of *Erysipelotrichi* and *Bacilli*) and *Proteobacteria* but a decrease in the *Bacteroidetes* (Hildebrandt et al. 2009; Turnbaugh et al. 2009). Studying the impact of long-term dietary patterns on gut microbial enterotypes, Wu et al. (2009) have found high-fibre diets to promote the bacterial phyla *Bacteroidetes* and *Actinobacteria* relative to the *Firmicutes* and *Proteobacteria.*

Both in invertebrates and vertebrates, the gut microbiota play a key role in host metabolism and essentially influence its physiology, health and well-being, and functionality and performance.

## 2.4 Plasmid-Encoded Functions

Plasmids, or mobile genetic elements, play an important role in the spread of antibiotic resistance genes in the environment (Jones et al. 2010). Horizontal and vertical transferable antibiotic resistance is crucial where selective pressure exists due to the use of antibiotics, either for therapeutic purposes, as feed additive for growth promotion, or by the antibiotic contamination of surface waters via wastewater from hospitals and private households.

Perhaps the best documentation of selective pressure (and its reversal) has been with regard to the use of the glycopeptide antibiotic, avoparcin, in animal feed as growth promoter during the 1990s. In countries such as the Netherlands, where vancomycin was scarcely used for therapeutic treatment, vancomycin-resistant enterococci (VRE) have been commonly detected in the commensal microbiota of food animals, on their meat and even in the commensal microbiota of healthy humans. However, at the same time, huge quantities of avoparcin were used as growth promoter for farm animals. Being structurally related to the therapeutically important vancomycin, selective pressure by avoparcin was suggested to be the cause of resistance transfer to enterococci associated with farm animals and beyond. This was supported by the fact that in countries where the use of avoparcin was forbidden, no VRE have been detected in food animals and neither in food of animal origin nor in healthy humans. The use of all antibiotics as growth promoters was banned in Sweden in 1986, where no VRE have been detected in the faecal microbiota of farm animals (Van den Bogaard and Stobberingh 1999). Following the ban of avoparcin use for poultry in Denmark in 1995, the prevalence of VRE decreased from 80% in 1995 to 5% in 1998. Similar observations have been made in other European countries, e.g. Germany, subsequently to banning the use of avoparcin as feed additive for farm animals (Bager et al. 1999; Van den Bogaard et al. 2000).

## 2.5 Herbivores

Herbivores consume only plant material of which a large part of the biomass, especially cellulose, cannot be digested by their own enzymes. All vertebrates are characterised by the absence of cellulase. However, in contrast to carnivores and omnivores, the conversion of complex carbohydrates such as cellulose into monomers is essential for herbivores due to their plant-based diet. Compared to carnivores, the digestive tract of herbivores is relatively long, thus enabling the digestion/breakdown of large amounts of fibre by a specialised microbial population. Based on the site and organ for fermentation, two types of herbivores are distinguished, i.e. the foregut fermenters and the hindgut fermenters. The fermentation chamber is characterised by anaerobic conditions, a regulated pH and temperature, sufficient fluid and a steady nitrogen supply.

Ruminants are the major group among the *foregut fermenters* and are represented by approximately 150 species, including cattle, goats, sheep, giraffes, yaks, deer, antelope and buffalo. Characterised by a four-chambered stomach (rumen, reticulum, omasum and abomasum), typical ruminants differ from pseudo-ruminants (e.g. camel and hippopotamus) by the three compartments of their stomach. Some rodents, marsupials, colombine monkeys and sloths also rely on foregut fermentation of their feed. In the first two chambers of the ruminant, the rumen and the reticulum, the food is mixed with saliva with subsequent separation into two layers of solid and liquid material (Fowler and Bravo 2010; Reece 2013). The water buffalo has significant economic importance in South and Southeast Asia and includes the river buffalo with distribution from South Asia to the Balkans to Italy, and with slight morphological differences from the swamp buffalo, more typical of India and the Yangtze valley of China. A major difference between the rumen of the water buffalo and that of other ruminants lies in a larger population of (mainly cellulolytic) bacteria and lower numbers of protozoa. Compared to cattle, the rumen of the water buffalo contains higher concentrations of ammonia nitrogen with a relatively higher pH (Wanapat et al. 2000).

Soon after birth, an early microbial population develops in the rumen of the dairy calf but rapidly develops through weaning towards an adult rumen microbiome (Jami et al. 2013). The intake of dietary complex carbohydrates induces a shift in the rumen microbial population towards plant carbohydrate fermentors, thereby leading to the production of volatile fatty acids. These short-chain fatty acids are of vital importance in the development and growth of rumen papillae and increase of their capacity for nutrient absorption and metabolism (Loor et al. 2016). Diet, microbiome and the host interact in a complex manner, while diverse factors, including age and nutrition, may influence the composition of the rumen microbiome in the calf (Tajima et al. 2001). Using next-generation sequencing, Li et al. (2012) have found age-dependent changes in the calf microbiome between 2 and 6 weeks of age, during which the abundance major groups shifted from of *Prevotella* (at 2 weeks) to domination of *Bacteroidetes* with a 1.5-fold increase (Loor et al. 2016).

Postpartum shifting of dairy cows to a high-grain, high-energy diet can result in a reduced rumen pH with a drastic effect on the microbial population and may cause subacute ruminal acidosis, often with consequences for the animal (Nagaraja and Titgemeyer 2007). Population numbers of bacterial genera involved in starch fermentation, e.g. *Ruminococcus*, *Bifidobacterium* and *Atopobium*, were increased during subacute ruminal acidosis, while the sensitivity of fibrolytic microorganisms (e.g. *Prevotella*, *Treponema*, *Anaeroplasma*, *Papillibacter* and *Acinetobacter*) to low rumen pH resulted in population decreases. It appears that Gram-positive bacteria such as the starch and sugar fermenting *Firmicutes* may selectively be promoted by a low rumen pH, while Gram-negative bacteria such as the *Bacteroidetes* will be reduced (Loor et al. 2016).

The digestive tract of *hindgut fermenters*, also categorised as non-ruminant herbivores (e.g. rabbits, horses, zebras and rhinoceroses), has mono- and polygastric functions. Starch, proteins, vitamins and minerals are digested in the stomach and small intestine. Thanks to a specialised microbiota, the fibrous part of the feed is degraded by fermentation in the hindgut after passage through the small intestine. Zebras and rhinoceroses exhibit lower feeding economics by expelling large amounts of undigested food. To obtain necessary nutrients, they must continually graze and eat huge quantities of food.

As a mammalian herbivore, the elephant has a relatively simple digestive system with features typical of a hindgut fermenter. The diet of the elephant is primarily made up of vegetation, and for conversion of the diverse complex dietary carbohydrates, its digestion system is strongly dependent of cellulose and carbohydrate breakdown by bacterial fermentation. The caecum (with its subdivisions) comprises the major location of fermentation and absorption (up to 44% of the consumed feed) and is located at the junction of the small and large intestines, the total length of which can reach of up to 19 m.

Secondary plant metabolites (SPMs) defend plants against herbivores; yet, indigenous mammalian herbivores seem not to be harmed by the ingestion of toxic plants. Studying the gut microbiota of desert woodrats (*Neotoma lepida*) specialising on the highly toxic creosote bush (*Larrea tridentata*), Kohl et al. (2014) have confirmed the vital importance of gut microbes for herbivores in the consumption of toxic plants.

Phenolics and terpenoids are major groups of SPMs known for their antimicrobial activity. Meta-analysis of 36 studies comprising 185 ruminant treatments with terpenoid compounds, also comprising essential oils and saponins, showed this group to exert the strongest negative effects. Relative to the controls, inhibition of plant cell-wall digestibility amounted to 23% for essential oils, 11% for saponins and 3% for tannins. Moreover, a range of essential oils have shown strong inhibitory activity against several bacterial genera and even against protozoa, thereby in particular inhibiting fermentation and decreasing the rate of bacterial deamination of protein in the lumen (Karasov and Douglas 2013).

## 2.6 Avians

The extreme diversity within the avian group of animals makes it difficult to generalise on their gut microbiota and is also hampered by the limited information available on the numerous bird types. Their varied and diverse diets include insects, seeds, nectar, fruit, plants and even small animals including other birds and rodents (Koutsos et al. 2001). By the absence of teeth, the GIT of birds is adapted for processing of unmasticated food components swallowed whole. Due to their economic importance, the gut of poultry such as the chicken and turkey has probably been studied most extensively within the avian group. Considered to be germ-free at hatching, the mature GIT of the chicken—comprising the crop, small intestine and caecum—rapidly develops to harbour a diverse bacterial population of more than 900 species within 6 weeks, dominated by *Firmicutes* (70%), and with lower numbers of *Bacteroidetes* and *Proteobacteria* (Apajalahti and Vienola 2016). In the ageing chicken, differences in bacterial communities between the ileum and caecum become more distinctive, with a more diverse population in the caecum. Nutrient absorption and bacterial colonisation appear to be key factors determining these differences (Shaufi et al. 2015). The chicken microbiota comprise commensal and pathogenic bacterial species, the latter of which may impact the health of either chickens (*E. coli* and *Clostridium*) or humans (*Salmonella* and *Campylobacter*). Factors such as litter management, diet and feed additives may influence the gut microbiota of the chicken. Moreover, maintenance of a healthy chicken gut microbiota may be achieved by application of pre- and probiotics (as alternative to in-feed subtherapeutic antibiotic administration) and thereby result in the improvement of animal health (Oakley et al. 2014).

Characteristic pathogens of chickens and humans include *E. coli* and *Clostridium* (chickens) and *Salmonella* and *Campylobacter* (humans). A stable gut microbial population fulfils an essential role in the health and general fitness of the host. It presents a strong barrier against the colonisation of pathogens and plays a multiple role for ensuring a well-functioning host physiology and immunity. The chicken gut microbiota provide specific enzymes that are not produced by the chicken itself and thereby enable the depolymerisation of dietary polysaccharides. The short-chain fatty acids (SCFAs) formed in this process play an important role in the maintenance of a healthy gut, with, e.g. butyrate as the primary energy source of colonic epithelia, also improving and/or promoting homeostasis of colonocytes, gut villus morphology development, growth performance and carcass quality characteristics. Moreover, SCFAs have been shown to regulate intestinal blood flow, stimulate enterocyte growth and proliferation, regulate mucin production and affect intestinal immune responses (Chambers and Gong 2011; Panda et al. 2009; Yeoman et al. 2014). The struggle for the elimination of *Salmonella* from poultry has been long and in part still continues. Preceding the currently accepted approach of faecal microbiota transplantation, the principle of competitive exclusion (of *Salmonella*) was first applied as the Nurmi concept in the 1970s. This involved administration of the gut contents of adult chickens (with a 'mature' microbial population) as a saline suspension to newly hatched chicks with a deficient microbiota (Nurmi and Rantala 1973; Mead 2000). Impressive success has been achieved in Finland towards the establishment of *Salmonella*-free chicken flocks; yet, this has not been the case for most other regions and countries. The simultaneous need for the reduction/ban of (in-feed) antibiotics has opened special challenges for the development of new strategies targeted at the chicken gut and beneficial modulation of its microbiota. These include the use of probiotics, prebiotics (including mannan- and xylo-oligosaccharides), phytobiotics and phage therapy, resulting in various levels of success (Yang et al. 2009; Chambers and Gong 2011; Pourabedin 2015).

## 2.7 Invertebrates

### 2.7.1 Diversity of Symbiotic Relationships

The invertebrates represent the majority among the animal species, while, within this group, the insects predominate. Insects rely on diverse and countless symbiotic relationships with beneficial intestinal microorganisms playing a key role in the general fitness and survival of the insect host. Functions include the upgrading of nutrient-poor diets, supporting the digestion of resistant food components. The gut microbiota provide support to the insect host against predators and against invasion by parasites and pathogens while also contributing to intraspecific communication, influencing efficiency as disease vectors and govern mating and reproductive systems. These and numerous other aspects related to the complex relationships between insects and their endosymbionts are adequately reviewed by Engel and Morgan (2013).

Two main categories of nutritional symbioses are recognised in insects, comprising:

- Intracellular associations, typical of the arthropods, with low richness in symbionts, and specialising on restricted diets (blood and plant sap)
- Extracellular associations, typical of most metazoans, represented by complex and diverse endosymbiont communities closely associated with the gut lumen

A major nutritional function of the microbial symbionts is to provide sap-feeding insects with essential amino acids, contributing to cellulose digestion in some termites and wood-feeding insects and supplying nutrients essential for viability and fertility (Douglas 2009; Feldhaar 2011; Pernice et al. 2014). Microbial symbionts may also play a key role in pathogen defence (Dillon et al. 2005) and insecticide resistance (Kikuchi et al. 2012). The overall fitness of the holobiont (host plus all symbionts) will be influenced both by the environment and the association between the host and its symbionts (Feldhaar 2011).

The complex microbial community in the hindgut of soil invertebrates can reach levels of $10^{11}$ cells/mL. The gut microbiota is essential for the digestion of food and plays an ecological role in the global carbon cycle. All three domains (*Bacteria*, *Archaea* and *Eukarya*) are found in the gut of soil invertebrates, with the major bacterial phyla represented by the *Proteobacteria*, *Firmicutes*, *Actinobacteria*, *Bacteroides*, *Flavobacterium* and *Spirochaetes*. The bacteria contribute to the redox status of the gut while fulfilling various metabolic functions in the intestine such as nitrogen fixation and the degradation of cellulose, hemicellulose and aromatic compounds (König 2006).

### 2.7.2 Honey Bee

Probably the best studied microbiota of bees are those associated with the worker caste, that differ from the reproductive castes suggest in the composition of their microbiota, and with naive workers harbouring no or very few bacteria. Honey bee microbial symbionts have been shown to be functionally distinct from that of bumble bees, suggesting the gut symbionts to play a decisive in biological differences among bee species.

Genomic, metagenomic and metatranscriptomic data have revealed interesting insights into metabolic functions, also supported by experimental verification of bacterial physiological activities using culture techniques.

The bee gut microbiota is simple compared with that of most vertebrates. Using deep sampling of gut bacterial communities of individual honey bees, and applying 454 pyrotags for diagnostic regions with amplification from the 16S rRNA gene, Moran et al. (2012) found eight species, with strains sharing >97% 16SrRNA identity, to represent >95% of the gut bacteria in adult honey bee (*Apis mellifera*) workers. The bacterial species *Snodgrassella alvi* and *Gilliamella apicola* appear to be unique to the eusocial honey bees (*Apis* spp.) and the bumble bees (*Bombus* spp.) and are the most prominent Gram-negative members of the intestinal microbial community, each comprising up to 30–39% of the total gut population (Moran et al. 2012).

Accumulating evidence points to *G. apicola* and *S. alvi* as mutualistic symbionts playing an important role in both pathogen defence and nutrition. With sufficient nutritional supply via the host, *S. alvi*, however, was shown to thrive without other gut bacteria and thus not to be dependent of *G. apicola* for survival. In addition to *S. alvi*, *Lactobacillus* and *Bifidobacterium* are also recognised as important fermentative members of the bee gut microbiota (Kwong et al. 2014; Kwong and Moran 2016).

### 2.7.3 Mediterranean Fruit Flies ('Medfly') and Animal Models

With lower microbial diversity than vertebrates, insects in particular are considered as potential models for studying gut–host interactions. The model probably most frequently used for research is the common fruit fly, *Drosophila melanogaster.* Only up to 20 species represent the total gut microbiota of the fruit fly, showing high taxonomic diversity at the species level with the genera *Lactobacillus* and *Acetobacter/Gluconobacter* as the most abundant and consistently found in hosts ranging from laboratory reared to wild-caught flies (Pernice et al. 2014).

In comparison, the Mediterranean fruit fly (*Ceratitis capitata*) shows a much more complex picture. The influence of the microbiota in the medfly has been studied with the aim of reducing its population. The medfly is a crop pest responsible for global crop devastation. In order to control their population, a sterile insect technique (SIT) has been used. This technique involves the release of mass populations of laboratory-bred sterile populations into the wild. When released, the sterile populations compete with wild populations for mating opportunities. As a result, the number of fertile matings is reduced and causes a decline in the population (Dyck et al. 2005). Analysis of the microbiota and subsequent enrichment of species with probiotic potential has been applied subsequently showing the ability to increase SIT effectiveness (Ben Ami et al. 2010; Gavriel et al. 2011; Hamden et al. 2013).

Culture-dependent and culture-independent studies have shown the medfly to have a stable community consisting largely of *Enterobacteriaceae* including the genera *Klebsiella, Enterobacter, Providencia, Pectobacterium, Pantoea, Morganella* and *Citrobacter* (Gavriel et al. 2011; Hamden et al. 2013; Aharon et al. 2012; Behar et al. 2005). Notably *Klebsiella oxytoca* and *Pectobacterium cypripedii* are vertically transmitted through the female by the inoculation of these bacteria during oviposition. Although not vertically transmitted, *Enterobacter* species are also detected ubiquitously within the medfly gut. These bacteria remain with the medfly throughout its life (Behar et al. 2008b). However, the abundance of these species is determined by the population of origin and developmental stage. Apart from a few species such as *Klebsiella oxytoca* and *Citrobacter freundii*, remaining stably colonised within the medfly gut, fluctuating patterns of microbial composition are observed in medflies from different areas (Behar et al. 2008a). Also, 16S rRNA pyrosequencing data of flies in different stages of ontogeny showed separate clustering at 98% OTU similarity. It should be noted that when samples are analysed with 97% similarity value, no clustering is observed, indicating that the diversity occurs within the *Enterobacteriaceae* (Aharon et al. 2013).

By identifying the stable populations, Behar et al. (2008b) suggested the medfly microbiota to be linked to the fitness of the host. They highlighted the stable species had either diazotrophic or pectinolytic capabilities. They suggest these functions aid in the growth of the larvae within the fruit it resides. The diazotrophic function of the bacteria supporting the nitrogen fixation process is believed to be required for the larvae as it grows within the high C/N environment of fruits. In addition, the pectinolytic capacity has a possible role in increasing the supply of metabolisable carbohydrates from the fruit. This is further supported by the fact that these pectinolytic species show a decline in populations during the adult stage, where they feed on honeydew, nectar and fruits (Yuval and Hendrichs 2000). Concurrently, it was reported the microbiota granted advantages in reproductivity as well as longevity upon restricted diet. Maintenance on the nutritionally limited sugar-only diet resulted in greater copulating success by nonantibiotic-treated flies compared to antibiotic treated flies (Ben-Yosef et al. 2008a, b). Further, the nonantibiotic flies had a greater life span on the same sugar-only diet. The same experiment with nutritionally complete results yielded no difference between the groups (Behar et al. 2008c).

In light of this, the sterility inducing irradiation treatment used in SIT was shown to have a significant impact on the structure of the microbiota. Despite the dominance of *Enterobacteriaceae* in both non-irradiated and irradiated populations, the irradiated populations presented with decreased *Klebsiella* species and a resultant increase in *Pseudomonas* species. As *Pseudomonas* colonisation within medflies is known to reduce life span, irradiation was seen to have induced dysbiosis. In addition, Lauzon and Potter (2012) presented electron microscopy evidence that SIT-related irradiation was related to reduced observation of attached bacteria in midgut sections. This reduction is the result of gut tissue damage and malformed peritrophic membrane. However, when these dysbiotic irradiated adult flies were fed *Klebsiella oxytoca*-containing diets, the *Klebsiella* species levels were raised while reducing the levels of *Pseudomonas* species. The alleviation of the dysbiotic state leads to improved mating success by the male flies (Ben Ami et al. 2010).

Recently, the application of probiotics in the SIT medflies has been further investigated. A probiotic mixture of *Klebsiella pneumoniae*, *Citrobacter freundii* and *Enterobacter* species were fed to medfly larvae. The administration in the larval stage bypassed the difficulties of supplementing probiotics to adult flies. The probiotic treatment significantly increased pupal weight, longevity, adult size, flight ability and adult emergence (Hamden et al. 2013). Also, Augustinos et al. (2015) conducted isolation of gut-associated *Enterobacter* species from non-irradiated adult medflies. These species were then diet supplemented as either live or heat-killed form to irradiation sterilised larvae. Both live and heat-killed supplementation groups had improvements in pupal and adult productivity (feed energy uptake) and shorter rearing duration. In particular, the live bacteria supplementation had a more pronounced effect.

## Conclusions

Research over the last decade has firmly established the key role of a healthy gut microbiota in the overall health of the host. *Bacteroidetes* and *Firmicutes* have been considered as the two major phyla dominating the 'normal' gut microbiota. Yet, phyla such as the *Actinobacteria* and *Proteobacteria* also contribute to the well-being of the host, although in diverse ways, and the balance among these and other 'minor' phyla may be influenced by several factors including the diet, the mode of delivery (vaginal or caesarean), the use of antibiotics, stress and the environment. Metagenomics have opened new doors towards studying and understanding the microbiota, their functions and their decisive role in host–microbe interactions. A healthy gut microbiota may serve as an 'insurance' for beneficial host metabolic functions, protection against pathogens (e.g. by maintenance of structural integrity of the gut mucosal barrier) and immunomodulation. Improved understanding of interactive mechanisms supporting a 'normal' gut microbiota has now opened new ways to the scientifically targeted modulation of the gut microbiota, both as a potential therapeutic option for treatment of gastrointestinal disorders and also for improvement of performance (e.g. feed conversion ratio in commercial animals) and the general well-being.

## References


Abe F, Ishibashi N, Shimamura S (1995) Effect of administration of bifidobacteria and lactic acid bacteria to newborn calves and piglets. J Dairy Sci 78:2838–2846. https://doi.org/10.3168/jds.S0022-0302(95)76914-4

Adamczyk-Sowa M, Medrek A, Madej P, Michlicka W, Dobrakowski P (2017) Does the gut microbiota influence immunity and inflammation in multiple sclerosis pathophysiology? J Immunol Res 2017:7904821., 14 p. https://doi.org/10.1155/2017/7904821

Adams MC, Luo J, Rayward D, King S, Gibson R, Moghaddam GH (2008) Selection of a novel direct-fed microbial to enhance weight gain in intensively reared calves. Anim Feed Sci Technol 145:41–52. https://doi.org/10.1016/j.anifeedsci.2007.05.035

Aharon Y, Pasternak Z, Ben Yosef M, Behar A, Lauzon C, Yuval B et al (2012) Phylogenetic, metabolic, and taxonomic diversities shape Mediterranean fruit fly microbiotas during ontogeny. Appl Environ Microbiol 79:303–313. https://doi.org/10.1128/AEM.02761-12. pmid:23104413

Aharon Y, Pasternak Z, Ben Yosef M, Behar A, Lauzon C, Yuval B, Jurkevitch E (2013) Phylogenetic, metabolic, and taxonomic diversities shape Mediterranean fruit fly microbiotas during ontogeny. Appl Environ Microbiol 79:303–313

Akbari MR, Haghighi HR, Chambers JR, Brisbin J, Read LR, Sharif S (2008) Expression of antimicrobial peptides in cecal tonsils of chickens treated with probiotics and infected with *Salmonella enterica* serovar typhimurium. Clin Vaccine Immunol 15:1689–1693. https://doi.org/10.1128/CVI.00242-08

Aly SM, Ahmed YA, Ghareeb AA, Mohamed MF (2008) Studies on Bacillus subtilis and *Lactobacillus acidophilus*, as potential probiotics, on the immune response and resistance of tilapia nilotica (Oreochromis niloticus) to challenge infections. Fish Shellfish Immunol 25:128–136

Apajalahti J, Vienola K (2016) Interaction between chicken intestinal microbiota and protein digestion. Anim Feed Sci Tech 221:323–330. https://doi.org/10.1016/j.anifeedsci.2016.05.004

Asgari F, Madjd Z, Falak R, Bahar MA, Nasrabadi MH, Raiani M, Shekarabi M (2016) Probiotic feeding affects T cell populations in blood and lymphoid organs in chickens. Benef Microbes 28:1–8. https://doi.org/10.3920/BM2016.0014

Augustinos AA, Kyritsis GA, Papadopoulos NT, Abd-Alla AM, Cáceres C, Bourtzis K (2015) Exploitation of the medfly gut microbiota for the enhancement of sterile insect technique: use of *Enterobacter* sp. in larval diet-based probiotic applications. PLoS One 10(9):e0136459. https://doi.org/10.1371/journal.pone.0136459

Bager F, Aarestrup FM, Madsen M, Wegener HC (1999) Glycopeptide resistance in Enterococcus faecium from broilers and pigs following discontinued use of avoparcin. Microb Drug Resist 5:53–56

Balcázar JL, de Blas I, Ruiz-Zarzuela I, Vendrell D, Calvo AC, Márquez I, Gironés O, Muzquiz JL (2007a) Changes in intestinal microbiota and humoral immune response following probiotic administration in brown trout (Salmo trutta). Br J Nutr 97:522–527. https://doi.org/10.1017/S0007114507432986

Balcázar JL, de Blas I, Ruiz-Zarzuela I, Vendrell D, Gironés O, Muzquiz JL (2007b) Enhancement of the immune response and protection induced by probiotic lactic acid bacteria against furunculosis in rainbow trout (Oncorhynchus mykiss). FEMS Immunol Med Microbiol 51:185–193. https://doi.org/10.1111/j.1574-695X.2007.00294.x

Bauer KC, Huus KE, Finlay BB (2016) Microbes and the mind: emerging hallmarks of the gut microbiota-brain axis. Cell Microbiol 18(5):632–644. https://doi.org/10.1111/cmi.12585

Behar A, Yuval B, Jurkevitch E (2005) Enterobacteria-mediated nitrogen fixation in natural populations of the fruit fly *Ceratitis capitata*. Mol Ecol 14:2637–2643. https://doi.org/10.1111/j.1365-294X.2005.02615.x

Behar A, Jurkevitch E, Yuval B (2008a) Bringing back the fruit into fruit fly-bacteria interactions. Mol Ecol 17:1375–1386. https://doi.org/10.1111/j.1365-294X.2008.03674.x

Behar A, Yuval B, Jurkevitch E (2008b) Community structure of the Mediterranean fruit fly microbiota: seasonal and spatial sources of variation. Israel J Ecol Evol 54(2):181–191. https://doi.org/10.1080/15659801.2008.10639612

Behar A, Yuval B, Jurkevitch E (2008c) Gut bacterial communities in the Mediterranean fruit fly (*Ceratitis capitata*) and their impact on host longevity. J Insect Physiol 54:1377–1383. https://doi.org/10.1016/j.jinsphys.2008.07.011

Belkan Y, Hand T (2014) Role of the microbiota in immunity and inflammation. Cell 157(1):121–141. https://doi.org/10.1016/j.cell.2014.03.011

Ben Ami E, Yuval B, Jurkevitch E (2010) Manipulation of the microbiota of mass-reared Mediterranean fruit flies *Ceratitis capitata* (Diptera: Tephritidae) improves sterile male sexual performance. ISME J 4:28–37. https://doi.org/10.1038/ismej.2009.82

Benson AK, Kelly SA, Legge R, Ma F, Low SJ, Kim J, Zhang M, Oh PL, Nehrenberg D, Hua K, Kachman SD, Moriyama EN, Walter J, Peterson DA, Pomp D (2010) Individuality in gut microbiota composition is a complex polygenic trait shaped by multiple environmental and host genetic factors. PNAS 107(44):18933–18938. www.pnas.org/cgi/doi/10.1073/pnas.1007028107

Ben-Yosef M, Jurkevitch E, Yuval B (2008a) Effect of bacteria on nutritional status and reproductive success of the Mediterranean fruitfly *Ceratitis capitata*. Physiol Entomol 33:145–154

Ben-Yosef M, Behar A, Jurkevitch E, Yuval B (2008b) Bacteria-diet interactions affect longevity in the medfly *Ceratitis capitata*. J Appl Entomol 132:690–694

Bhandari SK, Opapeju FO, Krause DO, Nyachoti CM (2010) Dietary protein level and probiotic supplementation effects on piglet response to *Escherichia coli* K88 challenge: performance and gut microbial population. Livest Sci 133:185–188. https://doi.org/10.1016/j.livsci.2010.06.060

Borie C, Albala I, Sánchez P, Sánchez ML, Ramírez S, Navarro C, Morales MA, Retamales AJ, Robeson J (2008) Bacteriophage treatment reduces *Salmonella* colonization of infected chickens. Avian Dis 52:64–67. https://doi.org/10.1637/8091-082007-Reg

Borie C, Sánchez ML, Navarro C, Ramírez S, Morales MA, Retamales J, Robeson J (2009) Aerosol spray treatment with bacteriophages and competitive exclusion reduces *Salmonella enteritidis* infection in chickens. Avian Dis 53:250–254. https://doi.org/10.1637/8406-071008-Reg.1

Brisbin JT, Haghighi HR, Kulkarni R, Sharif S (2007) Regulation of the intestinal immune system by commensal microbiota in avian and ruminant species. CAB Rev Perspect Agric Nutr Nat Resour 2(8). https://doi.org/10.1079/PAVSNNR20072008. http://www.cababstractsplus.org/cabreviews

Brisbin JT, Gong J, Orouji S, Esufali J, Mallick AI, Parvizi P, Shewen PE, Sharif S (2011) Oral treatment of chickens with lactobacilli influences elicitation of immune responses. Clin Vaccine Immunol 18:1447–1455. https://doi.org/10.1128/CVI.05100-11

Brown R, Clarke TB (2016) The regulation of host defences to infection by the microbiota. Immunology 150:1–6. https://doi.org/10.1111/imm.12634

Browne HP, Foster SC, Anonye BO, Kumar N, Neville BAS, Stares MD, Goulding D, Lawley TD (2016) Culturing of 'unculturable' human microbiota reveals novel taxa and extensive sporulation. Nature 533:543–546. https://doi.org/10.1038/nature17645

Brunt J, Austin B (2005) Use of a probiotic to control lactococcosis and streptococcosis in rainbow trout, *Oncorhynchus mykiss* (Walbaum). J Fish Dis 28:693–701. https://doi.org/10.1111/j.1365-2761.2005.00672.x

Buchon N, Broderick AN, Lemaitre B (2013) The gut microbiota—masters of host development and physiology. Nat Rev Microbiol 10:1038
Cerf-Bensussan N, Gaboriau-Routhiau V (2010) The immune system and the gut microbiota: friends or foes? Nat Rev Immunol 10(10):735–744
Chambers JR, Gong J (2011) The intestinal microbiota and its modulation for *Salmonella* control in chickens. Food Res Int 44:3149–3159
Chaucheyras-Durand F, Durand H (2010) Probiotics in animal nutrition and health. Benef Microbes 1:3–9. https://doi.org/10.3920/BM2008.1002
Chiquette J, Allison MJ, Rasmussen MA (2008) Prevotella bryantii 25A used as a probiotic in early-lactation dairy cows: effect on ruminal fermentation characteristics, milk production, and milk composition. J Dairy Sci 91:3536–3543. https://doi.org/10.3168/jds.2007-0849
Clarke G, Stilling RM, Kennedy PJ, Stanton C, Cryan JF, Dinan TG (2014) Gut microbiota: the neglected endocrine organ. Mol Endocrinol 28:1221–1238. https://doi.org/10.1210/me.2014-1108
Collier CT, Carroll JA, Ballou MA, Starkey JD, Sparks JC (2011) Oral administration of *Saccharomyces cerevisiae boulardii* reduces mortality associated with immune and cortisol responses to Escherichia coli endotoxin in pigs. J Anim Sci 89:52–58. https://doi.org/10.2527/jas.2010-2944
Collington GK, Parker DS, Armstrong DG (1990) The influence of inclusion of either an antibiotic or a probiotic in the diet on the development of digestive enzyme activity in the pig. Br J Nutr 64:59–70. https://doi.org/10.1079/BJN19900009
Cruywagen CW, Jordaan I, Venter L (1996) Effect of *Lactobacillus acidophilus* supplementation of milk replacer on preweaning performance of calves. J Dairy Sci 79:483–486. https://doi.org/10.3168/jds.S0022-0302(96)76389-0
D'Argenio V, Salvatore F (2015) The role of the gut microbiome in the healthy adult status. Clin Chim Acta 451:97–102. https://doi.org/10.1016/j.cca.2015.01.003
Dillon RJ, Vennard CT, Buckling A, Charnley AK (2005) Diversity of locust gut bacteria protects against pathogen invasion. Ecol Lett 8:1291–1298
Dinan TG, Cryan JF (2017) Gut–brain axis in 2016: brain–gut–microbiota axis—mood, metabolism and behaviour. Nat Rev Gastroenterol Hepatol 14:69–70. https://doi.org/10.1038/nrgastro.2016.200
Douglas AE (2009) The microbial dimension in insect nutritional ecology. Funct Ecol 23:38–47. https://doi.org/10.1111/j.1365-2435.2008.01442.x
Duncker SC, Lorentz A, Schroeder B, Breves G, Bischoff SC (2006) Effect of orally administered probiotic *E. coli* strain Nissle 1917 on intestinal mucosal immune cells of healthy young pigs. Vet Immunol Immunopathol 111:239–250. https://doi.org/10.1016/j.vetimm.2006.01.017
du Toit M, Franz CM, Dicks LM, Schillinger U, Haberer P, Warlies B, Ahrens F, Holzapfel WH (1998) Characterisation and selection of probiotic lactobacilli for a preliminary minipig feeding trial and their effect on serum cholesterol levels, faeces pH and faeces moisture content. Int J Food Microbiol 40:93–104. https://doi.org/10.1016/S0168-1605(98)00024-5
Dyck VA, Hendrichs JP, Robinson AS (eds) (2005) The sterile insect technique: principles and practice in area-wide integrated pest management. Springer, Dordrecht
Eckburg PB, Bik EM, Bernstein CN, Purdom E, Dethlefsen L, Sargent M, Gill SR, Nelson KE, Relman DA (2005) Diversity of the human intestinal microbial flora. Science 308:1635–1638
Elam NA, Gleghorn JF, Rivera JD, Galyean ML, Defoor PJ, Brashears MM, Younts-Dahl SM (2003) Effects of live cultures of *Lactobacillus acidophilus* (strains NP45 and NP51) and *Propionibacterium freudenreichii* on performance, carcass, and intestinal characteristics, and Escherichia coli strain O157 shedding of finishing beef steers. J Anim Sci 81:2686–2698
Engel P, Morgan NA (2013) The gut microbiota of insects—diversity in structure and Function. FEMS Microbiol Rev 37:699–735. https://doi.org/10.1111/1574-6976.12025
Fasano A (2017) Celiac disease, gut-brain axis, and behavior: cause, consequence, or merely epiphenomenon? Pediatrics 139(3):e20164323

Feldhaar H (2011) Bacterial symbionts as mediators of ecologically important traits of insect hosts. Ecol Entomol 36:533–543. https://doi.org/10.1111/j.1365-2311.2011.01318.x
Fischbach MA, Segre JA (2016) Signaling in host-associated microbial communities. Cell 164:1288–1300. https://doi.org/10.1016/j.cell.2016.02.037
Fowler ME, Bravo WP (2010) Chapter 1: General biology and evolution. In: Medicine and surgery of camelids, 3rd edn. Wiley-Blackwell, Ames
Frizzo LS, Soto LP, Zbrun MV, Bertozzi E, Sequeira G, Armesto RR, Rosmini MR (2010) Lactic acid bacteria to improve growth performance in young calves fed milk replacer and spray-dried whey powder. Anim Feed Sci Technol 157:159–167. https://doi.org/10.1016/j.anifeedsci.2010.03.005
Gavriel S, Jurkevitch E, Gazit Y, Yuval B (2011) Bacterially enriched diet improves sexual performance of sterile male Mediterranean fruit flies. J Appl Entomol 135:564–573. https://doi.org/10.1111/j.1439-0418.2010.01605.x
Ghorbani GR, Morgavi DP, Beauchemin KA, Leedle JAZ (2002) Effects of bacterial direct-fed microbials on ruminal fermentation, blood variables, and the microbial populations of feedlot cattle. J Anim Sci 80:1977–1985. https://doi.org/10.2527/2002.8071977x
Giri SS, Sukumaran V, Oviya M (2013) Potential probiotic Lactobacillus plantarum VSG3 improves the growth, immunity, and disease resistance of tropical freshwater fish, Labeo rohita. Fish Shellfish Immunol 34:660–666. https://doi.org/10.1016/j.fsi.2012.12.008
Godon J-J, Aulazhagan P, Steyer J-P, Hamelin J (2016) Vertebrate bacterial gut diversity: size also matters. BMC Ecol 16:12. https://doi.org/10.1186/s12898-016-0071-2
Gopalakannan A, Arul V (2011) Inhibitory activity of probiotic *Enterococcus faecium* MC13 against Aeromonas hydrophila confers protection against hemorrhagic septicemia in common carp Cyprinus carpio. Aquacult Int 19:973–985. https://doi.org/10.1007/s10499-011-9415-2
Gordon JI (2012) Honor thy gut symbionts redux. Science 336:1251–1253
Ha CW, Lam YY, Holmes AJ (2014) Mechanistic links between gut microbial community dynamics, microbial functions and metabolic health. World J Gastroenterol 20:16498–16517
Hamden H, Guerfali MM, Fadhl S, Saidi M, Chevrier C (2013) Fitness improvement of mass-reared sterile males of *Ceratitis capitata* (Vienna 8 strain) (Diptera: Tephritidae) after gut enrichment with probiotics. J Econ Entomol 106:641–647. https://doi.org/10.1603/EC12362. pmid:23786049
Harikrishnan R, Kim MC, Kim JS, Balasundaram C, Heo MS (2011) Protective effect of herbal and probiotics enriched diet on haematological and immunity status of *Oplegnathus fasciatus* (Temminck & Schlegel) against Edwardsiella tarda. Fish Shellfish Immunol 30:886–893. https://doi.org/10.1016/j.fsi.2011.01.013
Hayakawa T, Masuda T, Kurosawa D, Tsukahara T (2016) Dietary administration of probiotics to sows and/or their neonates improves the reproductive performance, incidence of post-weaning diarrhea and histopathological parameters in the intestine of weaned piglets. Anim Sci J 87(12):1501–1510. https://doi.org/10.1111/asj.12565
He S, Zhou Z, Meng K, Zhao H, Yao B, Ringø E, Yoon I (2011) Effects of dietary antibiotic growth promoter and Saccharomyces cerevisiae fermentation product on production, intestinal bacterial community, and nonspecific immunity of hybrid tilapia (*Oreochromis niloticus female* × *Oreochromis aureus male*). J Anim Sci 89:84–92. https://doi.org/10.2527/jas.2010-3032
Hildebrandt MA, Hoffmann C, Sherrill-Mix SA, Keilbaugh SA, Hamady M, Chen Y-Y, Knight R, Ahima RS, Bushman F, Wu GD (2009) High-fat diet determines the composition of the murine gut microbiome independently of obesity. Gastroenterology 137:1716–1724
Ivanov I, Honda K (2012) Intestinal commensal microbes as immune modulators. Cell Host Microbe 12:496–506
Jami E, Israel A, Kotser A, Mizrahi I (2013) Exploring the bovine rumen bacterial community from birth to adulthood. ISME J 7:1069–1079. https://doi.org/10.1038/ismej.2013.2
Jatkauskas J, Vrotniakiene V (2010) Effects of probiotic dietary supplementation on diarrhoea patterns, faecal microbiota and performance of early-weaned calves. Vet Med Czech 55:494–503

Jones BV, Fung S, Marchesi JR (2010) Comparative metagenomic analysis of plasmid encoded functions in the human gut microbiome. BMC Genomics 11:46. http://www.biomedcentral.com/1471-2164/11/46

Karasov WH, Douglas AE (2013) Comparative digestive physiology. Compr Physiol 3(2):741–783. https://doi.org/10.1002/cphy.c110054

Kikuchi Y, Hayatsu M, Hosokawa T, Nagayama A, Tago K, Fukatsu T (2012) Symbiont-mediated insecticide resistance. Proc Natl Acad Sci U S A 109:8618–8622. www.pnas.org/cgi/doi/10.1073/pnas.1200231109

Kim D, Beck BR, Heo SB, Kim J, Kim HD, Lee SM, Kim Y, Oh SY, Lee K, Do H, Lee K, Holzapfel WH, Song SK (2013) Lactococcus lactis BFE920 activates the innate immune system of olive flounder (*Paralichthys olivaceus*), resulting in protection against *Streptococcus* iniae infection and enhancing feed efficiency and weight gain in large-scale field studies. Fish Shellfish Immunol 35:1585–1590. https://doi.org/10.1016/j.fsi.2013.09.008

Kohl KD, Weiss RB, Cox J, Dale C, Dearing D (2014) Gut microbes of mammalian herbivores facilitate intake of plant toxins. Ecol Lett 17:1238–1246

König H (2006) *Bacillus* species in the intestine of termites and other soil invertebrates. J Appl Microbiol 101(3):620–627. https://doi.org/10.1111/j.1365-2672.2006.02914.x

Koutsos EA, Matson KD, Klasing KC (2001) Nutrition of birds in the order of Psittaciformes: a review. J Avian Med Surg 15(4):257–275. https://doi.org/10.1647/1082-6742(2001)015[0257:NOBITO]2.0.CO;2

Kritas SK, Govaris A, Christodoulopoulos G, Burriel AR (2006) Effect of Bacillus licheniformis and *Bacillus subtilis* supplementation of ewe's feed on sheep milk production and young lamb mortality. J Vet Med A Physiol Pathol Clin Med 53:170–173. https://doi.org/10.1111/j.1439-0442.2006.00815.x

Kwong WK, Moran NA (2016) Gut microbial communities of social bees. Nat Rev Microbiol 14:374–384. https://doi.org/10.1038/nrmicro.2016.43

Kwong KW, Engel P, Koch H, Moran NA (2014) Genomics and host specialization of honey bee and bumble bee gut symbionts. PNAS 111:11509–11514. www.pnas.org/cgi/doi/10.1073/pnas.1405838111

Lagier JC, Hugon P, Khelaifia S, Fournier PE, La Scola B, Raoult D (2015) The rebirth of culture in microbiology through the example of culturomics to study human gut microbiota. Clin Microbiol Rev 28:237–264. https://doi.org/10.1128/CMR.00014-14

Lagier JC, Khelaifia S, Tidjani M et al (2016) Culture of previously uncultured members of the human gut microbiota by culturomics. Nat Microbiol 1:16203. https://doi.org/10.1038/NMICROBIOL.2016.203

Lamari F, Mahdhi A, Chakroun I, Esteban MA, Mazurais D, Amina B, Gatesoupe FJ (2016) Interactions between candidate probiotics and the immune and antioxidative responses of European sea bass (*Dicentrarchus labrax*) larvae. J Fish Dis 39:1421–1432. https://doi.org/10.1111/jfd.12479

Lau JT, Whelan FJ, Herath I, Lee CH, Collins SM, Bercik P, Surette MG (2016) Capturing the diversity of the human gut microbiota through culture-enriched molecular profiling. Genome Med 8:72. https://doi.org/10.1186/s13073-016-0327-7

Lauzon CR, Potter SE (2012) Description of the irradiated and nonirradiated midgut of *Ceratitis capitata* Wiedemann (Diptera: Tephritidae) used for sterile insect technique. J Pest Sci 85(2):217–226

Lee IK, Yoon Chul Kye YC, Kim G, Kim HW, Gu MJ, Umboh J, Maaruf K, Kim SW, Yun C-H (2016) Stress, nutrition, and intestinal immune responses in pigs—a review. Asian Australas J Anim Sci 29(8):1075–1082. https://doi.org/10.5713/ajas.16.0118

Lee SH, Hosseindoust AR, Kim JS, Choi YH, Lee JH, Kwon IK, Chae BJ (2016) Bacteriophages as a promising anti-pathogenic option in creep-feed for suckling piglets: targeted to control *Clostridium spp*. and coliforms faecal shedding. Livest Sci 191:161–164. https://doi.org/10.1016/j.livsci.2016.08.003

Li RW, Connor EE, Li C, Baldwin Vi RL, Sparks ME (2012) Characterization of the rumen microbiota of pre-ruminant calves using metagenomic tools. Environ Microbiol 14:129–139. https://doi.org/10.1111/j.1462-2920.2011.02543.x

Loor JJ, Elolimy AA, McCann JC (2016) Dietary impacts on rumen microbiota in beef and dairy production. Anim Front 6(3):22–29

Lozupone CA, Stombaugh JI, Gordon JI, Jansson JK, Knight R (2012) Diversity, stability and resilience of the human gut microbiota. Nature 489:220–230. https://doi.org/10.1038/nature11550

Malik R, Bandla S (2010) Effect of source and dose of probiotics and exogenous fibrolytic enzymes (EFE) on intake, feed efficiency, and growth of male buffalo (*Bubalus bubalis*) calves. Trop Anim Health Prod 42:1263–1269. https://doi.org/10.1007/s11250-010-9559-5

Maltz MA, Weiss BL, O'Neill M, Wu Y, Aksoy S (2012) OmpA-mediated biofilm formation is essential for the commensal bacterium *Sodalis glossinidius* to colonize the tsetse fly gut. Appl Environ Microbiol 78(21):7760–7768

Mann ER, Landy JD, Bernardo D, Peake STC, Hart AL, Al-Hassi HO, Knight S (2013) Intestinal dendritic cells: their role in intestinal inflammation, manipulation by the gut microbiota and differences between mice and men. Immunol Lett 150:30–40. https://doi.org/10.1016/j.imlet.2013.01.007

Mappley LJ, Tchórzewska MA, Nunez A, Woodward MJ, Bramley PM, La Ragione RM (2013) Oral treatment of chickens with *Lactobacillus reuteri* LM1 reduces *Brachyspira pilosicoli*-induced pathology. J Med Microbiol 62:287–296. https://doi.org/10.1099/jmm.0.051862-0

McFall-Ngaia M, Hadfield MG, Bosch TGC, Carey HV, Domazet-Loso T et al (2013) Animals in a bacterial world, a new imperative for the life sciences. Proc Natl Acad Sci U S A 110:3229–3236. https://doi.org/10.1073/pnas.1218525110

McFarland LV (2014) Use of probiotics to correct dysbiosis of normal microbiota following disease or disruptive events: a systematic review. Br Med J 10:1136

Mead GC (2000) Prospects for 'competitive exclusion' treatment to control salmonellas and other food-borne pathogens in poultry. Vet J 159:111–123

Miller MB, Bassler BL (2001) Quorum sensing in bacteria. Annu Rev Microbiol 55:165–199. https://doi.org/10.1146/annurev.micro.55.1.165

Mohapatra S, Chakraborty T, Kumar V, DeBoeck G, Mohanta KN (2012) J Animal Phys Anim Nutr 97:405–430. https://doi.org/10.1111/j.1439-0396.2012.01301.x

Montalban AA, De Schryver P, Bossier P, Gorkiewicz G, Mulero V, Monroe DG, Galindo-Villegas J (2015) Selective manipulation of the gut microbiota improves immune status in vertebrates. Front Immunol 6:512. https://doi.org/10.3389/fimmu.2015.00512

Moran NA, Hansen AK, Powell JE, Sabree ZL (2012) Distinctive gut microbiota of honey bees assessed using deep sampling from individual worker bees. PLoS One 7(4):e36393

Morgan XC, Huttenhower C (2012) Chapter 12: Human microbiome analysis. PLoS Comput Biol 8(12):e1002808. https://doi.org/10.1371/journal.pcbi.1002808

Mortha A, Chudnovskiy A, Hashimoto D, Bogunovic M, Spencer SP, Belkaid Y, Merad M (2014) Microbiota-dependent crosstalk between macrophages and ILC3 promotes intestinal homeostasis. Science 343(6178):1249288. https://doi.org/10.1126/science.1249288

Mountzouris KC, Tsirtsikos P, Kalamara E, Nitsch S, Schatzmayr G, Fegeros K (2007) Evaluation of the efficacy of a probiotic containing *Lactobacillus, Bifidobacterium, Enterococcus*, and *Pediococcus* strains in promoting broiler performance and modulating cecal microflora composition and metabolic activities. Poult Sci 86:309–317. https://doi.org/10.1093/ps/86.2.309

Muegge BD, Kuczynski J, Knights D, Clemente JC, González A, Fontana L, Henrissat B, Knight R, Gordon JI (2011) Diet drives convergence in gut microbiome functions across mammalian phylogeny and within humans. Science 332(6032):970–974

Nagaraja TG, Titgemeyer EC (2007) Ruminal acidosis in beef cattle: the current microbiological and nutritional outlook. J Dairy Sci 90(Suppl 1):E17–E38. https://doi.org/10.3168/jds.2006-478

Nayak SK, Swain P, Mukherjee SC (2007) Effect of dietary supplementation of probiotic and vitamin C on the immune response of Indian major carp, *Labeo rohita* (Ham.) Fish Shellfish Immunol 23:892–896. https://doi.org/10.1016/j.fsi.2007.02.008

Nikoskelainen S, Ouwehand AC, Bylund G, Salminen S, Lilius EM (2003) Immune enhancement in rainbow trout (*Oncorhynchus mykiss*) by potential probiotic bacteria (*Lactobacillus rhamnosus*). Fish Shellfish Immunol 15:443–452. https://doi.org/10.1016/S1050-4648(03)00023-8

Nurmi E, Rantala M (1973) New aspects of *Salmonella* infection in broiler production. Nature 241:210–211

Oakley BB, Lillehoj HS, Kogut MH, Kim WK, Maurer JJ et al (2014) The chicken gastrointestinal microbiome. FEMS Microbiol Lett 360:100–112

Ohland LC, Jobin C (2015) Microbial activities and intestinal homeostasis: a delicate balance between health and disease. Cell Mol Gastroenterol Hepatol 1:28–40

Panda AK, Rama Rao SV, Raju MVLN, Shyam Sunder G (2009) Effect of butyric acid on performance, gastrointestinal tract health and carcass characteristics in broiler chickens. Asian Aust J Anim Sci 22:1026–1031

Panigrahi A, Kiron V, Satoh S, Hirono I, Kobayashi T, Sugita H, Puangkaew J, Aoki T (2007) Immune modulation and expression of cytokine genes in rainbow trout *Oncorhynchus mykiss* upon probiotic feeding. Dev Comp Immunol 31:372–382. https://doi.org/10.1016/j.dci.2006.07.004

Park SC, Shimamura I, Fukunaga M, Mori KI, Nakai T (2000) Isolation of bacteriophages specific to a fish pathogen, *Pseudomonas plecoglossicida*, as a candidate for disease control. Appl Environ Microbiol 66:1416–1422. https://doi.org/10.1128/AEM.66.4.1416-1422.2000

Penry DL, Jumars PA (1987) Modeling animal guts as chemical reactors. Am Nat 129:69–96

Pernice M, Simpson SJ, Ponton F (2014) Towards an integrated understanding of gut microbiota using insects as model systems. J Insect Physiol 69:12–18. https://doi.org/10.1016/j.jinsphys.2014.05.016

Pollmann M, Nordhoff M, Pospischil A, Tedin K, Wieler LH (2005) Effects of a probiotic strain of *Enterococcus faecium* on the rate of natural chlamydia infection in swine. Infect Immun 73:4346–4353. https://doi.org/10.1128/IAI.73.7.4346-4353.2005

Pourabedin M (2015) Effects of mannan-oligosaccharides and xylo-oligosaccharides on the chicken gut microbiota. PhD-Thesis, McGill University, Montreal

Qin J, Li R, Raes J, Arumugam M, Burgdorf KS et al (2010) A human gut microbial gene catalogue established by metagenomic sequencing. Nature 464:59–65

Rajilic-Stojanovic M, De Vos WM (2014) The first 1000 cultured species of the human gastrointestinal microbiota. FEMS Microbiol Rev 8:996–1047

Ran C, Huang L, Hu J, Tacon P, He S, Li Z, Wang Y, Liu Z, Xu L, Yang Y, Zhou Z (2016) Effects of dietary live and heat-inactive baker's yeast on growth, gut health, and disease resistance of Nile tilapia under high rearing density. Fish Shellfish Immunol 56:263–271. https://doi.org/10.1016/j.fsi.2016.07.001

Reece WO (2013) Functional anatomy and physiology of domestic animals, 4th edn. Wiley, Hoboken, p 359. ISBN 111868589X, 9781118685891

Ringø E, Vadstein O (1998) Colonization of *Vibrio pelagius* and *Aeromonas caviae* in early developing turbot (*Scophthalmus maximus* L.) larvae. J Appl Microbiol 84:227–233. https://doi.org/10.1046/j.1365-2672.1998.00333.x

Roos TB, Tabeleão VC, Dümmer LA, Schwegler E, Goulart MA, Moura SV, Corrêa MN, Leite FPL, Gil-Turnes C (2010) Effect of *Bacillus cereus* var. Toyoi and *Saccharomyces boulardii* on the immune response of sheep to vaccines. Food Agric Immunol 21:113–118. https://doi.org/10.1080/09540100903443691

Sanders JG, Powell S, Kronauer DJ, Vasconcelos HL, Frederickson ME, Pierce NE (2014) Stability and phylogenetic correlation in gut microbiota: lessons from ants and apes. Mol Ecol 23(6):1268–1283. https://doi.org/10.1111/mec.12611

Sandhu KV, Sherwin E, Schellekens H, Stanton H, Dinan TG, Cryan JF (2017) Feeding the microbiota-gut-brain axis: diet, microbiome, and neuropsychiatry. Transl Res 179:223–244. https://doi.org/10.1016/j.trsl.2016.10.002

Sekirov I, Russell SL, Antunes LC, Finlay BB (2010) Gut microbiota in health and disease. Physiol Rev 90:859–904. https://doi.org/10.1152/physrev.00045.2009

Sen S, Ingale SL, Kim YW, Kim JS, Lohakare JD, Kim EK, Kim HS, Ryu MH, Kwon IK, Chae BL (2012) Effect of supplementation of *Bacillus subtilis* LS 1-2 to broiler diets on growth performance, nutrient retention, caecal microbiology and small intestinal morphology. Res Vet Sci 93:264–268. https://doi.org/10.1016/j.rvsc.2011.05.021

Shaufi MAM, Sieo CC, Chong CW, Gan HM, Ho YW (2015) Deciphering chicken gut microbial dynamics based on high-throughput 16S rRNA metagenomics analyses. Gut Pathogens 7:4. https://doi.org/10.1186/s13099-015-0051-7

Smith PA (2015) Brain, meet gut. Nature 526:312–314

Sommer F, Bäckhed F (2013) The gut microbiota—masters of host development and physiology. Nat Rev Microbiol 10:1038

Spanggaard B, Huber I, Nielsen J, Sick EB, Pipper CB, Martinussen T, Slierendrecht WJ, Gram L (2001) The probiotic potential against vibriosis of the indigenous microflora of rainbow trout. Environ Microbiol 3:755–765. https://doi.org/10.1046/j.1462-2920.2001.00240.x

Sudo N (2014) Microbiome, HPA axis and production of endocrine hormones in the gut. Adv Exp Med Biol 817:177–194. https://doi.org/10.1007/978-1-4939-0897-4_8. In: Mark L, Cryan JF (eds) Microbial endocrinology: the microbiota-gut-brain axis in health and disease. Springer, New York, Heidelberg

Taheri HR, Moravej H, Tabandeh F, Zaghari M, Shivazad M (2010) Efficacy of combined or single use of *Lactobacillus crispatus* LT116 and *L. johnsonii* LT171 on broiler performance. Br Poult Sci 51:580–585. https://doi.org/10.1080/00071668.2010.508491

Tajima K, Aminov RJ, Nagamine T, Matsui H, Nakamura M, Benno Y (2001) Diet-dependent shifts in the bacterial population of the rumen revealed with realtime PCR. Appl Environ Microbiol 67:2766–2774. https://doi.org/10.1128/AEM.67.6.2766-2774.2001

Taoka Y, Maeda H, Jo JY, Jeon MJ, Bai SC, Lee WJ, Yuge K, Koshio S (2006) Growth, stress tolerance and nonspecific immune response of Japanese flounder *Paralichthys olivaceus* to probiotics in a closed recirculating system. Fish Sci 72:310–321. https://doi.org/10.1111/j.1444-2906.2006.01152.x

Thaiss CA, Zmora N, Levy M, Elinav E (2016) The microbiome and innate immunity. Nature 535:65–74. https://doi.org/10.1038/nature18847

Toro H, Price SB, McKee AS, Hoerr FJ, Krehling J, Perdue M, Bauermeister L (2005) Use of bacteriophages in combination with competitive exclusion to reduce *Salmonella* from infected chickens. Avian Dis 49:118–124. https://doi.org/10.1637/7286-100404R

Turnbaugh PJ, Ley RE, Hamady M, Fraser-Liggett CM, Knight R, Gordon JI (2007) The human microbiome project. Nature 449(7164):804–810

Turnbaugh PJ, Ridaura VK, Faith JJ, Rey FE, Knight R, Gordon JI (2009) The effect of diet on the human gut microbiome: a metagenomic analysis in humanized gnotobiotic mice. Sci Transl Med 1(6):6ra14. https://doi.org/10.1126/scitranslmed.3000322

Van den Bogaard AE, Stobberingh EE (1999) Antibiotic usage in animals: impact on bacterial resistance and public health. Drugs 58:589–607

Van den Bogaard AE, Bruinsma N, Stobberingh EE (2000) The effect of banning avoparcin on VRE carriage in The Netherlands. J Antimicrob Chemother 46:146–148

Villena J, Saavedra L, Hebert EM, Suda Y, Masumizu Y, Albarracin L, Clua P, Ikeda-Ohtsubo W, Kitazawa H (2017) Draft genome sequence of *L. plantarum* MPL16, a wakame-utilizing immunobiotic strain isolated from swine feces. Genome Announc 5:e00006–e00017. https://doi.org/10.1128/genomeA.00006-17

Walter J, Britton RA, Roos S (2011) Host-microbial symbiosis in the vertebrate gastrointestinal tract and the *Lactobacillus reuteri* paradigm. Proc Natl Acad Sci U S A 108(Suppl 1):4645–4652. www.pnas.org/cgi/doi/10.1073/pnas.1000099107

Wanapat M, Ngarmsang A, Korkhuntot S, Nontaso N, Wachirapakorn C, Beakes G, Rowlinson P (2000) A comparative study on the rumen microbial population of cattle and swamp buffalo raised under traditional village conditions in the northeast of Thailand. Asian Australas J Anim Sci 13(7):918–921

Wang F, Roy S (2016) Gut homeostasis, microbial dysbiosis, and opioids. Toxicol Pathol 45:150–156

Wang X, Farnell YZ, Peebles ED, Kiess AS, Wamsley KG, Zhai W (2016) Effects of prebiotics, probiotics, and their combination on growth performance, small intestine morphology, and resident *Lactobacillus* of male broilers. Poult Sci 95:1332–1340. https://doi.org/10.3382/ps/pew030

Williams BB, Van Benschoten AH, Cimermancic P, Donia MS, Zimmermann M, Taketani M, Ishihara A, Kashyap PC, Fraser JS, Fischbach MS (2014) Discovery and characterization of gut microbiota decarboxylases that can produce the neurotransmitter tryptamine. Cell Host Microbe 16(4):495–503. https://doi.org/10.1016/j.chom.2014.09.001

Wu GD, Chen J, Hoffmann C et al (2009) Linking long-term dietary patterns with gut microbial enterotypes. Science 334(6052):105–108. https://doi.org/10.1126/science.1208344

Yang Y, Iji PA, Choct M (2009) Dietary modulation of gut microflora in broiler chickens; a review of the role of six kinds of alternatives to in-feed antibiotics. Worlds Poult Sci J 65:97–114. https://doi.org/10.1017/S0043933909000008

Yang F, Hou C, Zeng X, Qiao S (2015) The use of lactic acid bacteria as a probiotic in swine diets. Pathogens 4:34–45. https://doi.org/10.3390/pathogens4010034

Yeo S, Park H, Ji Y, Park S, Yang Y, Lee J, Mathara JM, Shin HK, Holzapfel W (2015) Influence of gastrointestinal stress on autoinducer-2 activity of two *Lactobacillus* species. FEMS Microbiol Ecol 91:fiv065. https://doi.org/10.1093/femsec/fiv065

Yeo S, Lee S, Park H, Shin H, Holzapfel W, Huh CS (2016) Development of putative probiotics as feed additives: validation in a porcine-specific gastrointestinal tract model. Appl Microbiol Biotechnol 100:10043–10054. https://doi.org/10.1007/s00253-016-7812-1

Yeoman CJ, Chia N, Jeraldo P, Sipos M (2014) The microbiome of the chicken gastrointestinal tract. Anim Health Rev 13(1):89–99. https://doi.org/10.1017/S1466252312000138

Yu HF, Wang AN, Li XJ, Qiao SY (2008) Effect of viable *Lactobacillus fermentum* on the growth performance, nutrient digestibility and immunity of weaned pigs. J Anim Feed Sci 17:61–69

Yuval B, Hendrichs J (2000) Behavior of flies in the genus *Ceratitis* (Dacinae: Ceratidini). In: Aluja M, Norrbom A (eds) Fruit flies (Tehpritidae): phylogeny and evolution of behavior. CRC Press, Boca Raton, pp 429–457

Zhang ZF, Kim IH (2014) Effects of multistrain probiotics on growth performance, apparent ileal nutrient digestibility, blood characteristics, cecal microbial shedding, and excreta odor contents in broilers. Poult Sci 93:364–370. https://doi.org/10.3382/ps.2013-03314

# Protective Cultures for the Safety of Animal-Derived Foods

3

Jordi Rovira and Beatriz Melero

## 3.1 Definitions

Nowadays, consumers demand fresh, "natural" with no chemical additives, minimally processed foods that are safe with a long shelf life and more recently with some functional properties that make these foods to improve their health or prevent the appearance of some illness. All is due to drastic changes in the traditional way of eating, due to the new lifestyle that implies less time for cooking. The demand of more time for leisure after a long and stressing working day has raised the concepts of "convenience" or "ready-to-eat" foods that allow consumers to reduce the preparation time of foods for eating at home. Hence, there is a growing interest in foods that are both "fresh" and "convenient." This approach is described by technological terms as "minimal processing" and "hurdle technology" (Leistner and Gorris 1995; Leistner 2000). The concept behind these approaches implies the combination of different mild technologies, high-pressure processing, pulsed electric fields, or modified atmosphere packaging being some of them, allowing to keep "fresh" the sensory properties of foods and to substitute the more drastic interventions like heat treatment or the use or chemical additives. In this view, the use of food grade microbial cultures has played for a long time a relevant role, whereas new promising applications are appearing. According to their intended use in foods, "microbial food cultures" can be classified as starter cultures, probiotics, or protective cultures.

### 3.1.1 Starter Cultures

Traditionally microbial food cultures have been used, since the onset of mankind, to produce fermented foods. According to the definition given by Campbell-Platt (1994), a fermented food is the one which "has been subjected to the action of

J. Rovira (✉) • B. Melero
Department of Biotechnology and Food Science, University of Burgos, Burgos, Spain
e-mail: jrovira@ubu.es



D. Di Gioia, B. Biavati (eds.), *Probiotics and Prebiotics in Animal Health and Food Safety*, https://doi.org/10.1007/978-3-319-71950-4_3

microorganisms or enzymes so that desirable biochemical changes cause significant modification of the food." In fact, the processing of fermented foods consists on to give the conditions that allow the optimal growth of fermenting microorganisms that transform the raw material by its metabolic activity in the final desirable fermented food. The relation between fermentation and the metabolic activity of different microorganisms was established at the end of the nineteeth century by Louis Pasteur. Since then, the addition of selected food cultures in different kinds of food matrices to obtain the corresponding fermented food has become a usual practice. These selected food cultures receive the name of starter cultures, because when added in an appropriate amount (around $10^6$–$10^7$ CFU $g^{-1}$ or $mL^{-1}$), they allow to start the fermentation process quicker in comparison with a natural fermentation process. Nowadays, the main role of starter cultures is technological, due to their ability to behave as a preservation hurdle. In that sense, starter cultures contribute to an improvement in hygienic safety, sensory attractiveness, and high and constant levels of quality and shelf life (Hammes and Knauf 1994).

### 3.1.2 Probiotics

According to the definition put forward by FDA and WHO jointly, probiotics are "Live microorganisms which when administered in adequate amounts conferring a health benefit to the host." This concept emphasizes the idea that food is not only vital for living but also plays a role in the prevention and reduction of risk factors for several diseases and is also capable of enhancing certain vital physiological functions. In that sense, foods were those cultures were added, either singly or in combinations, with this aim are considered as functional foods.

Probiotics have been receiving growing attention in recent years for their ability to modulate physiological functions such as nutrition and metabolism, immunity, gut-brain axis, and pathological phenomena such as infections, cancer, inflammation, allergies, autoimmune diseases, oxidative stress, cardiovascular problems, and psychiatric disorders which affect health and life quality (Mangiapane et al. 2015). For more detailed information about food cultures used as probiotics, readers are invited to read the rest of the chapters of this book.

### 3.1.3 Protective Cultures

Protective cultures are those microbial food cultures that are only added with the specific aim to inhibit pathogens and/or to extend the shelf life of the product, while changing as little as possible, its sensory properties. This concept was suggested in 1994 by Lücke in his classical work about fermented meat products, where he proposed the selection of lactic meat starter cultures with the aim to be used as possible "biological preservatives" for non-fermented meats. One year later, Holzapfel et al. (1995) proposed as well the use of "biological" or "milder" preservation approaches by using the so-called protective cultures or their metabolites, notably enzymes and bacteriocins. In the same work, these authors stated that this biological preservation has the aim of reducing the health risks without changing the sensory quality of the product. One year later, in 1996, Stiles defined the terms *biopreservation* or *biocontrol* as "the use of natural or controlled microbiota, or its antibacterial products to extend the shelf life and enhance the safety of foods." More recently, Vignolo et al. (2015) have defined *bioprotection* mixing the former definitions as "the use of antagonistic microorganisms and/or their metabolic products to inhibit undesirable organisms in order to enhance food safety and extend shelf-life without significantly altering the sensory properties of the product."

A distinction is sometimes made between starter cultures and protective cultures, although in reality it may be the same culture applied for different purposes under different conditions. For instance, a metabolic activity in a starter culture such as acid production has a technological importance, while antimicrobial action may constitute a secondary effect. On the contrary, for a protective culture, the functional objectives are the inverse (Holzapfel et al. 1995). In that sense, according to Elsser-Gravesen and Elsser-Gravesen (2014), all starter cultures are per se also protective cultures, but not all protective cultures are also starter cultures. A clear distinction between starter cultures and bioprotective cultures is therefore neither possible nor meaningful.

In summary, we can say in a simple way that starter cultures have a technological and sensory purpose and probiotic cultures have a functional aim in the host, whereas the protective cultures have the objective to improve safety and shelf life of foods (Table 3.1). However, the line between these different functions is really very thin, and in many occasions, the same strains or a combination of different strains

**Table 3.1** Bioprotection due to microbial food cultures

| Bioprotection | |
|---|---|
| Food cultures (microorganisms directly) | Metabolic products produced by microorganisms externally (indirect) |
| Source: same ecological niche where the action should be exerted | • Organic acids<br>• Diacetyl<br>• Reuterin/reutericyclin<br>• Bacteriocins<br>• Bacteriolysins<br>• Antifungal substances |
| Starter cultures<br>• Technological and sensory purpose<br>• Fermented foods | |
| Probiotics<br>• Influence in host functionality<br>• Functional foods and some fermented foods (dairy) | |
| Source: could be different from the one where the action should be exerted | |
| Protective cultures<br>• Only added with a protection aim against pathogens and spoilers<br>• Fermented and non-fermented foods | |
| Legal status<br>• History of use<br>• USA: generally recognized as safe (GRAS)<br>• EU: qualified presumption of safety (QPS) | Legal status<br>• Requires specific regulatory approval as preservatives |

**Table 3.2** Diversity of microbial food cultures with beneficial use

| Phylum | Genus |
|---|---|
| *Actinobacteria* | *Bifidobacterium* (8)[a], *Brevibacterium (3), Corynebacterium (4), Brachybacterium (2), Microbacterium (1), Arthrobacter (4), Kocuria (2), Micrococcus (2), Propionibacterium (5), Streptomyces (1)* |
| *Firmicutes* | *Bacillus (3), Carnobacterium (3), Enterococcus (3), Tetragenococcus (2), Lactobacillus (84), Pediococcus (3), Leuconostoc (12), Oenococcus (1), Weissella (9), Macrococcus (1), Staphylococcus (15), Lactococcus (3), Streptococcus (3)* |
| *Proteobacteria* | *Acetobacter (9), Gluconobacter (9), Hafnia (1), Halomonas (1), Zynomonas (1)* |
| *Ascomycota* | *Lecanicillium (1), Geotrichum (1), Yarrowia (1), Galactomyces (1), Scopulariopsis (1), Fusarium (2), Candida (10), Cyberlindnera (2), Debaryomyces (1), Dekkera (1), Hanseniaspora (3), Kazachstania (2), Kluyveromyces (1), Lachancea (2), Metschnikowia (1), Pichia (4), Saccharomyces (4), Schwanniomyces (1), Starmerella (1), Trigonopsis (1), Wickerhamomyces (1), Zygosaccharomyces (1), Zygotorulaspora (1), Kluyveromyces (1), Torulaspora (1), Schizosaccharomyces (1), Neurospora (1), Aspergillus (4), Penicillium (7)* |
| *Basidiomycota* | *Cystofilobasidium (1), Guehomyces (1)* |
| *Zygomycota* | *Mucor (4), Rhizopus (4)* |

Based on "Inventory of Microbial Food Cultures" (Bourdichon et al. 2012)
[a]Number of species of that genus

applied together can exert more than one action in the food or even do in the host, in case of using probiotic strains in those combinations. What seems clear is that all three types of microbial food cultures have a common trait which is their ability to play an important role in *biopreservation* either directly, by the action of native or added microbial cultures in the food matrix, or indirectly by the production of antagonistic substances produced ex situ and afterward added in the food matrix as a preservatives (Table 3.2). The addition of these protective cultures in food matrices matches with the fresh, natural, and additive-free concept of the consumer and can be used as "green preservatives" and clean label.

## 3.2 Source of Protective Cultures

Several families, genera, and species can be considered as potential protective cultures. Table 3.2 shows a summary of an inventory of microbial food cultures with a documented use in food, established as a result of a joint project between the International Dairy Federation (IDF) and the European Food and Feed Cultures Association (EFFCA) in 2002 and recently reviewed by Bourdichon et al. (2012). This inventory includes 195 bacterial species and 69 species of yeasts and mold with desirable contributions to the food fermentation. In this list, probiotic species were only included if they take part of cultures used in a food fermentation process. In a broad sense, we can consider that all those species can exert a protective effect on food matrices at least as a competitive microorganism in the ecological niche (food matrix) where these microorganisms grow, although they do not produce other antimicrobial substances. Moreover, most of them are susceptible to be used as starter cultures for certain kind of foods.

As it is shown in Table 3.2, a high number of these species belong to lactic acid bacteria (LAB), which represents around 66% of the total bacteria and almost 50% of all microorganisms listed on the table. Historically, LAB have been used as biopreservative microorganisms, playing a key role in the diversity of fermented foods including milk, meats, fish, vegetables, and sourdoughs by producing a quick acidification of the different raw material. With increasing consumer pressure, toward more natural food preservatives, LAB have become ideal candidates for commercial exploitation. Their GRAS (generally regarded as safe) and qualified presumption of safety (QPS) status in the EU has enjoyed consistent and growing interest and, consequently, the scientific exploration of their potential as biocontrol agents. Moreover, LAB apart from their preserving qualities are also associated with health-promoting/probiotic properties (Crowley et al. 2013). Nowadays, LAB comprises a complex group including species from 17 genera: *Aerococcus*, *Alloiococcus*, *Carnobacterium*, *Dolosigranulum*, *Enterococcus*, *Globicatella*, *Lactobacillus*, *Lactococcus*, *Lactosphaera*, *Leuconostoc*, *Melissococcus*, *Oenococcus*, *Pediococcus*, *Streptococcus*, *Tetragenococcus*, *Vagococcus*, and *Weissella*. Although genotypically *Bifidobacterium* are completely different from LAB, some authors, from the practical point of view, have included them in this bacterial group due to their ability to produce also lactic acid as final metabolite. Among all of them, species from *Lactobacillus* and *Leuconostoc* genera are the most used as source of protective cultures.

In general, candidate species and strains used as potential starter cultures or probiotics are commonly isolated from the same ecological niche (food matrix) where it is aimed to be applied. In that sense, starter cultures employed in fermented products are usually isolated from the same type of fermented foods, whereas probiotics from the digestive tract of animals and humans are where the beneficial function should be exerted. However, in the case of protective cultures, the original source not always is related to the food they will be applied for its further preservation. In this respect, the source of these protective cultures is very variable (Table 3.3), because the important trait here is the ability of the protective strain to exert an antimicrobial effect against the spoilage and pathogen bacteria and this feature is common to all microbial cultures as it was mentioned above. In the literature it is possible to find three different situations taking into account the source and the application of protective cultures:

1. Same food, same product: this situation reflects the idea that the protective strains have been isolated from the same ecological niche (food) where these strains will exert their potential protective action, for example, in non-fermented products like vacuum-packed meat products, LAB often causes spoilage; however, they can also keep the meat product sensory fresh throughout the storage period (Bredholt et al. 1999), and these meat products are eminent sources for cultures, which can be used for biopreservation purposes.

**Table 3.3** Some examples of different sources of protective culture strains and their potential applications

| Protective culture | Source | Application | References |
|---|---|---|---|
| *Leuconostoc pseudomesenteroides PCK18* | Maasai milk (Kenya) | Fresh suckling-lamb meat packaged under modified atmosphere (MAP) | Osés et al. (2015) |
| *Lactobacillus harbinensis* K.V9.3.1Np and *Lactobacillus rhamnosus* K.C8.3.1I | Cow and goat milks | Yogurt | Delavenne et al. (2015) |
| *Lactobacillus plantarum* | 897 (LAB) isolated from: different herbs, fruits, and vegetables | Cheese | Cheong et al. (2014) |
| *Lactococcus lactis* subsp. *lactis* | 16 commercial LAB or bifidobacteria strains | Vacuum-packed raw salmon | Ibrahim and Vesterlund (2014) |
| *Leuconostoc pseudomesenteroides* PCK18 *Bifidobacterium longum* subsp. *longum* PCB133 | Maasai milk (Kenya) Newborn infant | Chicken products packaged under modified atmosphere (MAP) | Melero et al. (2013) |
| *Enterococcus faecium* PCD71 and *Lactobacillus fermentum* ACA-DC179 | 635 LAB strains Different food products and human origin | Ground raw chicken meat | Maragkoudakis et al. (2009) |
| *Lactobacillus sakei* and *Lactococcus lactis* | 181 chilled meat and processed meat | Vacuum-packaged lamb and beef | Jones et al. (2009) |
| *Carnobacterium divergens* and *Carnobacterium maltaromaticum* | 120 strains: Meat (beef and pork products) Culture collections Sea foods (cod, halibut, salmon, shrimps, and roe products) | Meat and fish products | Laursen et al. (2005) |
| *Lactobacillus sakei* subsp. *carnosus* | 91 strains Different meat products | Cooked cured ham | Vermeiren et al. (2004) |
| *Leuconostoc carnosum* 4010 | Vacuum-packed meat products | Vacuum-packed meats | Budde et al. (2003) |

(continued)

**Table 3.3** (continued)

| Protective culture | Source | Application | References |
|---|---|---|---|
| *Leuconostoc mesenteroides* L124 and *Lactobacillus curvatus* L442 | Dry-fermented sausages | Vacuum or modified atmosphere-packaged sliced cooked cured pork | Mataragas et al. (2003) |
| *Lactococcus lactis* subsp. *lactis* | Fresh cheese | Raw sausages (Merguez) | Benkerroum et al. (2003) |
| *Lactobacillus sakei* | Vacuum-packaged ham and servelat sausage | Cooked, sliced, vacuum-packaged meats | Bredholt et al. (2001) |

2. Same food, different product: in this case the protective strain has been isolated from the same type of food, but their application will be done in another product, for instance, protective strains isolated from fermented meats will be applied in non-fermented meats.
3. Different food or non-food origin: in this situation the ecological niche of the isolated protective strains is completely different from the one which will be applied, for instance, protective strains coming from milk or human origin applied in raw meats (Carlini et al. 2010; Melero et al. 2013; Osés et al. 2015).

In Table 3.3, some examples from the literature that shows the variability in the source of protective cultures are reported.

The selection of protective cultures has been described through different approaches. Bredholt et al. (1999) investigated the use of indigenous lactic acid bacteria (LAB) as protective cultures in cooked meat products, inoculating *Listeria monocytogenes* to commercial cooked meat products and monitoring the growth of the pathogen and the LAB indigenous population. Further, LAB were isolated from samples where *L. monocytogenes* failed to grow. *Lactobacillus sakei* was the identified species and further selected because of its appropriate antimicrobial, sensorial, physiological characteristics together with satisfying growing behavior in a cooked meat product without affecting its sensory properties. Vermeiren et al. (2006a) based the selection of potential protective cultures for cooked meat applications considering LAB strains that are homofermentative, salt tolerant, psychrotrophic, and adapted to meat-based substrates, with antibacterial capacities against spoiling and pathogen bacteria, and furthermore do not influence the sensory properties of the meat products on which they are applied.

In addition to LAB, in some fermented meat products such as dry sausages or dry-cured meats and in some cheeses, it is normal to use food cultures of white mycelia molds or yeasts like *Penicillium nalgiovense*, *Penicillium chrysogenum*, or *Debaryomyces hansenii* to exclude the growth in the surface of other toxigenic molds and to obtain a more attractive external appearance of the product for consumer. Acosta et al. (2009) have selected several *Penicillium* spp. strains producing antifungal proteins that could be useful to prevent hazards due to the growth of mycotoxigenic molds such as *P. echinulatum*, *P. commune*, and *Aspergillus niger* in the surface of dry-cured meats. Moreover, Rodríguez et al. (2015) inoculated on the surface of dry-cured Iberian hams two protective cultures consisting of the antifungal protein-producing *P. chrysogenum* RP42C and a mix of selected autochthonous nontoxigenic molds, limiting the growth of ochratoxin A-producing molds and its accumulation in this product throughout the processing period. In the same way, Nuñez et al. (2015) isolated from the surface of dry-cured meats two autochthonous *D. hansenii* strains that reduced significantly the growth of the ochratoxigenic *P. verrucosum*, keeping its counts under the level considered as hazardous for the mycotoxin presence. These types of microbial strains are normally used in the food industry and can be found commercially for these purposes.

## 3.3 Mode of Action

The aim of this section is to give a summary and a clear glance over how protective cultures exert their protective action against food spoilers and pathogens. As it has been shown in Tables 3.2 and 3.3, LAB are the most relevant group of bacteria associated with this protective role, probably because they are well adapted to grow in different ecological niches, as ferment foods, in non-fermented foods playing a role of spoilers or even do in the digestive tract as probiotics. They are good competitors in those complex ecological niches by using several antagonistic mechanisms. For this reason, many LAB can be used as protective cultures, although depending on the food matrix to protect, other types of microorganisms can be used as well. However, due to its relevant role in bioprotection from now onward, LAB will be the main focus of this section. There are many excellent reviews in the literature dealing with biopreservation mechanisms of LAB in general (Deegan et al. 2006; García et al. 2010; Nes et al. 2012; Elsser-Gravesen and Elsser-Gravesen 2014; Gálvez et al. 2014) and in some specific cases or applications such as dairy (Beshkova and Frengova 2012; Arqués et al. 2015), meat (Lücke 2000; Työppönen et al. 2003a; Aymerich et al. 2011; Vignolo et al. 2015), and aquatic products (Calo-Mata et al. 2008) that can be revised for further and more detailed information about this issue.

Food-grade microorganisms can produce a multitude of different substances that are inhibitory to other microorganisms. These mechanisms are part of the natural balance in complex microbial ecosystems. By exploiting this ability, it is possible to use protective cultures to design "natural" preservation systems that ensure an adequate safety and shelf life of foods while maintaining the desired quality of the product. The strategy to defend a population territory by the release of antimicrobial substances that inhibit growth or even kill competitors is known as amensalism (Gálvez et al. 2014).

In the last two or three decades, substantial research activities have aimed to develop protective cultures to be applied in different food matrices, especially those from animal origin and mainly in raw meats and sea food where no other preservatives can be added as well as in ready-to-eat (RTE) foods with a special focus on the control of *L. monocytogenes*.

In that sense, it is possible to distinguish between three different types of protective cultures which through antagonistic mechanisms exert this protective action: (1) bacteriocinogenic cultures acting as antagonistic by producing one or more bacteriocins or bacteriocin-like compounds, (2) non-bacteriocinogenic cultures using another antagonistic strategy, and (3) protective cultures producing antifungal substances (Vermeiren et al. 2006a; Elsser-Gravesen and Elsser-Gravesen 2014). Table 3.4 shows a summary of different antagonistic activities described for LAB.

**Table 3.4** Antimicrobial substances produced by LAB that can exert bioprotection action against food spoilers and pathogens

| Antimicrobial | Substances | Mode of action | Target |
|---|---|---|---|
| Organic acids | Lactic acid | Decrease the pH | Broad spectrum against non-acidophilic microorganisms |
| | Acetic acid (heterofermentative LAB) | Undissociated hydrophobic form | |
| | Propionic acid (in traces amounts) | Disruption of the cytoplasmic membrane and interference with membrane potential | |
| | *Carboxylic acids*: | Reduction in intracellular pH | |
| | Cinnamic acid derivatives | | |
| | D-glucuronic acid | | |
| | Salicylic acid | | |
| | Benzoic | | |
| | Hydroxybenzoic acids | | |
| $CO_2$ | Heterofermentative LAB | Creates an anaerobic environment | Aerobic bacteria |
| | | Carbonic acid, when it dissolves in water | |
| Diacetyl | 2,3-Butanedione | Gram-negative bacteria are generally more sensitive | Gram-negative bacteria are more sensitive than Gram-positive bacteria |
| | Produced during citrate fermentation by some strains | | |
| Hydrogen peroxide | Produced by flavoprotein oxidases in presence of oxygen | Oxidative damage of proteins | Antimicrobial |
| | | Increase membrane permeability | |
| Reuterin | 3-Hydroxypropionaldehyde (3-HPA) | Inhibit DNA synthesis | Broad spectrum including Gram-positive and Gram-negative bacteria, viruses, fungi, and protozoa |
| | | Oxidative stress to the cell by reaction between aldehyde group of reuterin with thiol groups of small molecules and proteins | |
| Reutericyclin | *N*-acylated tetramic acid, negatively charged | Dissipation of the proton motive force | Gram-positive bacteria |

(continued)

**Table 3.4** (continued)

| Antimicrobial | Substances | Mode of action | Target |
|---|---|---|---|
| Bacteriocins | Class I. posttranslationally modified bacteriocins (<15 kDa) | | Antibacterial |
| | • Type A: linear, cationic | Pore forming, cationic | Antilisterial |
| | – AI: modification by two enzymes | | |
| | – AII: modification by single enzyme | | |
| | • Type B: non-cationic, globular | Enzyme inhibition, non-cationic | |
| | • Type C: two peptides | | |
| | Class II. unmodified bacteriocins (<15 kDa) | Potent antilisterial activity | |
| | • IIa: pediocin-like | Cause leakage of the membrane | |
| | • IIb: two peptides | | |
| | • IIc: doesn't fit | | |
| | • IId: leaderless bacteriocins | | |
| | • IIe: cyclic bacteriocins | | |
| Bacteriolysins | Large peptides (<30 kDa), termolabile Class III bacteriocins | Cell lysis by cell wall hydrolysis | |
| Peptides with antifungal activity (Fungicides) | Medium length peptide, TV35b, from *Lactobacillus pentosus* | Not clear | Antifungal activity |
| | Small-peptide (3 kDa) *Lactobacillus coryniformis* subsp. *coryniformis* strain Si3 | Some, similar to Class II bacteriocins | |
| | Peptide (43 kDa) from *Lactobacillus paracasei* subsp. *paracasei* M3 | Defensin-like protein found in pear | |
| | | Antihypersensitive and antimicrobial-like peptides contained in caseins | |
| Fatty acids | Long-chain hydroxylated fatty acids (C8–C12) | Not clear | Antibacterial and antifungal activity against a broad spectrum of yeasts and molds |
| | 5-Oxododecanoic acid | Partition of the lipid bilayers of fungal membranes resulting in loss of membrane integrity | |
| | 3-Hydroxy decanoic acid | | |
| | 3-Hydroxy-5-dodecenoic acid | | |

**Table 3.4** (continued)

| Antimicrobial | Substances | Mode of action | Target |
|---|---|---|---|
| Phenyllactic acid | Phenyllactic acid (PLA) and 4-hydroxyphenyllactic acid (OH-PLA) | | Broad spectrum antibacterial against Gram-positive and Gram-negative bacteria |
| | | | Antifungal action, active against yeasts and molds at mg $mL^{-1}$ concentrations |
| Cyclic dipeptides | Low molecular weight | Not defined yet | Antimicrobial, antitumoral, and antifungal activities |
| | 2,5-dioxopiperazines | | Also involved in quorum sensing processes |
| | Cyclo(glycyl-L-leucyl) | | |
| | Cyclo(Phe-Pro) | | |
| | Cyclo(Phe-OH-Pro) | | |
| Lactones | Tetrahydro-4-hydroxy-4-methyl-2H-pyran-2-one (mevalonolactone) | | Antibacterial, antifungal, and antiviral activities |
| | δ-Dodecalactone | | |

### 3.3.1 Bacteriocinogenic Cultures

Bacteriocinogenic cultures are those that are able to produce bacteriocins, in most of the cases ribosomally synthesized peptides or proteins with antimicrobial activity. Nowadays, this term is mostly used to describe the small, heat-stable cationic peptides synthesized by Gram-positive bacteria, especially LAB, which display a wider spectrum of inhibition (García et al. 2010). However, it has been speculated that all members of the Eubacteria and also of the Archaea, when freshly isolated from their natural ecosystems, are probably equipped with the capability of expressing bacteriocins. Gram-positive bacteriocins, and in particular LAB bacteriocins, comprise a very heterogeneous group regarding their primary structure, composition, and physicochemical properties, and their classification is still a matter of discussion and disagreement. Several attempts have been done in that sense, but consensus is difficult due to the increasing rate of new bacteriocins that appear and their heterologous composition, structure, and mode of action. In Table 3.5 there is a summary of the different classification schemes proposed by several authors (Klaenhammer 1993; Cotter et al. 2005; Heng et al. 2007; Nes et al. 2012; Gálvez

**Table 3.5** Evolution of classification of Gram-positive bacteriocins

| Klaenhammer (1993) | Cotter et al. (2005) | Heng et al. (2007) | Nes et al. (2012) | Gálvez et al. (2014) | Examples |
|---|---|---|---|---|---|
| *Class I. lantibiotics* | *Class I. lantibiotics* | *Class I. lantibiotics* | *Class I. lantibiotics* | *Class I. lantibiotics* | |
| No subtypes | Up to 11 subclasses | Type A (Ia): linear | Type A (Ia): Linear | Type A: Linear | |
| | | Type AI | Type AI | | Nisin A/U/Z |
| | | Type AII | Type AII | | Lacticin 481, salivaricin A |
| | | Type B (Ib): globular | Type B (Ib): globular | Type B: globular | Mersacidin, cinnamycin |
| | | Type C (Ic): multicomponent | Type C (Ic): Multicomponent | | Lacticin 3147, plantaricin W |
| | | | | Labyrinthopeptides (contain labionin) | Labyrinthopeptins A1, A2 |
| | | | | Sactibiotics | Subtilosin A, thuricin CD |
| *Class II: unmodified peptides* | *Class II: unmodified peptides* | *Class II: unmodified peptides* | *Class II: nonlantibiotics* | *Class II: nonlantibiotics* | |
| IIa: pediocin-like | IIa: pediocin-like | IIa: pediocin-like | IIa: pediocin-like Four subgroups | IIa: pediocin-like | Pediocin PA-1, plantaricin C19, sakacin P, penocin A |
| IIb: two components | IIb: two components | IIb: two components | IIb: two components | IIb: two components | Enterocin L50, leucocin H, plantaricin E/F |
| IIc: thiol activated | IIc: cyclic peptides | IIc: miscellaneous | IIc: miscellaneous | IIc: cyclic peptides | |
| | IId: miscellaneous | | IId: leaderless bacteriocins | IId: other single-peptide, nonpediocin molecules | Lactococcins A, leucocin B and Q |
| | | | IIe: Cyclic peptides | | Enterocin AS-48, leucocyclicin Q |
| *Class III: large heat-labile proteins* | Eliminated | *Class III: large heat-labile proteins* | Not clearly state | *Class III (bacteriolysins)* | |
| | | IIIa: lysins | | | Lysostaphin, enterolysin A, Helveticin J |
| | | IIIb: non-lytic | | | |
| *Class IV: complex (+ lipids or carbohydrates)* | Eliminated | *Class IV: cyclic peptides* | | | |

et al. 2014). In general, there is a consensus for the two major classes of bacteriocins: Class I (lantibiotics) and Class II (nonlantibiotics).

Class I bacteriocins or "lantibiotics" are small peptides that undergo extensive posttranslational modification and contain lanthionine and β-methyl lanthionine residues, as well as dehydrated amino acids (Gálvez et al. 2014). Traditionally lantibiotics have been divided in two types A and B. Type A are linear, amphiphilic, cationic peptides up to 34 residues long, and they act by forming pores in target bacteria membranes causing leakage of low molecular weight compounds that eventually leads to cell death. More recently, Type A lantibiotics have been split in two subtypes according to the number of enzymes involved in their posttranslational modification. Type AI needs two enzymes, while Type AII only needs one enzyme (Nes et al. 2012). Type B lantibiotics include shorter, up to 19 residues, with a globular structure, and some of them act by inhibiting enzyme activities that have impact on cell wall synthesis. Other authors include in this class a third type of bacteriocins named Type C or Ic that comprises two peptides lantibiotic bacteriocins or multicomponent lantibiotics (Heng and Tagg 2006; García et al. 2010). The well-known bacteriocin nisin belongs to Class I Type AI and was identified by the first time in 1929 from a strain of *Lactococcus lactis* and now approved for use as a food additive in around 50 countries. Nisin has been used safely in the food industry as a preservative (E234) for over 40 years without the appearance of significant bacterial resistance. Nisin dissipates the proton motive force of the target cell by forming a pore through the cytoplasmic membrane which causes the flux of essential energy (ATP) and different ions from the cell.

Class II bacteriocins are small peptides (4–6 kDa) and heat-stable, which contrary to lantibiotics do not undergo extensive posttranslational modification, except for cleavage of a leader peptide (when present) during transport out of the cell. These Class II bacteriocins have been divided into several subclasses that, again, grouped the different types of bacteriocins depending on the authors. There is no discussion that Class IIa includes the pediocin-like bacteriocins characterized by their potent antilisterial activity. This subclass IIa has been divided into three or four different groups on the basis of sequence similarities and differences in the more variable C-terminal domain (Nes et al. 2012). The bactericidal mode of action of these subclass IIa bacteriocins such as pediocin PA-1 appears to involve three basic steps: (1) binding to the cytoplasmic membrane of target cell in a mannose-specific phosphotransferase (PTS) system, (2) insertion of the bacteriocin molecules into the membrane, and (3) formation of pores that permeabilizes the membrane disrupting the proton motive force and leading to cell death (Heng et al. 2007). Subclass IIb has been assigned the multicomponent bacteriocins that need the participation of two peptides to exert its action (Cotter et al. 2005; Heng et al. 2007; Nes et al. 2012). It was observed that when both peptides are combined, a strong synergistic activity has been shown. The target of the two-peptide bacteriocins was the membrane, where they caused leakage of monovalent cations, depletion of the ATP pool, dissipation of the membrane potential, and eventually death of the target microorganism (Nes et al. 2012). In subclass IIc Cotter et al. (2005) include the cyclic bacteriocins, characterized by a peptide bond between the C- and

N-terminus, whereas other authors include in this subclass a miscellaneous group of bacteriocins that does not fit with the previous subclasses IIa and IIb (Heng et al. 2007). The subclass IIc comprises various modes of action such as membrane permeabilization, specific inhibition of septum formation, and pheromone activity (Héchard and Sahl 2002). Moreover, Nes et al. (2012) added two more subclasses: subclass IId that includes the leaderless bacteriocins and subclass IIe comprising all circular or cyclic bacteriocins. This latter has been proposed as Class IV bacteriocins by Heng et al. (2007).

Another discrepancy has risen with the Class III bacteriocins that include large and heat-labile proteins with a distinct mechanism of action from other Gram-positive bacteriocins, also called nonbacteriocin lytic proteins and termed as bacteriolysins. According to Cotter et al. (2005), these peptides are completely different from bacteriocins and are out of their classification. On the contrary, Heng et al. (2007) retained these large bacteriocin groups as Class III, and they are subdivided into IIIa (bacteriolysins) and IIIb (non-lytic proteins).

As it has been seen above, based on their cationic and their hydrophobic nature, most bacteriocins act as membrane permeabilizers. Pore formation, which seems to be target-mediated, leads to the total or partial dissipation of the proton motive force, ultimately causing cell death of related Gram-positive bacteria. However, Gram-negative bacteria are intrinsically resistant due to the protective role of the external membrane. Nevertheless, some bacteriocins can become active in combination with other outer membrane-destabilizing agents such as EDTA (Heng et al. 2007).

It is quite frequent that the same species can produce different types of bacteriocins, for instance, *Lactobacillus plantarum* can produce a plantaricin C (Class I, Type A); plantaricin W (Class I, two peptides lantibiotic); plantaricin S, plantaricin 423, and plantaricin C19 (Class IIa); plantaricin EF (PlnE/F); and plantaricin JK (PlnJ/K) (Class IIb). In the same way, many *Leuconostoc* species also produce multiple bacteriocins, for example, *Ln. pseudomesenteroides* QU 15 produces one Class IIa (leucocin A) and two Class IId bacteriocins (leucocin Q and N), while *Ln. mesenteroides* TA33a has been reported to produce three bacteriocins: leucocins A and C (Class IIa) and leucocin B (Class IId) (Wan et al. 2015).

Bacteriocin-producing strains can be applied as the main starter cultures in fermented foods offering technological properties required for the fermentation or as an adjunct protective culture in combination with bacteriocin-resistant starter strains. They can also be applied as protective cultures in non-fermented foods without any adverse effects on the sensory properties of food. Bacteriocins also contribute to probiotic functionality of some LAB acting as colonizing peptides that facilitate the introduction or dominance of the bacteriocin-producing strain into the GIT niche. They may act as antimicrobial peptides directly killing other bacteria, as signaling peptides through quorum sensing and cross talk with bacterial communities or as signaling cells of the host immune system (Arqués et al. 2015).

Several Class I bacteriocinogenic-producing protective cultures have been applied in different food products. Nisin-producing *Lc. lactis* strains from Spanish fermented sausages were effective in inhibiting closely related LAB, *L. monocytogenes*, *Clostridium perfringens*, *Bacillus cereus*, and *Staphylococcus aureus* (Rodríguez et al. 1995).

Moreover, *Lactobacillus sakei* L45 isolated from Norwegian dry sausages and *Lb. sakei* 148 from Spanish fermented sausages secrete the lantibiotic lactocin S, whose moderate spectrum of activity comprises LAB and *Clostridium* (Aymerich et al. 1998).

As it was shown above, Class IIa bacteriocins have a narrow spectrum of activity but display a high specific activity against *L. monocytogenes.* Significant inhibition of *L. monocytogenes* growth in German-type fresh Mettwurst has been observed for *Lb. sakei* Lb 706 strain-producing sakacin A (Schillinger et al. 1991). Other Class II bacteriocins produced by a *Lb. curvatus* CRL705 strain that synthesize lactocin 705 a two-peptide (Class IIb) and lactocin AL705 pediocin-like (Class IIa) were used as a bioprotective culture in meat discs, and it was effective in preventing the growth of *Listeria innocua*, while *Brochothrix thermosphacta* experienced a reduction of 1.5 log along the storage period (Castellano et al. 2008). Moreover, Hugas (1998) found also that *Lb. sakei* CTC494, a sakacin K producer, exerted a bacteriostatic effect on vacuum-packaged fresh meat products. Similarly, Benkerroum et al. (2003) indicated that during fermentation of merguez, a red, spicy mutton or beef-based fresh sausage, reduction in *L. monocytogenes* counts was greater in samples fermented with the bacteriocinogenic *Lactococcus lactis* subsp. *lactis* strain than in those fermented with the bacteriocin-negative culture. On the other hand, Jacobsen et al. (2003) showed in sliced meat products that the bacteriocinogenic culture of *Leuconostoc carnosum* 4010 was more effective than leucocins inhibiting the growth of *L. monocytogenes*. More examples can be found in the literature showing the protective action of bacteriocinogenic cultures in several food matrices.

### 3.3.2 Non-bacteriocinogenic Cultures

Not all microbial food cultures used as protective cultures are able to produce bacteriocins. However, they are also able to inhibit the growth of other microorganisms in a food matrix in some specific conditions, as it will be shown later on in this section. This means that protective cultures use other strategies to exert their protective action. Table 3.4 shows different antagonistic substances used by LAB in addition to the production of bacteriocins. Among these substances, it is possible to find active antagonistic metabolites such as organic acids (lactic, acetic, formic, propionic, butyric) and diverse antagonistic compounds (carbon dioxide, ethanol, hydrogen peroxide, fatty acids, acetoin, diacetyl, reuterin, reutericyclin). Moreover, another hypothesis to explain the protective action of non-bacteriocinogenic cultures is directed toward competition for nutrients (Buchanan and Bagi 1997). In that sense, Nilsson et al. (2005) showed that a non-bacteriocinogenic *Carnobacterium piscicola* strain reduced the growth of *L. monocytogenes* partly by glucose depletion in vitro. Probably, a more complex combined effect of production of antimicrobials and competition for or depletion of specific nutrients might explain the protective effect of these cultures (Devlieghere et al. 2004).

Bredholt et al. (2001) have reported that a non-bacteriocin-producing strain of *Lb. sakei* was able to inhibit the growth of *L. monocytogenes* inoculated on cooked ham and stored at 4 and 8 °C. These authors pointed out that fast growth rate and greater competitiveness for micronutrients give a selective advantage of *Lb. sakei* over slower growing competitors. They also suggested as contributory factors the lower pH (acidification) and the bacteriostatic action of undissociated lactic acid. Similarly, Juven et al. (1998) attributed to lactic acid (ca. 50 mM), produced by a commercial psychotropic strain *Lb. alimentarius* (Flora Carn L-2), the reduction of around 4 log CFU $g^{-1}$ of *L. monocytogenes* in vacuum-packaged ground beef stored for 9 weeks at 4 °C. Other authors have compared the effect against spoilage or pathogen microbial population in different food matrices of a bacteriocinogenic vs. non-bacteriocinogenic protective cultures. Alves et al. (2006) have reported that the growth of two strains of *L. monocytogenes* was significantly suppressed in sliced cooked vacuum-packaged ham when the samples were co-inoculated with either bacteriocin-producing or non-producing *Lb. sakei* strains. Similarly, Zanette et al. (2015) have not found significant differences in using two *Lb. plantarum* starter cultures (bacteriocinogenic and bacteriocin-negative strains) in reducing the levels of *L. monocytogenes* during sausages maturation. Kaban et al. (2010) showed in sliced beef Bologna-type sausages inoculated with *L. monocytogenes* and packaged with vacuum and MAP that the pathogen strain increased only in the vacuum-packaged samples inoculated with the non-bacteriocinogenic strain *Lb. sakei* Lb 706b. However, in MAP, both the bacteriocinogenic strain *Lb. sakei* Lb 706 and *Lb. sakei* Lb 706b suppressed the growth of *L. monocytogenes* completely, probably due to the synergic effect with $CO_2$. On the contrary, Jones et al. (2009) applying the same *Lb. sakei* strains in vacuum-packaged lamb meat did not find significant differences in the reduction of *L. monocytogenes* population between samples inoculated with both strains. These authors reported that inhibition due a protective strain in a certain food did not always correlate with inhibition observed in earlier media-based studies. They found that *Lb. sakei* 27, which was non-inhibitory in agar-based studies, was associated with reduced recovery of *C. jejuni*, while strain *Lb. sakei* 63, which was inhibitory in agar-based studies, did not affect the target population in vacuum-packaged beef (Jones et al. 2009). This supports the conclusion that because antimicrobial effects of LAB can differ under different substrate and storage conditions, simple laboratory screening methods, while convenient, may not detect LAB strains with inhibitory properties in other environments. Moreover, Saraoui et al. (2016) have shown that the inhibition mechanism of *L. monocytogenes* by the protective strain of *Lactococcus piscium* CNCM I-4031 is cell-to-cell contact dependent.

### 3.3.3 Protective Cultures with Antifungal Activity

Molds in foods can be considered as bacteria, beneficial, spoilers, or toxigenic depending on the role of some specific species and strains in different foods where they grow. In some cases, fungal spoilage of foods represents a major cause of concern for food manufacturers. However, in other foods such as mold-ripened cheeses or some dry-cured and fermented meat products, molds play an important role from the technological and sensory point of view. In that sense, in some occasions it will be necessary to avoid their growth, while in others, it will be important to favor the growth and colonization of nontoxigenic molds, contributing in uniqueness of the final product.

The use of protective cultures against the occurrence of molds in some dry and fermented meat products and cheeses has been used commercially since many years ago with sensory, technological, and safety purposes. In dry-cured and fermented meat products, the ripening process conditions favor the growth of a large fungal population on the surface which is essential for the flavor development and more attractive appearance of the product. However, some mold strains growing in these products are able to produce mycotoxins. Among them, ochratoxin A (OTA) is the mycotoxin most frequently found in dry-cured ham and other types of ripened meat products (Rodríguez et al. 2015). The use of protective strains of white mycelia molds such *Penicillium nalgiovense* and *P. chrysogenum* or yeasts like *Debaryomyces hansenii* are well known to avoid the growth, by competitive exclusion, of toxigenic molds with black or green mycelia on the surface of those products.

As it was mentioned above, some foodborne fungi, both yeasts and molds, cause serious spoilage in stored food, and some of them may also produce health-damaging mycotoxins such as aflatoxins or ochratoxin A among others. Moreover, consumer demands for minimally processed foods and reduced use of chemical preservatives stimulating the research on antifungal lactic acid bacteria as biopreservatives. In the last decade, there has been an increase in the literature reporting the production of these antifungal substances by LAB and the use of these active strains as protective cultures as well (Schnürer and Magnusson 2005; Crowley et al. 2013). Among these substances, it is possible to find some organic acids such as acetic, formic, propionic, butyric, caproic, and *n*-valeric acid; other carboxylic acids have also been described including cinnamic acid derivatives, D-glucuronic acid, and salicylic acid. Phenyllactic acid (PLA) and its derivate 4-hydroxyphenyllactic acid (OH-PLA) are perhaps the most extensively studied antifungal organic acids from LAB. They possess a broad antibacterial and antifungal spectrum and usually play a synergistic role with other metabolites. Cyclic dipeptides, known also as 2,5-dioxopiperazines, have been also identified by its antifungal activity from several LAB strains together with some proteinaceous compounds and 3-hydroxylated fatty acids (Tables 3.1 and 3.4). To get a deeper insight in these kinds of antifungal substances, we recommend to read excellent reviews such as those from Schnürer and Magnusson (2005) and Crowley et al. (2013). Although these compounds have more application in cereal derivate foods such as bakery products, fruits, vegetables, and animal feed, some interesting applications have been also described in dairy products including cheeses and yogurt, which are susceptible to fungal contamination (Garcha and Natt 2011; Delavenne et al. 2012).

Another factor that influences the mode of action of protective culture application is the inoculum level. It plays an important role in the protective cultures' efficacy, and a compromise must be reached between the protective effect and the modification of the sensory properties due to their own metabolism. A minimum inoculation level is necessary to be competitive and assure the protective action, especially in meat and seafood products where it is not possible to eliminate or reduce the indigenous bacterial population of raw material. Several authors have reported the influence of the inoculum level. Budde et al. (2003) showed in a vacuum-packaged meat sausage stored at 5 °C that the addition of *Leuconostoc carnosum* 4010 at $10^7$ CFU $g^{-1}$ immediately reduced the number of viable *L. monocytogenes* cells to a level below the detection limit, while at a level of $10^5$ CFU $g^{-1}$, no inhibitory effect against *L. monocytogenes* was observed during the first week of storage. Similarly, Vermeiren et al. (2006a) studied the influence of the inoculum concentration on the protective effect, comparing two inoculum levels, $10^5$ vs. $10^6$ CFU $g^{-1}$, of *Lb. sakei*. The lower amount of inoculum failed to prevent the growth of *L. monocytogenes* at 7 °C, whereas the highest inoculum was successful in the prevention. Delavenne et al. (2015) also found that to show an effective antifungal activity against *Yarrowia lipolytica* in yogurt, *Lb. harbinensis* K.V9.3.1Np must, at the time of contamination, reach a concentration greater than $2.5 \times 10^6$ CFU $g^{-1}$, because lower concentrations were unable to control the growth of that spoilage yeast. Moreover, Osés et al. (2015) established a difference between the protective culture *Leuconostoc pseudomesenteroides* PCK 18 and *L. monocytogenes* inocula higher than 2 log CFU $g^{-1}$ to exert an effective control of the pathogen grow. At the same time, a lower concentration of protective culture was not enough to control its growth in vacuum and MAP suckling lamb.

Antagonistic substances such as some purified bacteriocins like nisin can be directly applied to the food matrix. However, this strategy has some drawbacks, for instance, bacteriocins may bind to the food fat or protein particles, and some food additives, natural proteases, or other inhibitors may inactivate them. In addition, the effect may be seen only in a narrow pH range, which excludes their utilization in many food products (Työppönen et al. 2003a). Moreover, they should be declared in the label as additives, while the uses of protective cultures directly to the food matrix do not require regulatory approval or label declarations as they are considered GRAS of QPS and are frequently considered a more attractive strategy to incorporate bacteriocins in foods (Table 3.1).

## 3.4 Application of Protective Cultures in Animal Derivate Foods

In this section a brief review will be done to point out the application of protective cultures directly in different animal derivate food matrices, such as dairy, meat, and aquatic products. Only the real food application of protective cultures will be considered along this section, so in vitro experiments or the direct application of antimicrobial substances such as bacteriocins alone will not be discussed.

### 3.4.1 Dairy Products

Dairy products present several advantages in relation with other animal derivate food matrices. Pasteurization of milk before manufacture of dairy products is often required or recommended. This represents that most of the vegetative bacterial indigenous population is eliminated prior to proceed with the fermentation step. That means that the food cultures added can grow with less competitive pressure. However, some traditional, highly appreciated fermented dairy foods are still made from raw milk. Moreover, although pasteurization destroys potential pathogenic microorganisms, postpasteurization processing can lead to the recontamination of dairy products mainly by *Listeria monocytogenes*, *Salmonella* spp., *Staphylococcus aureus*, and pathogenic *Escherichia coli*. Soft cheeses can support the growth of *Listeria* introduced after processing independently of the use of raw or pasteurized milk. Recalls of dairy products contaminated with the pathogen, especially those coming from soft cheeses, are relatively frequent (Arqués et al. 2015). In dairy products, in some occasions, starter, probiotic, and protective cultures can be the same bacterial species and strains. In that sense, the development of multistrain probiotic dairy products with good technological properties and with improved characteristics to those shown by individual strains has gained increased interest being able to act not only as protective cultures in foods but also as probiotics capable to exert a protective action against infections.

Several types of bacteriocins, such as nisin, lacticin, pediocin, piscicolin, or enterocins, have been applied successfully to prevent the growth or reduce the population of *L. monocytogenes* in different kinds of cheese, cream, cheese sauces, or yogurt. However, as direct addition of bacteriocins to food systems could result in some loss of the antimicrobial activity due to the diffusion into the food matrix or the interaction with food components, different strategies of incorporation have been considered such as microencapsulation in liposomes and nanovesicles from soy lecithin or attached to produce bioactive films for surface decontamination (Arqués et al. 2015). Nevertheless, bacteriocinogenic cultures as starter or adjunct cultures in cheese making allow the production in situ of bacteriocins, reduce the cost of the biopreservation, and do not require regulatory approval. Several authors have successfully used different bacteriocinogenic cultures to control spoilers and mainly *L. monocytogenes* in dairy products. This later is considered the main foodborne pathogen of concern in cheese and dairy products. Therefore, many different studies have been focused on the application of antilisterial starter or adjunct cultures for inhibition of these bacteria. In summary, bacteriocinogenic cultures are the most frequently used protective cultures in dairy products, although more recently antifungal protective cultures are also gaining consideration in their application against fungal spoilage of dairy products (Table 3.6).

#### Application of Bacteriocinogenic Protective Cultures in Dairy Products

These protective cultures are mainly used in dairy fermented products, where they can exert their action as starter or even do as probiotic cultures.

In cheese making, many *Lactococcus lactis* strains producing Class I bacteriocins such as nisin or lacticin 3147 or lacticin 481 have been successfully used to increase food safety or extend the shelf life of these kinds of products. However, some problems arise in using nisin-producing cultures in cheese making, the most obvious being a lack of compatibility between the bacteriocin-producing strain and

**Table 3.6** Application of protective cultures to dairy products (some examples)

| Product | Protective culture | Inoculum | Target microorganism | Inoculum | Effect | References |
|---|---|---|---|---|---|---|
| Yogurt | *Lactobacillus harbinensis* K.V9.3.1Np | $10^6$ CFU $mL^{-1}$ | *Yarrowia lipolytica* (spoiler yeast) | $10^2$–$10^3$ CFU per yogurt | A total inhibition of *Y. lipolytica* | Delavenne et al. (2015) |
| Cottage cheese | *Lactococcus lactis* nisin A | 1% of $10^8$ CFU $mL^{-1}$ ($10^6$ CFU $mL^{-1}$) | *L. monocytogenes* F2365 | $10^3$ CFU $mL^{-1}$ | No growth of *L. monocytogenes* in 7 days | Dal Bello et al. (2012) |
| Yogurt | *Streptococcus salivarius* subsp. *thermophilus* B (Bac +) and *Lactobacillus delbrueckii* subsp. *bulgaricus* CY (Bac−) | $10^6$ CFU $mL^{-1}$ | *L. monocytogenes*<br>*St. aureus* | $10^3$ CFU $mL^{-1}$<br>$10^6$ CFU $mL^{-1}$ | *L. monocytogenes* below the detection limit<br>*St. aureus* survive<br>5 days shelf life extension | Benkerroum et al. (2002) |
| Cottage cheese | *Lactococcus lactis* DPC4275 (lacticin 3147) | 1% cheese vat | *L. monocytogenes* Scott A | $10^4$ CFU $g^{-1}$ | Reduction of <10 CFU $g^{-1}$ in 5 days | McAuliffe et al. (1999) |
| Cheddar cheese | *Lactococcus lactis* DPC3147 (lacticin 3147) | Inoculum of 0.7% as part of starter culture in milk | NSLAB (nonstarter LAB) | Native | Reduction | Ryan et al. (1996) |
| Camembert cheese | *Lactococcus lactis* subsp. *lactis* (nisin + and nisin −) | Inoculum of 2% in milk<br>$10^7$ CFU $mL^{-1}$ | *L. monocytogenes* | $10^1$–$10^3$ CFU $mL^{-1}$ | 2.4 log CFU $g^{-1}$ difference between Nis + and Nis− after 6 weeks | Maisnier-Patin et al. (1992) |

other cultures required for the fermentation. Nisin-producing starters were assessed, but they lack the technological properties required for cheese making, as, for instance, poor acidification rates, inadequate proteolytic activity, and often with enhanced susceptibility to bacteriophage attack (Deegan et al. 2006). These drawbacks can be compensated with different strategies, including the ones where previously mentioned nisin-producing strains are used as adjunct cultures (co-culture) together with a bacteriocin-resistant starter or by producing transconjugant strains that gather the bacteriocin-producing capability and the desirable technological properties in the same strain (O'Sullivan et al. 2003).

Different nisin-producing cultures were effective against *L. monocytogenes* in Camembert (Maisnier-Patin et al. 1992) and in raw milk cheese (Rodríguez et al. 2001). Bouksaim et al. (2000) developed also with success a starter culture system that produced both acid and nisin at acceptable rates in Gouda cheese mixing a *Lactococcus lactis* subsp. *lactis* biovar. *diacetylactis* strain-producing nisin Z together with a commercial starter culture. Some strains of *Lc. lactis* are able to produce other lantibiotic bacteriocins such as lacticin 3147 or lacticin 481 that also reduce the counts of *L. monocytogenes* in Cheddar (Ryan et al. 1996) and cottage cheese (McAuliffe et al. 1999). Moreover, lacticin 3147 modified starters (transconjugant) successfully inhibited this pathogen on the surface of smear-ripened cheese (O'Sullivan et al. 2006).

Some of these lantibiotic-producing protective cultures have also been successfully used to control spoilage in cheese. For instance, lacticin 3147-producing starter cultures have been tested to control the nonstarter lactic acid bacteria population in Cheddar cheese (Ryan et al. 1996). Moreover, in most cheeses late blowing is an undesirable defect caused by gas production mainly due to the outgrowth of several *Clostridium* species spores surviving heat treatments applied to milk before processing such as *Cl. tyrobutyricum*, *Cl. sporogenes*, *Cl. beijerinckii*, or *Cl. butyricum* (Cocolin et al. 2004). This abnormal butyric fermentation originates texture and flavor defects in cheeses, causing important economic losses in the cheese industry (Gálvez et al. 2014). Application of the nisin producer strain of *Lc. lactis* subsp. *lactis* IPLA 729 reduced around 3 log the level of a *Cl. tyrobutyricum* spoilage strain inoculated in cheeses, in comparison with control cheeses inoculated with a commercial starter culture supplemented with nitrate (Rilla et al. 2003). Other *Lc. lactis* strains producing lacticin 3147, thermophilic streptococci, or some lactobacilli such as the bacteriocin producer strain *Lb. gasseri* K7 have also a potential for inhibition of *Cl. tyrobutyricum* in some cheeses (Gálvez et al. 2014).

Other bacteriocinogenic cultures producing Class II bacteriocins have also been described to be active against *Listeria*. Some cultures of *Lactobacillus plantarum* have been used by their ability to produce Class IIa pediocin-like bacteriocins such as plantaricin 423 and as adjunct to a nisin-producing starter. Bacteriocin-producing enterococci such as *Enterococcus faecium*, *Enterococcus faecalis*, and *Enterococcus mundtii* have been investigated as adjunct cultures for cheese making because of their robustness, natural presence in cheeses, and production of several bacteriocins with strong antilisterial activity (Gálvez et al. 2008). *Enterococcus faecalis* producing the cyclic bacteriocin enterocin AS-48 has been used as starter or co-culture together with a commercial lactic starter in the production of raw milk Manchego cheese (Mills et al. 2011). Other strains producing enterocin AS-48 showed strong inhibition of *L. monocytogenes*, as well as *Bacillus cereus* and *St. aureus* (Núñez et al. 1997, Muñoz et al. 2004, 2007).

In yogurt, thermophilic streptococci starter culture such as *Streptococcus salivarius* subsp. *thermophilus* B reduced counts of *L. monocytogenes* below detectable levels and was able to extend the product shelf life by 5 days (Benkerroum et al. 2002). Normally, pediococci are not well adapted to dairy substrates, due to their lack or very slow lactose fermentation activity. However, when they are added as co-culture with yogurt starter cultures, they grow at the expense of the excess sugar released from lactose hydrolysis by the starters, allowing the production of pediocin active against *L. monocytogenes* (Somkuti and Steinberg 2010). More information about the application of protective cultures in dairy products can be obtained from the excellent reviews of Gálvez et al. (2014) and Arqués et al. (2015).

#### Application of Antifungal Protective Cultures in Dairy Products

More recently, it has been demonstrated the ability of some LAB, routinely used as starter cultures in fermented dairy products, to reduce fungal contamination. Yogurts have been primarily targeted as they are liable to yeast growth due to their low pH, storage at refrigeration temperatures, and presence of fruit in certain products (Crowley et al. 2013). Delavenne et al. (2012) demonstrated the protective action of *Lactobacillus harbinensis* K.V9.3.1Np and *Lb. rhamnosus* K.C8.3.1I against a number of fungi including *Debaryomyces hansenii* and *Rhizopus mucilaginosa* in yogurts, while they maintain the technological properties of the starter cultures used and the final sensory characteristics of the product. In the same way, a co-culture of *Lb. paracasei* subsp. *paracasei* and *Propionibacterium jensenii* was found to retard the growth of various *Candida* species in yogurt and cheese surface (Schwenninger and Meile 2004). Moreover, Garcha and Natt (2011) obtained an improvement of processed cheese slices shelf life, after applying antifungal LAB. Furthermore, several *Lb. plantarum* strains isolated from various herbs, fruits, and vegetables with antifungal activity were found to prevent the visible growth of *Penicillium commune* on cottage cheese by between 14 and more than 25 days longer than cottage cheese, which contained either no added LAB or LAB that did not have antifungal activity (Cheong et al. 2014). All these protective cultures could offer a natural alternative to manufacturers instead of using chemical preservatives such as sodium benzoate, sorbic acids, and natamycin in yogurt and cheese production (Crowley et al. 2013).

### 3.4.2 Meat Products

Meat products are highly perishable food products and widely recognized as possible foodborne pathogen carriers. Some bacteria such as *Salmonella* spp., *Escherichia coli* O157:H7, and other enterohemorrhagic *E. coli* (EHEC), *Listeria monocytogenes*, *Staphylococcus aureus*, and *Campylobacter jejuni* are the pathogens of concern in pork and poultry raw meat and derived products and *Clostridium botulinum* in cured hams and fermented sausages. Indeed, many outbreaks have been linked with some of these pathogens and meat products (Jiménez et al. 2005; Mor-Mur and Yuste 2010; Cartwright et al. 2013; Baker et al. 2016). Depending on the type of meat product, these pathogens can be controlled by using different technological processes such as fermentation, curing, cooking, and the addition of diverse additives to improve their preservation. Additionally, food safety and shelf life can be enhanced with the use of vacuum and modified atmosphere packaging (MAP) especially in the case of raw meat. In that sense, the use of protective cultures has also been proposed as a natural way to extend the shelf life and improve the food safety of meat products. These strategies have been applied both in fresh meat products as well in ready-to-eat products, such as fermented/dry-cured or cooked meat products, especially those sliced.

**Application of Protective Cultures in Raw Meat and Poultry**
The food industry has used and developed hurdle strategies to improve the shelf life of raw meat since its composition makes this animal product a perfect substrate for bacterial growth. These strategies are mainly based on the combination of physical methods such as refrigeration and freezing with vacuum packaging or modified atmosphere packaging (MAP). In this manner, spoilage bacterial growth is delayed by the low temperature used, and the modification of the gas environment contributes in different ways to extend the shelf life. The gas composition selected for meat preservation is important from the point of view of the spoilage microorganism selection. Vacuum- and MAP-packaged meat is spoiled by Gram-positive bacteria, while Gram-negative bacteria are favored by aerobically packaging. Moreover, Nieminen et al. (2015) studied the effect of different gas composition on the selection of LAB inoculated in raw loin pork stored 14 days at 6 °C and observed that high-$CO_2$ percentage promotes the growth of *Lactobacillus sakei* and *Lactobacillus oligofermentans*. However, high-$O_2$ atmosphere favors the growth of *Leuconostoc gelidum* subsp. *gasicomutatum*, and *Lactococcus piscium* predominated in a high-$N_2$ atmosphere without oxygen. These results should be considered to select the most appropriate LAB genera for biopreservation. This is also observed in Table 3.7 where most of the studies performed in vacuum packaging are combined with *Lb. sakei* strains. Katikou et al. (2005) showed in their study that the counts of *Enterobacteriaceae*, *Pseudomonas*, *Brochothrix thermosphacta*, and yeast and molds were 1–2 log CFU $g^{-1}$ less in the batch inoculated with *Lb. sakei* 4808 than in the control batch in beef meat stored in vacuum packaging during 28 days at refrigeration temperatures. Nevertheless, a lower effect was observed with *Lb. curvatus* $904^T$. The authors attribute these results with the fact that *Lb. sakei* 4808 produce a bacteriocin with activity against *B. thermosphacta* and *Pseudomonas* spp., while no effect was observed for the bacteriocin produced by *Lb. curvatus* $904^T$. Furthermore, sensorial analysis indicated that the limit of acceptability for abnormal odors was achieved 2 days later with *Lb. sakei* 4808 indicating that the bioprotective strain did not affect negatively the evolution of the odor. Similarly, beef meat under two different atmospheres (70% $O_2$/20% $CO_2$/10% $N_2$ and 60% $O_2$/40% $CO_2$)

**Table 3.7** Some examples of application of protective cultures on meat products (some examples)

| Product | Protective culture | Inoculum | Target microorganism | Inoculum | Effect | References |
|---|---|---|---|---|---|---|
| Raw meat | | | | | | |
| MAP and vacuum-packaged lamb | *Leuconostoc pseudomesenteroides* PCK 18 | $10^6$ CFU $g^{-1}$<br>$10^4$ CFU $g^{-1}$ | *Listeria monocytogenes* | $10^5$ CFU $g^{-1}$<br>$10^3$ CFU $g^{-1}$ | 3 log difference with control ($10^6$)<br>Sensory modification | Osés et al. (2015) |
| Ground beef in vacuum and MAP | *Lactobacillus sakei* cocktails | $10^3$ CFU $g^{-1}$ | *Salmonella enterica typhimurium*, *Escherichia coli* O157:H7, and *Brochothrix thermosphacta* | $10^2$ CFU $g^{-1}$<br>$10^2$ and<br>$10^4$ CFU $g^{-1}$ | Reduction around 2.5 log in *S. typhimurium* in vacuum packaging<br>Reduction around 2.5 log in *E. coli* in MAP<br>No effect against *B. thermosphacta* | Chaillou et al. (2014) |
| Chicken burger meat in MAP<br>Chicken legs in MAP | *Leuconostoc pseudomesenteroides* PCK18<br>*Bifidobacterium longum* subsp. *longum* PCB133 | $10^5$ CFU $g^{-1}$<br>$10^4$ CFU $g^{-1}$ | *Listeria monocytogenes* cocktail<br>*Campylobacter jejuni* cocktail | $10^3$ CFU $g^{-1}$<br>$10^5$ CFU $g^{-1}$ | Reduction of 1.22 log CFU $g^{-1}$<br>Reduction of 1.16 log CFU $g^{-1}$ | Melero et al. (2013) |
| Chicken burger meat in MAP | *Leuconostoc pseudomesenteroides* PCK 18 | $10^6$ CFU $g^{-1}$ | *Listeria monocytogenes* cocktail<br>*Campylobacter jejuni* cocktail | $10^3$ CFU $g^{-1s}$<br>$10^2$ CFU $g^{-1}$ | Reduction of 0.90 log CFU $g^{-1}$<br>No effect | Melero et al. (2012) |
| Goat meat | *Pediococcus pentosaceus* GOAT 01 (Bac+)<br>*Lactobacillus plantarum* GOAT 012 | $10^6$ CFU $g^{-1}$ | *Listeria monocytogenes*<br>*Salmonella typhimurium* | $10^6$ CFU $g^{-1}$ | Below detection limit (<2 log) on day 2<br>2 log reduction | Olaoye et al. (2011) |

| Product | Protective culture | Inoculum | Target microorganism | Inoculum | Effect | References |
|---|---|---|---|---|---|---|
| Vacuum-packed chilled lamb | *L. sakei* Lb 706 (Bac+) and Lb 706b (Bac−)<br>*L lactis* 75 | $10^3$ and 3 CFU $g^{-2}$ | *L. monocytogenes*<br>*Brochothrix thermosphacta* | $10^2$ CFU $mL^{-1}$<br>$10^2$ CFU $mL^{-1}$ | Below detection limit (high inoculum level)<br>No difference with control (Lb 706b)<br>No difference with control | Jones et al. (2009) |
| Vacuum-packed chilled beef | *L sakei* strains 27, 44 and 63 | $10^3$ and 3 CFU $g^{-2}$ | *Campylobacter jejuni*<br>*Clostridium estertheticum* | $10^3$ CFU $mL^{-1}$<br>Spores containing $10^2$ CFU $mL^{-1}$ | Lower counts (high inoculum level)<br>Gas appears 1 week later (high inoculum level) | |
| Raw chicken ground meat aerobically stored | *Enterococcus faecium* PCD71<br>*Lactobacillus fermentum* ACA-DC179 | $10^7$ CFU $g^{-1}$<br>$10^7$ CFU $g^{-1}$ | *Listeria monocytogenes 21412*<br>*Salmonella enteritidis* | $10^5$ CFU $g^{-1}$<br>$10^5$ CFU $g^{-1}$ | 0.7 log CFU $g^{-1}$ reduction<br>1.2 log CFU $g^{-1}$ reduction | Maragkoudakis et al. (2009) |
| Raw vacuum-packed beef | *Lactobacillus sakei* CECT 4808<br>*Lactobacillus curvatus* CECT 904$^T$ | $10^8$ CFU $g^{-1}$ | Native spoilage microflora | | Reduction between 1 and 2 log CFU $g^{-1}$ | Katikou et al. (2005) |
| Ground beef | FloraCarn L-2 (*Lactobacillus alimentarius*) | $10^7$ CFU $g^{-1}$ | Two *Listeria monocytogenes* strains | $10^6$ CFU $g^{-1}$ | Final population reduction of 4.3 or 4.1 log CFU $g^{-1}$ | Juven et al. (1998) |

(continued)

**Table 3.7** (continued)

| Product | Protective culture | Inoculum | Target microorganism | Inoculum | Effect | References |
|---|---|---|---|---|---|---|
| Fermented meat products and dry-cured ham | | | | | | |
| Dry cured Iberian ham | *Penicillium chrysogenum* RP42C and a mix of four nontoxigenic *Penicillium* strains | $10^6$ conidia $mL^{-1}$ | Native OTA-producing molds | | No OTA contaminated hams when inoculated with *P. chrysogenum* RP42C | Rodríguez et al. (2015) |
| Dry-fermented sausages | *Lactococcus lactis* subsp. *lactis* LMG21206 (Bac+) and *Lactobacillus curvatus* LBPE (Bac+) | $10^7$ CFU $g^{-1}$ | Four different strains of *L. monocytogenes* | $10^2$–$10^3$ CFU $g^{-1}$ | No growth Below detection limit | Benkerroum et al. (2005) |
| Fermented Merguez sausage (fermentation period 24 h) | *Lactococcus lactis* subsp. *lactis* M (Bac+) | $10^7$ CFU $g^{-1}$ | *Listeria monocytogenes* ATCC 7644 | $10^6$ CFU $g^{-1}$ | Final difference between product with and without protective culture of around 5 log CFU $g^{-1}$ | Benkerroum et al. (2003) |
| North European-type dry-fermented sausages | *Lb. rhamnosus* E-97800<br>*Lb. rhamnosus* LC-705<br>*Lb. plantarum* ALC01<br>*P. pentosaceus* RM2000 (control) | $10^7$ CFU $g^{-1}$ | *Listeria monocytogenes* | $10^3$ CFU $g^{-1}$ | *L. monocytogenes* negative after 7 days of ripening, while 28 days in the control | Työppönen et al. (2003b) |
| Cooked meat products | | | | | | |
| RTE poultry meat | *Lactococcus lactis* subsp. *lactis* | $10^7$–$10^8$ CFU $g^{-1}$ | *Staphylococcus aureus* | $10^4$–$10^5$ CFU $g^{-1}$ | 1 log reduction | Akbar and Anal (2014) |

| Product | Protective culture | Inoculum | Target microorganism | Inoculum | Effect | References |
|---|---|---|---|---|---|---|
| Frankfurters | Three LAB strains (Lactiguard+): *Lactobacillus animalis* La51, *L. amylovorus* M35, and *Pediococcus acidilactici* D3<br>*Lactiguard® do not grow in refrigeration* | $10^9$ CFU mL$^{-1}$ | Six strains of *Listeria monocytogenes* | $10^6$ CFU mL$^{-1}$ | Reduction of 0.6 log<br>Reduction of 1.2 log with cell-free extract (CFS)<br>Reduction of 3.3 log with CFS + lactate/diacetate | Koo et al. (2012) |
| Several cooked meat products (CMP) | *Lactobacillus sakei* 10[a] (Bac−) | $10^6$ CFU g$^{-1}$ | *Listeria monocytogenes* | $10^2$ CFU g$^{-1}$ | Reduction between 2 and 4 log CFU g$^{-1}$ | Vermeiren et al. (2006c) |
| *Tsire*, a Nigerian cooked stick meat | *Pediococcus acidilactici* NCIMB 700993 (Bac+) | $10^6$ CFU g$^{-1}$ | *Listeria monocytogenes*<br>*Salmonella* typhimurium | $10^6$ CFU g$^{-1}$ | Reduction of 2 log CFU g$^{-1}$<br>Reduction of 5 log CFU g$^{-1}$ | Olaoye and Dodd (2010) |
| Vacuum-packed pork sausages | *Leuconostoc carnosum* 4010 (Bac+) | $1.2 \times 10^5$ CFU g$^{-1}$ and $6.3 \times 10^6$ CFU g$^{-1}$ | Cocktail of five *L. monocytogenes* | $10^4$ CFU g$^{-1}$ | 7 log reduction after 28 days | Budde et al. (2003) |
| Sliced cooked cured pork shoulder | *Leuconostoc mesenteroides* L124 (Bac+)<br>*Lactobacillus curvatus* L442 (Bac+) | $10^5$ CFU mL$^{-1}$ | *L. innocua* 7510 S1 | $10^3$ CFU mL$^{-1}$ | 1.5 log CFU g$^{-1}$ (around 6 log CFU g$^{-1}$ difference with the control at the end of storage period) | Mataragas et al. (2003) |

(continued)

**Table 3.7** (continued)

| Product | Protective culture | Inoculum | Target microorganism | Inoculum | Effect | References |
|---|---|---|---|---|---|---|
| Cooked ham and servelat sausage | *Lactobacillus sakei* TH1 (Bac−) | $10^5$–$10^6$ CFU $g^{-1}$ | Cocktail of *L. monocytogenes* (Strain 2230/92 serotype 1, strain 167 serotype 4b, strain 187 serotype 4b) | $10^3$ CFU $g^{-1}$ | No grow | Bredholt et al. (2001) |
| Cooked ham | Flora Carn L-2 (*Lactobacillus alimentarius*) | $10^{10}$ CFU $g^{-1}$ in brine prior to cook | Indigenous microbiota | | Shelf life was extended 1 week | Kotzekidou and Bloukas (1996) |

did not present any off-odor after 28 days at 1 °C when inoculated either with *Lb. sakei* CTC 372 (bacteriocinogenic strain) and *Lactobacillus* CTC 711 (non-characterized) (Djenane et al. 2006). In addition, *Lb. sakei* CTC 372 reduced the level of native *B. thermosphacta* and *Pseudomonas* in 2 log at the end of the study in both atmospheres, while *Lactobacillus* CTC 711 only achieved one log of reduction in 20% $CO_2$. Likewise, the presence of *Lb. curvatus* CRL705 (bacteriocinogenic) decreased 2.5 log CFU $g^{-1}$ the population of native *B. thermosphacta* in raw vacuum-packed beef, and additionally no substantial sensorial and structural changes were reported (Castellano et al. 2010).

Moreover, the effect of *Lb. sakei* strains has also been tested against different foodborne pathogens. *Salmonella typhimurium* and *E. coli* O157:H7 counts were reduced around 2.5 log CFU $g^{-1}$ in minced beef meat under vacuum packaging and MAP, respectively, inoculated with a different mixture of three *Lb. sakei* strains ($10^3$ CFU $g^{-1}$) (Chaillou et al. 2014). This study also reveals the importance of using genetically diverse strain mixture rather than a single strain and performing previous studies to select the best combination of strains to enhance food safety. However, the combination of *Lb. plantarum* GOAT 012 and *Pediococcus pentosaceus* GOAT 01 (pediocin producer) did not have a synergic effect against *S. typhimurium* in goat meat reaching the same reduction (2 log CFU $g^{-1}$) than *P. pentosaceus* GOAT 01 alone on the third day of study (Olaoye et al. 2011). The used of *Lb. fermentum* ACA-DC179 (bacteriocin-like inhibitory substances, BLIS, producing strain) in raw chicken ground meat reduced also the level of *S. enteritidis* PT4 in 1.2 log CFU $g^{-1}$ after 7 days (Maragkoudakis et al. 2009). These studies showed up the action of other mechanisms apart of bacteriocin producing of these protective cultures because it is well known that normally bacteriocins have no effect against Gram-negative bacteria (Albano et al. 2007; Castellano et al. 2008). The effect against other Gram-negative bacteria such as *Campylobacter jejuni* has been also proved with *Bifidobacterium longum* subsp. *longum* PCB133 and *Lb. sakei* strains 27, 44, and 63 in chicken legs in MAP and vacuum-packed beef, respectively (Jones et al. 2009; Melero et al. 2013).

*Leuconostoc pseudomesenteroides* PCK18 used for biopreservation in chicken burger meat showed an effective antilisteria action (0.90 log CFU $g^{-1}$ reduction) in combination with MAP and freeze stress (−18 °C for 48 h) (Melero et al. 2012). In addition, in their study Osés et al. (2015) obtained a reduction of 3 log CFU $g^{-1}$ in *L. monocytogenes* when inoculated $10^5$ CFU $g^{-1}$ in suckling lamb in MAP with *Lc. pseudomesenteroides* PCK18 (6.89 log CFU $g^{-1}$) after 18 days in refrigeration. However, the product was rejected by day 4 due to the lactic acid odors produced by the protective culture.

## Application of Protective Cultures in Fermented Meat Products and Dry-Cured Ham

Fermented meat products are characterized by the possibility of storage at room temperature without affecting its safety and sensorial properties due to the formulation used in their production. Salt, nitrites, and spices may act as antimicrobial additives as well as contribute to the technological and sensory characteristics of the final product. Sugar is needed as substrate for starter cultures, responsible of the fermentation process. Those commercial starter cultures have been selected, historically, from similar products naturally produced due to the specific characteristics provided to the product. Although fermented meat products are shelf stable, foodborne pathogens could survive during fermentation and storage period. Thus, starter cultures with protective action have been studied against diverse pathogens (Table 3.7). *L. monocytogenes* is once again one of the pathogens more studied in this type of products because fermentation and drying processes do not affect its survival. Työppönen et al. (2003b) obtained the deletion of *L. monocytogenes* (3 log CFU $g^{-1}$) after 7 days of fermentation in North European-type dry sausages produced with three starter cultures (*Lactobacillus rhamnosus* E-97800, *Lb. rhamnosus* LC-705, and *Lb. plantarum* ALC01) from different origins, while 28 days was needed to eliminate the pathogen in the control batch. The pH of the final product was the normal in this type of product. *Lactococcus lactis* subsp. *lactis* M, a bacteriocinogenic strain, reduced in 2.7 log CFU $g^{-1}$ the counts of *L. monocytogenes* inoculated in Merguez after 24 h of fermentation, while 1.6 log CFU $g^{-1}$ was reduced with a non-bacteriocinogenic strain (*Lc. lactis* J) when compared with the initial inoculum (Benkerroum et al. 2003). However, the effect of the bacteriocinogenic strain was slightly reduced when the sausages were prepared with nitrites (only 2.4 log CFU $g^{-1}$). The pH behavior was the same in all the batches independently on the use of starter cultures or nitrites. Moreover, no modifications on the sensory characteristics were detected. In their study, Rubio et al. (2013) combined protective cultures and HHP in the production of fuet (low-acid fermented sausage) against *L. monocytogenes* and *St. aureus*. *Enterococcus faecium* CTC8005 reduced the level of *L. monocytogenes* in approximately 2 log CFU $g^{-1}$ after stunning and maintained the level of the pathogen during 21 days. Additionally, counts were reduced below 2 log CFU $g^{-1}$ after a week of HHP treatment (600 MPa 5 min). None of the three protective cultures showed in vivo effect against *St. aureus*, and the combination with HHP had a slight effect against the pathogen comparing with the control batch without bioprotective cultures. This study also showed that the three starter cultures used controlled the growth of *Enterobacteriaceae* contributing to regulate the biogenic amines produced by microorganisms of this family. In this respect, the combination of two strains of *Lb. sakei* (CTC6469 and CTC6626) and two of *Staphylococcus xylosus* (CTC6013 and CTC6169) in the production of chorizo and fuet reduced the number of *Enterobacteriaceae* comparing with control batch. This result was reflected in the lower amount of biogenic amines (tyramine, cadaverine, and putrescine) normally produced during fermentation (Latorre-Moratalla et al. 2007). Other authors have also studied the combined effect of LAB and drying against another microorganism of concern in this type of products such as *E. coli* O157:H7 (Faith et al. 1998; Riordan et al. 1998). The use of *Lb. reuteri* in co-culture to meat starter cultures (*Pediococcus pentosaceus* and *St. carnosus*) reduced the level of *E. coli* O157:H7 in 3 log CFU $g^{-1}$ in dry-fermented sausages at the end of the drying. However, when the protective culture was applied micro-encapsulated in an alginate matrix, although no reduction of *E. coli* O157:H7 was obtained; *Lb. reuteri* strain was protected from low pH during the fermentation process (Muthukumarasamy and Holley 2007).

The application of biopreservation could be done also to the final product to prevent the growth of foodborne pathogens in case of cross-contamination during the slicing and packaging. Jácome et al. (2014) obtained a reduction of more than 2 log CFU $g^{-1}$ in *L. monocytogenes* inoculated in sliced Chouriço after 30 days at 5 °C when treated with *Lb. sakei* ST15 or a commercial mixed starter culture (BLC35, were *Lb. curvatus*, *St. xylosus*, and *Pediococcus acidilactici*; CHR Hansen). Moreover, consumer's acceptability perception was not affected as the panelist found few differences compared with the control product.

In the case of dry-cured ham, the presence of mycotoxins like ochratoxin A (OTA) should be avoided due to its harmful effect on humans. Dry-ham slices at different drying stages determined by $a_w$ (0.94 and 0.84) showed low load of *Penicillium nordicum* CBS 323.92 (OTA producer) in the presence of *Debaryomyces hansenii,* and in addition, a low amount of OTA was detected (Andrade et al. 2014). The effect was higher at $a_w$ 0.94 indicating that the use of the protective culture would be better at the beginning of drying period to avoid the growth of the toxigenic mold and, thus, the production of OTA. Regarding OTA presence in dry-cured Iberian ham, Rodríguez et al. (2015) showed that *P. chrysogenum* RP42C inoculated in the surface of this type of product did not present OTA after 9 months of processing. Moreover, the load of OTA-producing molds was significantly lower with the protective culture compared with the control batch without *P. chrysogenum* RP42C and the batch with four *Penicillium* wild-type nontoxigenic strains. Authors explain that the effect of the protective culture could be due to the production of the antifungal protein PgAFP.

#### Application of Protective Cultures in Cooked Meat Products

The last group of meat products where protective cultures have been applied is cooked meat products (Table 3.7) that belong to the category of ready-to-eat products. Normally, studies have tried to enhance the safety of this type of products due to their characteristics but also the improvement of the shelf life, and thus, their quality has been also the focus of other studies. Kotzekidou and Bloukas (1996) studied the effect of Flora Carn L-2 (*Lb. alimentarius*, $10^{10}$ CFU $g^{-1}$) on the preservation of cooked vacuum-packaged ham when added with the curing solution before the heat treatment. Authors found that the shelf life of the product with the protective culture was extended for a total of 4 weeks in comparison with 3 weeks obtained in the control batch due to the reduction in the growth of total aerobic bacteria, *B. thermosphacta*, *Staphylococcus*, and micrococci; however, no effect was observed against *Pseudomonas*. A recent study showed that *Lc. lactis* subsp. *lactis* (inoculated $10^3$ CFU $g^{-1}$) and *Lb. sakei* (inoculated $10^3$ CFU $g^{-1}$) inhibit the growth of *Ln. mesenteroides* (inoculated $10^3$ CFU $g^{-1}$), which is the principal responsible of the spoilage of cooked bacon stored under vacuum packaging and refrigeration temperatures (Comi et al. 2016).

Authors obtained a shelf life of 90 days, while normally spoilage, characterized by the appearance of greening color, slime, package inflation, off-odors, and off-flavors, produced by *Ln. mesenteroides* starts after 30 days. The growth of this spoilage microorganism was also reduced in cooked ham stored under vacuum packaging during 42 days at 7 °C, reaching 7 log CFU $g^{-1}$ 14 days later when co-inoculated with *Lb. sakei* subsp. *carnosus* in comparison with control batch (Vermeiren et al. 2006b).

*Listeria monocytogenes* is the pathogen of concern in this type of products; thus, higher scientific studies have been performed in comparison with other foodborne pathogens. Alves et al. (2006) showed the same antilisterial effect of *Lb. sakei* 1 (bacteriocinogenic, inoculated $10^6$ CFU $g^{-1}$) and *Lb. sakei* ATCC 15521 (non-bacteriocinogenic, inoculated $10^6$ CFU $g^{-1}$) reducing the final counts of *L. monocytogenes* ATCC 19115 (serotype 4b, inoculated $10^2$ CFU $g^{-1}$) and *L. monocytogenes* IAL 633 (serotype 1/2a, inoculated $10^2$ CFU $g^{-1}$) in 2 and 4 log CFU $g^{-1}$, respectively, in sliced cooked vacuum packed at 8 °C after 10 days. Kaban et al. (2010) reported around 3 log reduction of *L. monocytogenes* when bologna-type sausages were inoculated with *Lb. sakei* Lb 706 and the variant non-bacteriocinogenic (inoculated $10^3$–$10^4$ CFU $g^{-1}$) in comparison with the control batch when preserved under vacuum or MAP (50% $CO_2$/50% $N_2$) at 4 °C during 6 weeks. However, only the bacteriocinogenic strain was effective under vacuum inhibiting the growth of the pathogen. *Leuconostoc carnosum* 4010 (bacteriocinogenic strain) isolated from vacuum-packaged ham showed an antilisterial dose-dependent effect in slices of cooked vacuum-packaged ham (Budde et al. 2003). Authors reported that when *Lc. carnosum* 4010 was inoculated at a level of 5 log CFU $g^{-1}$, no effect was observed during the first week of treatment, whereas 6 log CFU $g^{-1}$ produced a slight reduction (less than one log) in the counts compared with the pathogen's inoculum (4 log CFU $g^{-1}$). Moreover, the level of *L. monocytogenes* was reduced below 10 CFU $g^{-1}$ by day 21 with the high level of protective culture and by day 28 with the low level of protective culture obtaining in both cases 7 log CFU $g^{-1}$ reduction in comparison with the control batch at the end of the study. *Lb. sakei* 10A (6 log CFU $g^{-1}$ of inoculum) exerted an antilisterial effect in cooked ham and cooked chicken fillet reducing the level of a cocktail of five strains of *L. monocytogenes* in 2 and 4.5 log CFU $g^{-1}$, respectively, after 14 days at 7 °C compared with control batch (Vermeiren et al. 2006c). Moreover, authors also reported that the growth of *B. thermosphacta* (2 log CFU $g^{-1}$ of inoculum) was suppressed in cooked sausage and cooked chicken fillet when the protective culture was present. Although no shelf life extension was reported for all the products, only cooked ham reached better acceptability with the protective culture because it could overgrow *Ln. mesenteroides* that spoil the product.

Olaoye and Dodd (2010) reported the applicability of *Pediococcus acidilactici* NCIMB 700993 (6 log CFU $g^{-1}$ of inoculum) in Nigerian cooked stick meat (*tsire*) preserved at 30 °C during 7 days (normal conditions in Nigeria) in combination with spices to reduce *L. monocytogenes* (6 log CFU $g^{-1}$ of inoculum). The counts in the control batch reached 11.29 log CFU $g^{-1}$ at the end of the study, while in the presence of the protective culture, the counts were below the detection limit (2 log) after day 2 with a reduction of 1.5 log CFU $g^{-1}$ during the first 24 h of treatment. Moreover, *Salmonella typhimurium* (6 log CFU $g^{-1}$ of inoculum) was reduced of around 5 log CFU $g^{-1}$ at the end of the study compared with the control batch with an initial reduction of 2 log on day 3. The quality of *tsire* was also improved with *Pc. acidilactici* obtaining lower numbers of natural spoilage microflora such as *Enterobacteriaceae* (more than 1 log reduction at the end of the study), *Staphylococcus* (below the detection limit from day 2 till the end of the study), total bacterial count (5 log CFU $g^{-1}$ reduction compared with control at day 7), and molds and yeast (3 log counts reduction at the end of the study). Akbar and Anal (2014) showed anti *St. aureus* activity of *Lc. lactis* subsp. *lactis* in ready-to-eat poultry products stored aerobically at 5 °C during 35 days with one log of reduction during the first week of study for both *St. aureus* strains. Moreover, after 25 days of study, 5 log CFU $g^{-1}$ reduction was achieved in the methicillin-resistant strain and 4 log in the other one.

### 3.4.3 Fish and Seafood Products

The microbiota of marine fish from temperate waters is usually composed of Gram-negative psychrotrophic bacteria, belonging mainly to the class *Gammaproteobacteria* such as *Pseudomonas*, *Shewanella*, *Acinetobacter*, *Aeromonas*, *Vibrio*, *Moraxella*, *Psychrobacter*, *Photobacterium*, etc. that are able to grow between 0 and 25 °C. Moreover, it is possible to find also some Gram-positive bacteria, such as *Micrococcus*, *Corynebacterium*, *Bacillus*, *Clostridium*, and *Lactobacillus*. Generally, LAB are found to form part of the intestinal microbiota of this type of fish, especially *Lb. plantarum* and species of the genera *Carnobacterium* such as *Cn. maltaromaticum* (previously *piscicola*), *Cn. divergens*, *Cn. gallinarum*, and *Cn. inhibens*. After fish death the raw material can be contaminated with endogenous microorganisms due to different postmortem operations. The production environment and human manipulations are also a source of post-contamination (Leroi 2010).

In fresh fish flesh, LAB have long been ignored because they are not currently present in seafood due to particular physicochemical characteristics of fish flesh that shows a high pH (above 6), low sugar concentration, and high content of low molecular weight nitrogenous molecules that stimulate the rapid growth of psychrotrophic pH-sensitive Gram-negative bacteria like *Pseudomonas* and *Shewanella*. Moreover, vacuum packaging, which could favor the development of LAB, does not slow the growth of these marine bacteria. Many of them can persist in these conditions especially *Shewanella putrefaciens*, *Photobacterium phosphoreum*, and *Vibrionaceae*, due to their ability to use trimethylamine oxide (TMAO), as a terminal electron acceptor for anaerobic respiration (Leroi 2010). The population of respiratory Gram-negative bacteria, like *Pseudomonas* and *Shewanella*, can be decreased by using MAP rich in $CO_2$. However, *Ph. phosphoreum*, a typical marine microorganism that is not present in meat and dairy products, is resistant to $CO_2$, and therefore it multiplies well in these conditions becoming the main spoilage bacterium of fresh MAP fish (Dalgaard et al. 1993; Dalgaard et al. 1997). In fish, MAP selects both *Ph. phosphoreum* and LAB, but the latter are less competitive so often play a minor role in spoilage.

Lightly preserved fish products (LPFP) are uncooked or mildly cooked ready-to-eat products that generally combine mild processing steps such as salting-drying-smoking. LPFP include carpaccio-type marinated fish, gravads, pickled fish, seafood in brine, cold-smoked fish, and peeled shrimp stored in MAP or in brine. These highly perishable products are characterized by a low level of preservatives, such as sorbate, benzoate, and $NO_2$, normally less than 6% of NaCl that correspond to $a_w$ less than 0.96 and a pH above 5 and are usually stored at chilled temperature under vacuum packaging or MAP to extend their shelf life. Those conditions inhibit the growth of some Gram-negative bacteria such as *Pseudomonas* spp.; decrease the growth of *Ph. phosphoreum*, *Aeromonas* spp., and *Sh. putrefaciens*; and let other more resistant microorganisms like psychrotrophic LAB grow. Many LAB strains are able to grow at refrigeration temperatures; tolerate MAP, low pH, high salt concentrations, and the presence of additives such as some preservatives and lactic acid, ethanol, or acetic acid; and comprise the dominant microbiota in many LPFP (Calo-Mata et al. 2008). Among LAB commonly found in seafood are often members of the genera *Lactobacillus*, *Leuconostoc*, and *Carnobacterium*. The latter are non-aciduric bacteria with low spoiling potential that have been extensively used as putative protective culture in LPFP without altering their final sensory properties.

The major microbial risks associated with LPFP are non-proteolytic *Clostridium botulinum* that produces neurotoxin-type E and *L. monocytogenes* (Huss et al. 2000; Løvdal 2015). The former is adequately controlled by combining salt content higher than 3.5% and low temperature (below 5 °C). However, *L. monocytogenes* can grow at chilled temperature (0 °C) and support low pH (4.5) and $a_w$ (0.92). The prevalence of *L. monocytogenes* in LPFP is highly variable from 0 to 80% depending on the kind of product being the smoked fish the one with a higher prevalence (Jami et al. 2014). This pathogen can survive the different processing steps followed by the manufacturing of these products. Therefore, the use of protective culture may be a useful hurdle to prevent *L. monocytogenes* development in the food matrix and thus control the safety risk (Leroi 2010).

Several authors have reported the use of different LAB as protective culture to delay the growth of some spoilage bacteria in fresh fish or inhibit the growth *L. monocytogenes* in LPFP. In fresh fish, Anacarso et al. (2014) reported in fresh salmon fillets stored at 4 °C the effectiveness of the bacteriocinogenic strain *Lactobacillus pentosus* 39 in reducing the population of *Aeromonas hydrophila* and *L. monocytogenes* in 2.1 and 3.6 log CFU $g^{-1}$, respectively. This reduction was higher in both cases 2.8 and 5.8 log CFU $g^{-1}$ under simulated cold-chain break conditions (30 °C for 12 h). Moreover, these authors found that in samples treated with the putative bacteriocin alone, a less marked decrease in both target bacteria population was observed. Similarly, Ibrahim and Vesterlund (2014) described an increase of 3 days of shelf life of vacuum-packed raw Atlantic salmon stored at refrigeration treated with a strain of *Lc. lactis* subsp. *lactis* in comparison with the control without changing the organoleptic and textural properties of the fish. Among the LPFP, the most studied product has been smoked salmon due to its importance from the commercial point of view and because of the risk of *L. monocytogenes* growing as it has been mentioned above. Again LAB and especially members of the genus *Carnobacterium* have been the most used as protective cultures in the literature due to they use to form part of the natural microbiota of these kinds of products and because they show a low acidification metabolism. Nilsson et al. (1999) showed that a bacteriocin-producing strain of *Cn. piscicola* (A9b) initially caused a 7-day lag phase of *L. monocytogenes*, followed by a reduction in numbers of *L. monocytogenes* from $10^3$ CFU $mL^{-1}$ to below 10 CFU $mL^{-1}$ after 32 days of incubation, coinciding with the detection of antilisterial compounds. The presence of a non-bacteriocin-producing strain of *Cn. piscicola* (A10a) prevented also the growth of *L. monocytogenes* during the 32-day incubation, although the reduction was 1 log lower than the former. On the contrary, Duffes et al. (1999) investigated the inhibition of *L. monocytogenes* by *Carnobacterium* strains on sterile and commercial vacuum-packed cold-smoked salmon stored at 4 and 8 °C. Different species and strains of *Carnobacterium* tested showed a bacteriostatic or bactericidal effect against the pathogen at chilling temperatures. However, *Listeria* growth was not affected by a non-bacteriocin-producing *Cn. piscicola* strain. Moreover, no product spoilage could be observed with the use of such bacteriocin-producing strains as demonstrated by good sensorial analyses and low biogenic amine production. Similarly, the *Cn. maltaromaticum* CS526 strain isolated from frozen surimi and identified as a bacteriocin producer showed a strong inhibitory activity against *L. monocytogenes*. The ability of this bacteriocinogenic strain, together with the non-bacteriocinogenic strain JCM5348, to inhibit the growth of the pathogen was examined in cold-smoked salmon stored at 4, 12, and 20 °C. *Cn. maltaromaticum* CS526 showed a bactericidal effect against *L. monocytogenes* within 21 and 12 days at 4 and 12 °C, respectively, and reduced its population by two and three log cycles, even at 20 °C. However, *Cn. maltaromaticum* JCM5348 did not prevent the growth of the pathogen, except at 4 °C. The presence of bacteriocin was detected in the samples co-inoculated with strain CS526 (Yamazaki et al. 2003). Brillet et al. (2004) studied the antilisterial capacity co-inoculating different *Carnobacterium* spp. and mixtures of *L. monocytogenes* strains in sterile blocks of cold-smoked salmon. They found that the strain *Cn. divergens* V41 was the most efficient, maintaining the level of *L. monocytogenes* at levels lower than 50 CFU $g^{-1}$ during the 4 weeks of vacuum storage at 4 and 8 °C. More recently, Leroi et al. (2015) have evaluated the effect of different indigenous and commercial LAB as protective cultures against four spoilage bacteria of vacuum-packaged cold-smoked salmon such as *Ph. phosphoreum*, *B. thermosphacta*, and

*Serratia proteamaculans* which released strong off-odors and the weaker spoiler *Cn. divergens*. The protective effect of the LAB differed from one specific spoilage organisms (SSO) to another and no correlation could be established between the sensory improvement, SSO inhibition, and the implantation or acidification of protective cultures (PCs). For more information about the application of protective cultures to seafood products, the reader can obtain further information in several nice reviews on this topic (Calo-Mata et al. 2008; Leroi 2010; Ghanbari et al. 2013; Gálvez et al. 2014).

### 3.4.4 Commercial Protective Cultures

Although there are a lot of studies dealing with the use of protective cultures in different food matrices, it seems that from the point of view of the real application, they are not used as was expected. There are several reasons that can justify this situation. Many studies have been done in lab conditions or food models that can be far from the real situation that these cultures will find when they will be applied in a real food. It is obvious that initial microbial population and variability of the target bacteria play an important role in the efficacy of the treatment, because not all strains from the same species have the same sensibility to the protective culture action as it was pointed out in previous sections of this chapter (Tolvanen et al. 2008). In addition, the increase in resistance by the target population to the different mechanisms used by the protective strain should be considered. In that sense, some authors proposed that the use of a bacteriocin-negative LAB may be more suited for practical use than a bioprotective agent against *L. monocytogenes* in ready-to-eat foods (Nilsson et al. 2004). Moreover, food matrix and composition are also factors to take in account in order to choose the proper protective culture and also define the method of application of such culture in order to obtain the desirable results.

Despite the drawbacks stated above, it is worthy to say that in some kind of products, the use of protective culture supposes the addition of another hurdle to spoilage or pathogen bacteria that can improve the shelf life and safety of the food product. In that sense, even though commercial protective cultures were first introduced around 20 years ago, now they have started to be well established in the food industry and recognized as an efficient tool to ensure the safety and durability of certain food products.

Commercial protective cultures are produced in the same way as starter cultures do: starting with batch fermentation, subsequent concentration by centrifugation, and final formulation as frozen pellets or freeze-dried powders. In Table 3.8 there is a list of different brands that produce commercial protective cultures available in this moment in the market. In this table, the reader will find some gaps mainly related to information about the real microorganisms that exert the protective action. In those cases the information is not easily accessible in open access from the different commercial brands that produce such type of cultures.

**Table 3.8** Available commercial protective cultures and their main applications

| Brand | Commercial name | Species | Benefit | Application |
|---|---|---|---|---|
| DuPont-Danisco | HOLDBAC® YM-C Plus<br>HOLDBAC® YM-B Plus | *Lactobacillus rhamnosus* (YM-B)<br>*Lactobacillus paracasei (YM-C)*<br>*Propionibacterium freudenreichii* subsp. *shermani* | Growth control of yeasts and molds and some heterofermentative lactic bacteria | Fresh fermented foods<br>White cheese |
| | HOLDBAC® LC | *Lactobacillus rhamnosus* | Growth control of leuconostoc, heterofermentative lactobacilli, and enterococci | Hard and semihard cheese |
| | HOLDBAC® Listeria | *Lactobacillus plantarum* | Growth control of *Listeria* | Soft and smear cheese, dry- and semidry-cured meats, cooked and fresh ground meats |
| | HOLDBAC® YM-XPK | | Growth control of yeast and mold spoilage | All cheese types |
| | HOLDBAC® YM-XPM | | Growth control of yeast and mold spoilage | Mild and very mild yogurt and fresh fermented dairy |
| Christian Hansen | SafePro® B-LC-007<br>SafePro® B-LC-20 | Multi-strain *Pediococcus* spp. | Growth control of *Listeria* | Salami and peperoni |
| | SafePro® ImPorous<br>SafePro® B-LC-78<br>SafePro® B-LC-48 | *Lactobacillus curvatus* | Reduce the risk of pore formations from spoilage bacteria<br>Enhance color and flavor development and reduce the risk of *Listeria monocytogenes*<br>Growth control of *Listeria* | Bacon<br>Raw bacon<br>Sliced cooked products |
| | SafePro® B-LC-20<br>SafePro® B-SF-77 | | Enhances freshness in beef applications, such as tartar, ground beef, cevapcici<br>Improved food safety of fresh sausages or products such as filet americain | Fresh meat |
| | SafePro® NovaLox | | Growth control of *Listeria* | Fish/salmon |

(continued)

**Table 3.8** Continued

| Brand | Commercial name | Species | Benefit | Application |
|---|---|---|---|---|
| Clerici-Sacco | Lyoflora FP-18<br>Lyoflora FP-77<br>Lyoflora FP-29 | *Carnobacterium* producing bacteriocins | Growth control of *Listeria* | Seafood products such as cold-smoked salmon |
| | Lyocarni BMX-37 | *Carnobacterium* producing bacteriocins<br>*Lactobacillus sakei*<br>*Staphylococcus carnosus* and *Staphylococcus xylosus* | Growth control of *Listeria*<br>Suppresses the growth of some spoilage bacteria<br>Enhance color and aroma formation | Fresh and dry-cured meat products |
| | Lyocarni BOM-13 | *Lactobacillus sakei* | Enhances quality and safety | Fresh and cooked meat products |
| | Lyocarni BOX-74 | *Carnobacterium producing bacteriocins*<br>*Lactobacillus sakei* | Growth control of *Listeria*<br>Suppresses the growth of some spoilage bacteria | Fresh and cooked meat products |
| | Lyocarni BXH-12<br>Lyocarni BXH-69 | *Lactobacillus sakei*<br>*Staphylococcus xylosus* | Suppresses the growth of some spoilage bacteria<br>Enhance color and aroma formation | Fresh and dry-cured meat products, spreadable pâté products, hamburgers and fresh sausages |
| | Lyofast CA 35 | *Lactobacillus paracasei* | Control of the secondary NSLAB | Dairy products cheese |
| | Lyofast FPR 2 | *Enterococcus faecium*<br>*Lactobacillus plantarum*<br>*Lactobacillus rhamnosus* | Inhibition of yeasts and molds | Dairy products cheese |
| | Lyofast LPR A | *Lactobacillus plantarum*<br>*Lactobacillus rhamnosus* | Inhibition of yeasts and molds | Dairy products cheese |
| | Lyofast LR B | *Lactobacillus rhamnosus* | Inhibition of yeasts and molds, propionic bacteria and heterofermentative LAB | Dairy products cheese |
| | Lyofast LPAL | *Lactobacillus plantarum* producing bacteriocin | Growth control of *Listeria* | Dairy products cheese |
| CSK food enrichment | Dairy Safe™ | | Protect cheeses from late blowing defects, inhibiting the growth of *Clostridium tyrobutyricum* | Dairy products organic cheese |

**Conclusion**

In this moment there are several protective cultures commercially available to be applied to different food products. It is important to take into account that protective cultures can be used as part of the concept of hurdle technology, which is based on the combination of different barriers acting in different ways on microbial cells, so that the cells have to activate different repair and adaptation mechanisms in order to survive and/or proliferate under the imposed selective conditions. However, their implementation should support, but not substitute, good manufacturing practices. In that sense, protective cultures can be very useful in ready-to-eat food products, such as mildly fermented meat products and sliced cooked meat products or cold-smoked fish or marinated seafood among others, where no intervention will be done before consumption by consumer to improve their food safety, especially against *Listeria monocytogenes*. They will be also very useful in raw foods to extent shelf life of raw meat and fish, where it is not allowed to add additives to extent shelf life a part of using physical preservation methods such as vacuum or MAP. In addition, new perspectives of using such cultures are rising in the area of fungal control in certain foods. Moreover, the scientific community and commercial brands must follow the research in this topic to get insight in the antagonistic mechanisms used by these cultures and improve the knowledge about possible resistance by target bacteria and to look for new and better applications.

## References

Acosta R, Rodríguez-Martín A, Martín A, Núñez F, Asensio MA (2009) Selection of antifungal protein-producing molds from dry-cured meat products. Int J Food Microbiol 135:39–46

Akbar A, Anal AK (2014) Occurrence of *Staphylococcus aureus* and evaluation of anti-staphylococcal activity of *Lactococcus lactis* subsp. *lactis* in ready-to-eat poultry meat. Ann Microbiol 64:131–138

Albano H, Todorov SD, Reenen CA, Hogg T, Dicks LMT, Teixeira P (2007) Characterization of two bacteriocins produced by *Pediococcus acidilactici* isolated from "Alheira", a fermented sausage traditionally produced in Portugal. Int J Food Microbiol 116:239–247

Alves VF, Martinez RCR, Lavrador MAS, De Martinis ECP (2006) Antilisterial activity of lactic acid bacteria inoculated on cooked ham. Meat Sci 74:623–627

Anacarso I, Messi P, Condò C, Iseppi R, Bondi M, Sabia C, de Niederhäusern S (2014) A bacteriocin-like substance produced from *Lactobacillus pentosus* 39 is a natural antagonist for the control of *Aeromonas hydrophila* and *Listeria monocytogenes* in fresh salmon fillets. LWT Food Sci Technol 55:604–611

Andrade MJ, Thorsen L, Rodríguez A, Córdoba JJ, Jespersen L (2014) Inhibition of ochratoxigenic moulds by *Debaryomyces hansenii* strains for biopreservation of dry-cured meat products. Int J Food Microbiol 170:70–77

Arqués JL, Rodríguez E, Langa S, Landete JM, Medina M (2015) Antimicrobial activity of lactic acid bacteria in dairy products and gut: effect on pathogens. Biomed Res Int 2015:584183

Aymerich MT, Hugas M, Monfort JM (1998) Review: bacteriocinogenic lactic acid bacteria associated with meat. Food Sci Technol Int 4:141–158

Aymerich T, Garriga M, Monfort J (2011) Applications of protective cultures, bacteriocins and bacteriophages in fermented meat products. In: Lacroix C (ed) Protective cultures, antimicrobial metabolites and bacteriophages for food and beverage biopreservation. Woodhead Publishing, Cambridge, pp 297–323

Baker CA, Rubinelli PM, Park SH, Carbonero F, Ricke SC (2016) Shiga toxin-producing *Escherichia coli* in food: incidence, ecology, and detection strategies. Food Control 59:407–419

Benkerroum N, Oubel H, Mimoun LB (2002) Behavior of *Listeria monocytogenes* and *Staphylococcus aureus* in yogurt fermented with a bacteriocin-producing thermophilic starter. J Food Prot 65:799–805
Benkerroum N, Daoudi A, Kamal M (2003) Behaviour of *Listeria monocytogenes* in raw sausages (merguez) in presence of a bacteriocin-producing lactococcal strain as a protective culture. Meat Sci 63:479–484
Benkerroum N, Daoudi A, Hamraoui T, Ghalfi H, Thiry C, Duroy M, Evrart P, Roblain D, Thonart P (2005) Lyophilized preparations of bacteriocinogenic *Lactobacillus curvatus* and *Lactococcus lactis* subsp. *lactis* as potential protective adjuncts to control *Listeria monocytogenes* in dry-fermented sausages. J Appl Microbiol 98:56–63
Beshkova D, Frengova G (2012) Bacteriocins from lactic acid bacteria: microorganisms of potential biotechnological importance for the dairy industry. Eng Life Sci 12:419–432
Bouksaim M, Lacroix C, Audet P, Simard RE (2000) Effects of mixed starter composition on nisin Z production by *Lactococcus lactis* subsp. *lactis* biovar. *diacetylactis* UL 719 during production and ripening of Gouda cheese. Int J Food Microbiol 59:141–156
Bourdichon F, Casaregola S, Farrokh C, Frisvad JC, Gerds ML, Hammes WP, Harnett J, Huys G, Laulund S, Ouwehand A, Powell IB, Prajapati JB, Seto Y, Schure ET, Boven AV, Vankerckhoven V, Zgoda A, Tuijtelaars S, Hansen EB (2012) Food fermentations: microorganisms with technological beneficial use. Int J Food Microbiol 154:87–97
Bredholt S, Nesbakken T, Holck A (1999) Protective cultures inhibit growth of *Listeria monocytogenes* and *Escherichia coli* O157:H7 in cooked, sliced, vacuum- and gas-packaged meat. Int J Food Microbiol 53:43–52
Bredholt S, Nesbakken T, Holck A (2001) Industrial application of an antilisterial strain of *Lactobacillus sakei* as a protective culture and its effect on the sensory acceptability of cooked, sliced, vacuum-packaged meats. Int J Food Microbiol 66:191–196
Brillet A, Pilet MF, Prevost H, Bouttefroy A, Leroi F (2004) Biodiversity of *Listeria monocytogenes* sensitivity to bacteriocin-producing *Carnobacterium* strains and application in sterile cold-smoked salmon. J Appl Microbiol 97:1029–1037
Buchanan RL, Bagi LK (1997) Microbial competition: effect of culture conditions on the suppression of *Listeria monocytogenes* Scott A by *Carnobacterium piscicola*. J Food Prot 60:254–261
Budde BB, Hornbæk T, Jacobsen T, Barkholt V, Koch AG (2003) *Leuconostoc carnosum* 4010 has the potential for use as a protective culture for vacuum-packed meats: culture isolation, bacteriocin identification, and meat application experiments. Int J Food Microbiol 83:171–184
Calo-Mata P, Arlindo S, Boehme K, de Miguel T, Pascoal A, Barros-Velazquez J (2008) Current applications and future trends of lactic acid bacteria and their bacteriocins for the biopreservation of aquatic food products. Food Bioproc Tech 1:43–63
Campbell-Platt G (1994) Fermented foods – a world perspective. Food Res Int 27:253–257
Carlini N, Mogna L, Strozzi GP (2010) Reducing foodborne pathogens in the food chain by the use of protective and probiotic cultures. In: Rovira J (ed) How to improve the food safety of my products? Instituto Tomás Pascual Sanz, Madrid, pp 73–88
Cartwright EJ, Jackson KA, Johnson SD, Graves LM, Silk BJ, Mahon BE (2013) Listeriosis outbreaks and associated food vehicles, United States, 1998–2008. Emerg Infect Dis 19:1–9
Castellano P, Belfiore C, Fadda S, Vignolo G (2008) A review of bacteriocinogenic lactic acid bacteria used as bioprotective cultures in fresh meat produced in Argentina. Meat Sci 79:483–499
Castellano P, González C, Carduza F, Vignolo G (2010) Protective action of *Lactobacillus curvatus* CRL705 on vacuum-packaged raw beef. Effect on sensory and structural characteristics. Meat Sci 85:394–401
Chaillou S, Christieans S, Rivollier M, Lucquin I, Champomier-Verges MC, Zagorec M (2014) Quantification and efficiency of *Lactobacillus sakei* strain mixtures used as protective cultures in ground beef. Meat Sci 97:332–338
Cheong EYL, Sandhu A, Jayabalan J, Le TTK, Nhiep NT, Ho HTM, Zwielehner J, Bansal N, Turner MS (2014) Isolation of lactic acid bacteria with antifungal activity against the common cheese spoilage mould *Penicillium commune* and their potential as biopreservatives in cheese. Food Control 46:91–97

Cocolin L, Innocente N, Biasutti M, Comi G (2004) The late blowing in cheese: a new molecular approach based on PCR and DGGE to study the microbial ecology of the alteration process. Int J Food Microbiol 90:83–91

Comi G, Andyanto D, Manzano M, Iacumin L (2016) *Lactococcus lactis* and *Lactobacillus sakei* as bio-protective culture to eliminate *Leuconostoc mesenteroides* spoilage and improve the shelf life and sensorial characteristics of commercial cooked bacon. Food Microbiol 58:16–22

Cotter PD, Hill C, Ross RP (2005) Bacteriocins: developing innate immunity for food. Nat Rev 3:777–788

Crowley S, Mahonya J, van Sinderena D (2013) Current perspectives on antifungal lactic acid bacteria as natural biopreservatives. Trends Food Sci Technol 33:93–109

Dal Bello B, Cocolin L, Zeppa G, Field D, Cotter PD, Hill C (2012) Technological characterization of bacteriocin producing *Lactococcus lactis* strains employed to control *Listeria monocytogenes* in cottage cheese. Int J Food Microbiol 153:58–65

Dalgaard P, Gram L, Huss HH (1993) Spoilage and shelf-life of cod fillets packed in vacuum or modified atmospheres. Int J Food Microbiol 19:283–294

Dalgaard P, Mejlholm O, Christiansen TJ, Huss HH (1997) Importance of *Photobacterium phosphoreum* in relation to spoilage of modified atmosphere-packed fish products. Lett Appl Microbiol 24:373–378

Deegan LH, Cotter PD, Hill C, Ross P (2006) Bacteriocins: biological tools for bio-preservation and shelf-life. Int Dairy J 16:1058–1071

Delavenne E, Ismail R, Pawtowski A, Mounier J, Barbier B (2012) Assessment of lactobacilli strains as yoghurt bioprotective cultures. Food Control 30:206–213

Delavenne E, Cliquet S, Trunet C, Barbier G, Mounier J, Le Blay G (2015) Characterization of the antifungal activity of *Lactobacillus harbinensis* K.V9.3.1Np and *Lactobacillus rhamnosus* K.C8.3.1I in yogurt. Food Microbiol 45:10–17

Devlieghere F, Vermeiren L, Debevere J (2004) New preservation technologies: possibilities and limitations. Int Dairy J 14:273–285

Djenane D, Martínez L, Sánchez-Escalante A, Montanes L, Blanco D, Yangüela J, Beltrán JA, Roncalés P (2006) Effect of lactic acid bacteria on beef steak microbial flora stored under modified atmosphere and on *Listeria monocytogenes* in broth cultures. Food Sci Technol Int 12:287–295

Duffes F, Corre C, Leroi F, Dousset X, Boyaval P (1999) Inhibition of *Listeria monocytogenes* by *in situ* produced and semipurified bacteriocins of *Carnobacterium* spp. on vacuum-packed, refrigerated cold-smoked. J Food Prot 62:1394–1403

Elsser-Gravesen D, Elsser-Gravesen A (2014) Biopreservatives. Adv Biochem Eng Biotechnol 143:29–49

Faith NG, Parniere N, Larson T, Lorang TD, Kaspar CW, Luchansky JB (1998) Viability of *Escherichia coli* O157:H7 in salami following conditioning of batter, fermentation and drying of sticks, and storage of slices. J Food Prot 61:377–382

Gálvez A, Lopez RL, Abriouel H, Valdivia E, Omar NB (2008) Application of bacteriocins in the control of foodborne pathogenic and spoilage bacteria. Crit Rev Biotechnol 28:125–152

Gálvez A, Grande-Burgos MJ, Lucas-López R, Pérez Pulido R (2014) Food biopreservation. Springer, London

Garcha S, Natt NK (2011) *In situ* control of food spoilage fungus using *Lactobacillus acidophilus* NCDC 291. J Food Sci Technol 49:643–648

García P, Rodríguez L, Rodríguez A, Martínez B (2010) Food biopreservation: promising strategies using bacteriocins, bacteriophages and endolysins. Trends Food Sci Technol 21:373–382

Ghanbari M, Jami M, Domig KJ, Kneifel W (2013) Seafood biopreservation by lactic acid bacteria: a review. LWT- Food Sci Technol 54:315–324

Hammes WP, Knauf HJ (1994) Starters in the processing of meat products. Meat Sci 36:155–168

Héchard Y, Sahl HG (2002) Mode of action of modified and unmodified bacteriocins from Gram-positive bacteria. Biochimie 84:545–557

Heng NCK, Tagg JR (2006) What is in a name? Class distinction for bacteriocins. Nat Rev Microbiol 4. https://doi.org/10.1038/nrmicro1273-c1

Heng NCK, Wescombe PA, Burton JP, Jack RW, Tagg JR (2007) The diversity of bacteriocins in Gram-positive bacteria. In: Riley MA, Chavan MA (eds) Bacteriocins: ecology and evolution. Springer, Berlin, pp 45–92

Holzapfel WH, Geisen R, Schillinger U (1995) Biological preservation of foods with reference to protective cultures, bacteriocins and food-grade enzymes. Int J Food Microbiol 24:343–362

Hugas M (1998) Bacteriocinogenic lactic acid bacteria for the biopreservation of meat and meat products. Meat Sci 49(suppl 1):S139–S150

Huss HH, Reilly A, Embarek PKB (2000) Prevention and control of safety hazards in cold smoked salmon production. Food Control 11:149–156

Ibrahim F, Vesterlund S (2014) *Lactococcus lactis ssp. lactis* as protective culture in vacuum-packed raw salmon (*Salmo salar*). J Aquat Food Prod Technol 23:601–607

Jacobsen T, Budde BB, Koch AG (2003) Application of *Leuconostoc carnosum* for biopreservation of cooked meat products. J Appl Microbiol 95:242–249

Jácome SL, Fonseca S, Pinheiro R, Todorov S, Noronha L, Silva J, Gomes A, Pintado M, Morais AMMB, Teixeira P, Vaz-Velho M (2014) Effect of lactic acid bacteria on quality and safety of ready-to-eat sliced cured/smoked meat products. Chem Eng Trans 38:403–408

Jami M, Ghanbari M, Zunabovic M (2014) *Listeria monocytogenes* in aquatic food products: a review. Compr Rev Food Sci Food Saf 13:798–813

Jiménez M, Soler P, Venanzi JD, Cante P, Varela C, Martínez NF (2005) An outbreak of *Campylobacter jejuni* enteritis in a school of Madrid, Spain. Euro Surveill 10:118–121

Jones R, Zagorec M, Brightwell G, Tagg JR (2009) Inhibition by *Lactobacillus sakei* of other species in the flora of vacuum packaged raw meats during prolonged storage. Food Microbiol 25:876–881

Juven BJ, Barefoot SF, Pierson MD, McCaskill LH, Smith B (1998) Growth and survival of *Listeria monocytogenes* in vacuum-packaged ground beef inoculated with *Lactobacillus alimentarius* FloraCarn L-2. J Food Prot 61:551–556

Kaban G, Kaya M, Lücke F (2010) The effect of *Lactobacillus sakei* on the behavior of *Listeria monocytogenes* on sliced bologna-type sausages. J Food Saf 30:889–901

Katikou P, Ambrosiadis I, Georgantelis D, Koidis P, Georgakis SA (2005) Effect of *Lactobacillus* protective cultures with bacteriocin like inhibitory substances' producing ability on microbiological, chemical and sensory changes during storage of refrigerated vacuum packaged sliced beef. J Appl Microbiol 99:1303–1313

Klaenhammer TR (1993) Genetics of bacteriocins produced by lactic acid bacteria. FEMS Microbiol Rev 12:39–85

Koo OK, Eggleton M, O'Bryan CA, Crandall PG, Ricke SC (2012) Antimicrobial activity of lactic acid bacteria against *Listeria monocytogenes* on frankfurters formulated with and without lactate/diacetate. Meat Sci 92:533–537

Kotzekidou P, Bloukas JG (1996) Effect of protective cultures and packaging film permeability on shelf-life of sliced vacuum-packed cooked ham. Meat Sci 42:333–345

Latorre-Moratalla ML, Bover-Cid S, Aymerich T, Marcos B, Vidal-Carou MC, Garriga M (2007) Aminogenesis control in fermented sausages manufactured with pressurized meat batter and starter culture. Meat Sci 75:460–469

Laursen BG, Bay L, Cleenwerck I, Vancanneyt M, Swings J, Dalgaard P, Leisner JJ (2005) *Carnobacterium divergens* and *Carnobacterium maltaromaticum* as spoilers or protective cultures in meat and seafood: phenotypic and genotypic characterization. Syst Appl Microbiol 28:151–164

Leistner L (2000) Basic aspects of food preservation by hurdle technology. Int J Food Microbiol 55:181–186

Leistner L, Gorris LGM (1995) Food preservation by hurdle technology. Trends Food Sci Technol 6:41–46

Leroi F (2010) Occurrence and role of lactic acid bacteria in seafood products. Food Microbiol 27:698–709

Leroi F, Cornet F, Chevalier F, Cardinal M, Coeuret G, Chaillou S, Joffraud JJ (2015) Selection of bioprotective cultures for preventing cold-smoked salmon spoilage. Int J Food Microbiol 213:79–87

Løvdal T (2015) The microbiology of cold smoked salmon. Food Control 54:360–373
Lücke FK (1994) Fermented meat products. Food Res Int 27:299–307
Lücke FK (2000) Utilization of microbes to process and preserve meat. Meat Sci 56:105–115
Maisnier-Patin S, Deschamps N, Tatini SR, Richard J (1992) Inhibition of *Listeria monocytogenes* in Camembert cheese made with a nisin-producing starter. Lait 72:249–263
Mangiapane E, Mazzoli R, Pessione A, Svenssonb B, Riedel K, Pessione E (2015) Ten years of subproteome investigations in lactic acid bacteria: a key for food starter and probiotic typing. J Proteomics 127:332–339
Maragkoudakis PA, Mountzouris KC, Psyrras D, Cremonese S, Fischer J, Cantor MD, Tsakalidou E (2009) Functional properties of novel protective lactic acid bacteria and application in raw chicken meat against *Listeria monocytogenes* and *Salmonella enteritidis*. Int J Food Microbiol 130:219–226
Mataragas M, Drosinos EH, Metaxopoulos J (2003) Antagonistic activity of lactic acid bacteria against *Listeria monocytogenes* in sliced cooked cured pork shoulder stored under vacuum or modified atmosphere at 4±2 °C. Food Microbiol 20:259–265
McAuliffe O, Hill C, Ross RP (1999) Inhibition of *Listeria monocytogenes* in cottage cheese manufactured with a lacticin 3147-producing starter culture. J Appl Microbiol 86:251–256
Melero B, Diez AM, Rajkovic A, Jaime I, Rovira J (2012) Behaviour of non-stressed and stressed *Listeria monocytogenes* and *Campylobacter jejuni* cells on fresh chicken burger meat packaged under modified atmosphere and inoculated with protective culture. Int J Food Microbiol 158:107–112
Melero B, Vinuesa R, Diez AM, Jaime I, Rovira J (2013) Application of protective cultures against *Listeria monocytogenes* and *Campylobacter jejuni* in chicken products packaged under modified atmosphere. Poult Sci 92:1108–1116
Mills S, Serrano LM, Griffin C, O'Connor PM, Schaad G, Bruining C, Hill C, Ross RP, Meijer WC (2011) Inhibitory activity of *Lactobacillus plantarum* LMG P-26358 against *Listeria innocua* when used as an adjunct starter in the manufacture of cheese. Microb Cell Fact 10(Suppl 1):S7
Mor-Mur M, Yuste J (2010) Emerging bacterial pathogens in meat and poultry: an overview. Food Bioproc Tech 3:24–35
Muñoz A, Maqueda M, Gálvez A, Martínez-Bueno M, Rodríguez A, Valdivia E (2004) Biocontrol of psychrotrophic enterotoxigenic *Bacillus cereus* in a non fat hard type cheese by an enterococcal strain-producing enterocin AS-48. J Food Prot 67:1517–1521
Muñoz A, Ananou S, Gálvez A, Martínez-Bueno M, Rodríguez A, Maqueda M, Valdivia E (2007) Inhibition of *Staphylococcus aureus* in dairy products by enterocin AS-48 produced in situ and ex situ: Bactericidal synergism through heat and AS-48. Int Dairy J 17:760–769
Muthukumarasamy P, Holley RA (2007) Survival of *Escherichia coli* O157: H7 in dry fermented sausages containing micro-encapsulated probiotic lactic acid bacteria. Food Microbiol 24:82–88
Nes IF, Kjos M, Bao Diep D (2012) Antimicrobial components of lactic acid bacteria. In: Sampo Lahtinen S, Ouwehand AC, Ssalminen S, von Wright A (eds) Lactic acid bacteria: microbial and functional aspects, 4th edn. CRC Press, Boca Ratón, pp 285–329
Nieminen TT, Nummela M, Björkroth J (2015) Packaging gas selects lactic acid bacterial communities on raw pork. J Appl Microbiol 119:1310–1316
Nilsson L, Gram L, Huss HH (1999) Growth control of *Listeria monocytogenes* on cold-smoked salmon using a competitive lactic acid bacteria flora. J Food Prot 62:336–342
Nilsson L, Ng YY, Christiansen JN, Jørgensen BL, Grótinum D, Gram L (2004) The contribution of bacteriocin to inhibition of *Listeria monocytogenes* by *Carnobacterium piscicola* strains in cold-smoked salmon systems. J Appl Microbiol 96:133–143
Nilsson L, Hansen TB, Garrido P, Buchrieser C, Glaser P, Knøchel S, Gram L, Gravesen A (2005) Growth inhibition of *Listeria monocytogenes* by a nonbacteriocinogenic *Carnobacterium piscicola*. J Appl Microbiol 98:172–183
Núñez M, Rodríguez JL, García E (1997) Inhibition of *Listeria monocytogenes* by enterocin 4 during the manufacture and ripening of Manchego cheese. J Appl Microbiol 83:671–677
Nuñez F, Lara MS, Peromingo B, Delgado J, Sánchez-Montero L, Andrade MJ (2015) Selection and evaluation of *Debaryomyces hansenii* isolates as potential bioprotective agents against toxigenic penicillia in dry-fermented sausages. Food Microbiol 46:114–120

O'Sullivan L, Ryan MP, Ross RP, Hill C (2003) Generation of food-grade lactococcal starters which produce the lantibiotics lacticin 3147 and lacticin 481. Appl Environ Microbiol 69:3681–3685
O'Sullivan L, O'Connor EB, Ross RP, Hill C (2006) Evaluation of live-culture-producing lacticin 3147 as a treatment for the control of *Listeria monocytogenes* on the surface of smear-ripened cheese. J Appl Microbiol 100:135–143
Olaoye OA, Dodd CER (2010) Evaluation of bacteriocinogenic *Pediococcus acidilactici* as protective culture in the preservation of tsire, a traditional Nigerian stick meat. J Food Saf 30:867–888
Olaoye OA, Onilude AA, Idowu OA (2011) Microbiological profile of goat meat inoculated with lactic acid bacteria cultures and stored at 30 °C for 7 days. Food Bioproc Tech 4:312–319
Osés SM, Diez AM, Gómez EM, Wilches-Pérez D, Luning PA, Jaime I, Rovira J (2015) Control of *Escherichia coli* and *Listeria monocytogenes* in suckling-lamb meat evaluated using microbial challenge tests. Meat Sci 110:262–269
Rilla N, Martínez B, Delgado T, Rodríguez A (2003) Inhibition of *Clostridium tyrobutyricum* in Vidiago cheese by *Lactococcus lactis* ssp. *lactis* IPLA 729, a nisin Z producer. Int J Food Microbiol 85:23–33
Riordan DCR, Duffy G, Sheridan JJ, Eblen BS, Whiting RC, Blair IS, McDowell DA (1998) Survival of *Escherichia coli* O157:H7 during the manufacture of pepperoni. J Food Prot 61:146–151
Rodríguez JM, Cintas LM, Casaus P, Horn N, Dodd HM, Hernandez PE, Gasson MJ (1995) Isolation of nisin-producing *Lactococcus lactis* strains from dry fermented sausages. J Appl Bacteriol 78:109–115
Rodríguez E, Arqués JL, Gaya P, Nuñez M, Medina M (2001) Control of *Listeria monocytogenes* by bacteriocins and monitoring of bacteriocin-producing lactic acid bacteria by colony hybridization in semi-hard raw milk cheese. J Dairy Res 68:131–137
Rodríguez A, Bernáldez V, Rodríguez M, Andrade MJ, Núñez F, Córdoba JJ (2015) Effect of selected protective cultures on ochratoxin A accumulation in dry-cured Iberian ham during its ripening process. LWT Food Sci Technol 60:923–928
Rubio R, Bover-Cid S, Martin B, Garriga M, Aymerich T (2013) Assessment of safe enterococci as bioprotective cultures in low-acid fermented sausages combined with high hydrostatic pressure. Food Microbiol 33:158–165
Ryan MP, Rea MC, Hill C, Ross RP (1996) An application in Cheddar cheese manufacture for a strain of *Lactococcus lactis* producing a novel broad-spectrum bacteriocin, lacticin 3147. Appl Environ Microbiol 62:612–619
Saraoui T, Fall PA, Leroi F, Antignac JP, Chéreau S, Pilet MF (2016) Inhibition mechanism of *Listeria monocytogenes* by a bioprotective bacteria *Lactococcus piscium* CNCM I-4031. Food Microbiol 53:70–78
Schillinger U, Kaya M, Lücke FK (1991) Behaviour of *Listeria monocytogenes* in meat and its control by a bacteriocin-producing strain of *Lactobacillus sake*. J Appl Bacteriol 70:473–478
Schnürer J, Magnusson J (2005) Antifungal lactic acid bacteria as biopreservatives. Trends Food Sci Technol 16:70–78
Schwenninger SM, Meile L (2004) A mixed culture of *Propionibacterium jensenii* and *Lactobacillus paracasei* subsp. *paracasei* inhibits food spoilage yeasts. Syst Appl Microbiol 27:229–237
Somkuti GA, Steinberg DH (2010) Pediocin production in milk by *Pediococcus acidilactici* in co-culture with *Streptococcus thermophilus* and *Lactobacillus delbrueckii* subsp. *bulgaricus*. J Ind Microbiol Biotechnol 37:65–69
Stiles M (1996) Biopreservation by lactic acid bacteria. Antonie Van Leeuwenhoek 70:331–345
Tolvanen R, Hellström S, Elsser D, Morgenstern H, Björkroth J, Korkeala H (2008) Document survival of *Listeria monocytogenes* strains in a dry sausage model. J Food Prot 71:1550–1555
Työppönen S, Petäjä E, Mattila-Sandholm T (2003a) Bioprotectives and probiotics for dry sausages. Int J Food Microbiol 83:233–244
Työppönen S, Markkula A, Petäjä E, Suihko ML, Mattila-Sandholm T (2003b) Survival of *Listeria monocytogenes* in North European type dry sausages fermented by bioprotective meat starter cultures. Food Control 14:181–185

Vermeiren L, Devlieghere F, Debevere J (2004) Evaluation of meat born lactic acid bacteria as protective cultures for the biopreservation of cooked meat products. Int J Food Microbiol 96:149–164

Vermeiren L, Devlieghere F, Vandekinderen I, Debevere J (2006a) The interaction of the non-bacteriocinogenic *Lactobacillus sakei* 10A and lactocin S producing *Lactobacillus sakei* 148 towards *Listeria monocytogenes* on a model cooked ham. Food Microbiol 23:511–518

Vermeiren L, Devlieghere F, Debevere J (2006b) Co-culture experiments demonstrate the usefulness of *Lactobacillus sakei* 10A to prolong the shelf-life of a model cooked ham. Int J Food Microbiol 108:68–77

Vermeiren L, Devlieghere F, Vandekinderen I, Rajtak U, Debevere J (2006c) The sensory acceptability of cooked meat products treated with a protective culture depends on glucose content and buffering capacity: a case study with *Lactobacillus sakei* 10A. Meat Sci 74:532–545

Vignolo G, Castellano P, Fadda S (2015) Bioprotective cultures. In: Toldrá F (ed) Handbook of fermented meat products and poultry, 2nd edn. Wiley Blackwell, Oxford, pp 129–137

Wan X, Saris PEJ, Takala TM (2015) Genetic characterization and expression of leucocin B, a class IId bacteriocin from *Leuconostoc carnosum* 4010. Res Microbiol 166:494–503

Yamazaki K, Suzuki M, Kawai Y, Inoue N, Montville TJ (2003) Inhibition of *Listeria monocytogenes* in cold-smoked salmon by *Carnobacterium piscicola* CS526 isolated from frozen surimi. J Food Prot 66:1420–1425

Zanette CM, Dalla Santa OR, Bersot LS (2015) Effect of *Lactobacillus plantarum* starter cultures on the behavior of *Listeria monocytogenes* during sausage maturation. Int Food Res J 22:844–848

# Probiotics and Prebiotics for the Health of Pigs and Horses

4

Spyridon K. Kritas

## 4.1 Introduction

Current and increasing demands for food, particularly animal origin food, have imposed mass intensified animal farming. In large populations, clinical or subclinical effects of microbial pathogens are magnified, and therefore administration of antibiotics at therapeutic or low subtherapeutic doses (antibiotic growth promoters, AGP) had been extensively practiced. For decades, antibiotics in subtherapeutic doses and for prolonged periods have been used as feed additives to minimize the impact of certain scours in farm animals and especially in monogastric farm animals (Dibner and Richards 2005). Furthermore, the use of antibiotics as growth promoters improved performance indicators like body weight gain and feed conversion ratio, through modifying the intestinal microbiota and supporting the general health status of the animals (Dibner and Richards 2005). At least four mechanisms have been proposed as explanations of antibiotic-mediated growth enhancement: (1) inhibition of subclinical infections, (2) reduction of growth-depressing microbial metabolites, (3) reduction of microbial use of nutrients, and (4) enhanced uptake and use of nutrients through the thinner intestinal wall associated with antibiotic-fed animals (Anderson et al. 1999; Visek 1978).

The use of chemotherapeutics in animal feeds was considered a threat by many scientists since they belong to classes of drugs used to treat human diseases (e.g., tylosin, spiramycin, bacitracin, virginiamycin, penicillins, tetracyclines), or were considered unacceptable occupational toxicity risks (olaquindox and carbadox). In many cases, a pathogen may develop resistance and then cause a human health problem that cannot be treated with the related classes of drugs. Considering that bacteria can transfer resistance to other types of bacterial species via R plasmids (bits of genetic material

S.K. Kritas, D.V.M., M.H.M., Ph.D.
Department of Microbiology and Infectious Diseases, Faculty of Veterinary Medicine, School of Health Sciences, Aristotle University of Thessaloniki, Macedonia, Greece
e-mail: skritas@vet.auth.gr



D. Di Gioia, B. Biavati (eds.), *Probiotics and Prebiotics in Animal Health and Food Safety*, https://doi.org/10.1007/978-3-319-71950-4_4

smaller than chromosomes that replicate autonomously in the cell cytoplasm), this threat is thought by many to be particularly important (Aiello 1998). A well-known example of a bacterium that has acquired resistance to multiple antibiotics is methicillin-resistant *Staphylococcus aureus* (MRSA). The awareness concerning the antibiotic resistance in the European Union (EU) and the possible spread of antibiotic resistance genes from bacteria of animal origin to humans has led to the prohibition of chemotherapeutics as feed additives since 2006. Another potential risk of agricultural use of antibiotics is the presence of chemotherapeutic residues in food. For these reasons, both in Europe and the USA, large supermarkets commonly ask suppliers to provide assurance that meat products were produced without growth-promoting feed additives. Almost all growth promoters have been banned in the European Union. A similar trend is beginning in the USA, and some producers who have become involved in both raising and marketing antibiotic-free products are receiving a premium price.

Furthermore, the use of certain feed additives, like zinc oxide, with beneficial effect against postweaning diarrhea in pigs, was restricted by the EU Regulation 1334/2003 as it can be harmful for the environment. Therefore, the removal of all these substances from animal feeding increases the pathogen pressure and risks in livestock. As pig and poultry are major food-producing animals, important efforts on antibiotic alternatives were performed in these animal species. Among such alternatives, probiotics and prebiotics can enhance animal resistance to pathogenic bacteria and improve their general health status (Choct 2009; Williams et al. 2001; Cheng et al. 2014).

## 4.2 Gut Microbiota

Gastrointestinal (GI) tract, and in particular intestine, (a) is one of the largest contact surfaces with the environment, (b) digests food and permits the absorption of nutrients, (c) blocks the entrance of many pathogens, and (d) contains one of the largest mucosal lymphoid tissue of the body (Tizard 2013). As intestine plays important role in nutrient absorption and in immunological response (e.g., more than 80% of the body's activated B cells are found in the intestine) (Tizard 2013), any malfunctions in the gut are not expected to be restricted only locally, but may indirectly affect remote organs (e.g., through lack of vitamins).

The intestinal tract is sterile at birth and becomes colonized in a series of successive steps (Dominguez-Bello et al. 2010; Koenig et al. 2011). Colonization is dependent on the microorganisms in the host animal's environment, the host's physiology, and the host animal's response to the early colonizers (Vondruskova et al. 2010). Usually the microbiota that colonizes the gastrointestinal (GI) tract consists of more than 800 different bacterial strains and contains a total count of $10^{14}$ bacteria (Vondruskova et al. 2010). They are divided in indigenous bacteria that tend to colonize the intestinal tract in a permanent way and transient bacteria which temporarily reside in the intestinal tract. In the healthy animal, the balance of microorganisms in the GI tract helps in efficient digestion and maximum absorption of nutrients. Moreover, it increases resistance to infectious diseases and is important for the normal development of gut morphology, maturation, and functionality (Zoetendal et al. 2004).

The presence of the gut microbiota provides constant antigenic stimulus and keeps intestinal lymphoid tissue in a constant state of activation, while immunity in the intestine is known to play a role in priming immunity even in remote areas such as the respiratory tract, urogenital tract, and mammary gland, e.g., by movement of IgA-positive B cells via intestinal lymphatics and bloodstream to these areas (Schachtschneider et al. 2013; Tizard 2013). Intestinal microbes are important for lymphocyte development such as B-cell class switching, Th17 effector T-cell development, and induction of regulatory T cells (Duerkop et al. 2009).

The commensal population has profound effects on the overall health and performance, and the development of organ, tissue, and immune system, in monogastric animals in particular, is something becoming even more obvious by experiments comparing conventionally reared versus sterile (germ-free) animals (Snel et al. 2002). In normal animals, if the microbiota becomes disrupted and a bacterial population prevails, health problems can occur (Fig. 4.1). During periods of stress, e.g., at weaning or after diet modifications, this balance can be altered, generally resulting in disturbance of the commensal flora. Beneficial bacteria (e.g., lactobacilli, bifidobacteria) have been shown to decrease when stress factors occur (Si et al. 2004). The resulting overgrowth of pathogenic bacteria can contribute to subclinical manifestations like decreased feed conversion ratio and production performance or cause clinical signs such as diarrhea (Gareau et al. 2009). For instance,

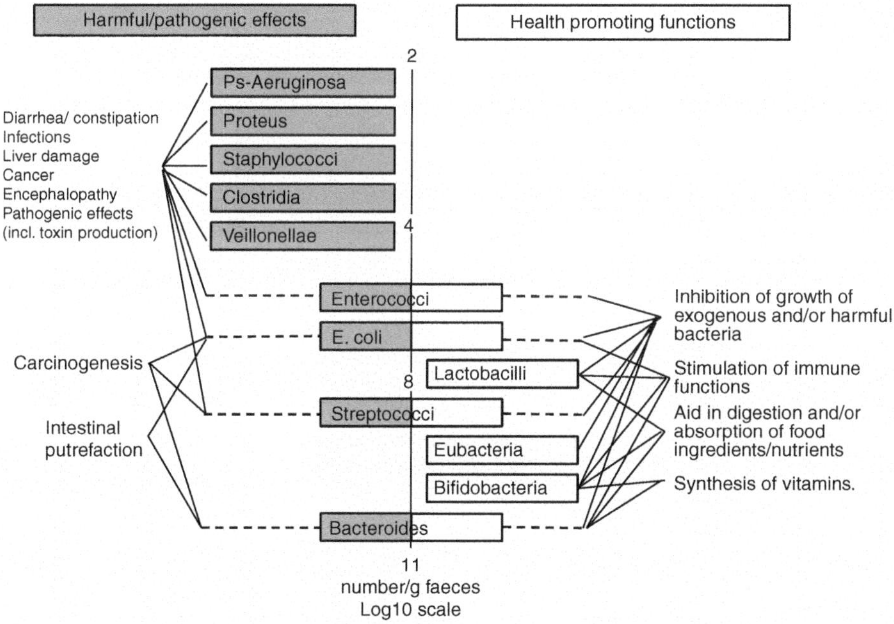


**Fig. 4.1** Generalized scheme of the composition and health effects of predominant human fecal bacteria. The figure shows approximate numbers of the different genera. The bacteria are generally split into those groups that have harmful or pathogenic influences on human health, those that have beneficial effects, and those that may have both. Potential reasons for the classification scheme are given (From Gibson and Roberfroid 1995, J. Nutr. 125:1401–1412)

in case of predominance of *Escherichia coli* in the gut lumen, diarrhea will be produced by most bacterial *E. coli* strains, while certain *E. coli* strains will cause edema disease due to exotoxin production, a nervous disease of weaned pigs (Vondruskova et al. 2010).

The microbiome of the gastrointestinal track can be considered as a metabolically active organ. The major commensal groups in monogastric animals (such as pig, chicken, rabbit, and human) are *Lactobacillus*, *Bifidobacterium*, *Bacteroides*, *Eubacterium*, *Streptococcus*, *Enterobacteriaceae*, *Clostridium*, *Fusobacterium*, *Propionibacterium*, and *Peptostreptococcus* (Vondruskova et al. 2010; Kenny et al. 2011).

## 4.3 Probiotics and Prebiotics

Probiotics (or direct-fed microbials, DFM) are harmless live microorganisms with beneficial effects on the host animal species. As presented in Fig. 4.2, their main mechanisms of actions include (Salminen et al. 1996; Mazmanian et al. 2008; Hooper et al. 2002; Timmerman et al. 2004; Salzman et al. 2003; Gill 2003):

- Production of antimicrobial substances such as organic acids (mostly lactic, acetic, and formic acid), bacteriocins, antibiotics, hydrogen peroxide, and other compounds that inhibit intestinal pathogens (Corcionivoschi et al. 2010; Murali and Kavitha 2010).

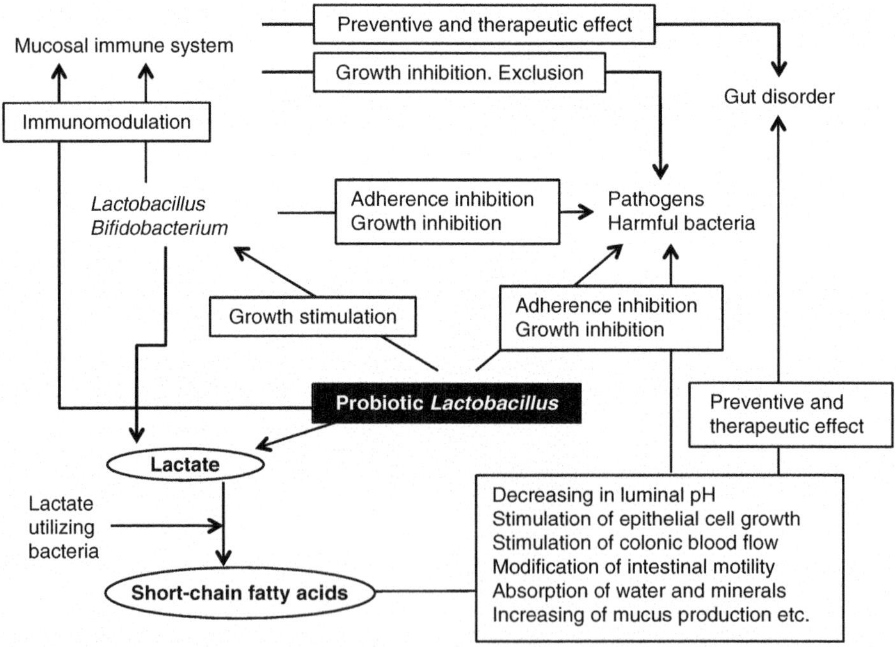


**Fig. 4.2** Overview of the health-beneficial effects of probiotic *Lactobacillus* (From Ohashi and Ushida 2009, *Animal Science Journal* 80, 361–371)

- Production of enzymes (e.g., proteases, amylases, lipases, and glycosidases) by the microbiota. *Bifidobacterium bifidum* produces a DNA polymerase that has been reported to be important in repairing damaged cells. This enzyme production may also explain improvements in feed efficiency that has been observed when certain DFMs are fed.
- Reductions of toxic amines, which are produced by some intestinal microbes, have irritating and toxic activity, or cause diarrhea.
- Competition for nutrients and/or for attachment sites (competitive exclusion) on the intestinal mucosa that potentially pathogenic bacteria use and thereby prevent them from colonizing the intestinal tract.
- Stimulation of the immune system. Probiotics may influence intestinal physiology either directly or indirectly through modulation of the commensal flora or the immune system. Thus, for example, they play a role in both specific and non-specific host immune responses and stimulation of production of pro- and anti-inflammatory cytokines(O'Hara and Shanahan 2006; Walsh et al. 2008; Wang et al. 2009). Some probiotic strains act as immunomodulators by enhancing macrophage activity, by increasing the local antibody levels, by inducing the production of interferon, and by activating killer cells (Yasui et al. 1989; Perdigon et al. 2001).

The administration of probiotics as feed additives has been used to correct these alterations and disturbances of the intestinal microbiota and establish its ideal synthesis. The application of probiotics may also be an alternative to the ongoing practice of subtherapeutic antibiotic use (Kritas and Morrison 2005).

Probiotics need to meet the following criteria (Fuller 1989):

1. Probiotic bacteria must be prepared in a viable manner and on a large scale.
2. They should remain viable and stable during use and under storage.
3. They should be able to survive in the intestinal tract.
4. The host should gain direct and indirect beneficial effects from the probiotics (improved intestinal microbiota).
5. Their safety should be evident.

Basic for livestock:

(a) Justified cost
(b) Ability to be used massively

## 4.4 The Effect of Probiotics and Prebiotics in the Pig

The main productive stages in pig farming are a) the reproduction cycle of the dam, e.g., estrus-conception, pregnancy, farrowing, and lactation and b) the growing of the pigs, e.g., suckling (0 to 3–4 weeks of age), nursery (from weaning at 3–4 weeks of age to 10 weeks of age), and growing-finishing periods (10–23 weeks of age).

The use of probiotics in farm animals has a main goal to restore the beneficial intestinal microbiota and the improvement of the animal general health. Another important aim is the increase of the productivity of farm animals. The establishment of a beneficial microbiota at the time of birth will result in healthier young animals, and this may be most likely achieved by treating dams, which represent an amplification step and will flood the neonatal pigs' environment with desirable bacterial strains. In contrast, at the time of weaning (e.g., the time of major crisis with instability and loss of certain commensal populations), it may be sufficient to simply provide a supportive, protective microbiota.

In pigs, there are studies indicating several possible mechanisms for the beneficial role of probiotics. These include an increase in intestinal lactic acid concentration and suppression of *E. coli* counts (Kovacs-Zomborszky et al. 1994), inhibition of adhesion of enterotoxigenic *E. coli* K88 to porcine small intestinal mucus, and increase of the number of ileal goblet cells and thickness of colonic mucosa (Jin et al. 2000). Lactobacilli and bifidobacteria administration, immediately after the birth of piglets, promotes the colonization of a healthy and beneficial commensal microbiota and limits mucosal atrophy, dysfunction, and pathogen load, thereby reducing the colonization of the intestine in neonatal piglets (Siggers et al. 2008).

Most probiotic species have increased fermentation activity and digestion capability (Ouwehand et al. 2002). Lactic acid bacteria (LAB) like *Lactobacillus* spp. produce lactic acid and proteolytic enzymes that improve digestibility (Yu et al. 2008). Meng et al. (2010) had observed better energy and crude protein digestibility in growing pigs fed a probiotic mixture containing *Bacillus subtilis* and *Clostridium butyricum* spores. In a study of Collado et al. (2007), in vitro adherence of *Salmonella* spp., *E. coli*, and *Clostridium* spp. to the porcine intestine had been reduced in the presence of *Bifidobacterium lactis* and *Lactobacillus rhamnosus*.

### 4.4.1 Sows and Suckling Piglets

During her life, the sow is subjected to many stressors, e.g., service, changes of housing, gestation in farrowing stalls, separation at weaning, etc. (Robertson and Tournout 1994). In addition, sow gains weight during pregnancy and subsequently loses it during lactation because her limited intake capacity cannot meet the nutritional requirements needed for milk production after parturition (Robertson and Tournout 1994; Whittemore 1998). Furthermore, the intake of nutrients by the sow during gestation and lactation influences the number of piglets born alive, the number of stillborn piglets, and the body weight of piglets at birth and weaning (English et al. 1984; Whittemore 1998).

From the side of the newborn, its intestinal flora starts to be established immediately after birth, while its balance is crucial for effective digestion and maximal absorption of nutrients, as well as for adequate body's resistance against infectious diseases (Dufresne 1998). Any stress factor that may disturb the equilibrium of commensals within the gut will adversely affect the growth and the health of the young animal. Despite the piglet being a separate individual, its viability up to the age of weaning greatly depends upon its mother in an indirect manner (colostrum and milk quantity and quality, diseases acquired by the dam, dam's aggressive behavior, etc.). It is therefore reasonable that improvement of health and productivity of the dam during pregnancy and lactation will be crucial for reducing mortality of the suckling pigs. To prevent or control both preweaning and postweaning illnesses, antibiotics in the creep feed or grower's ration have been used with good results (Kyriakis et al. 1992, 1995, 1997). Moreover, antibiotics incorporated in the feed for the pregnant sow, usually at a growth promoter dosage, appear to enhance their performance and to improve the viability of their progeny (Kantas et al. 1998; Giannakopoulos et al. 2001). The general consideration of removing antibiotics from sows and piglets and substituting them for probiotics applies also here.

Several probiotics have been examined on the sow and her offspring. Administration of a *Bacillus cereus* or a *B. licheniformis*/*B. subtilis* product in sow feed, 2 weeks prior to farrowing up to weaning, as well as in creep feed was enough to reduce suckling piglet mortality in farms even with mastitis problems (Alexopoulos et al. 2001, 2004a). Dams experienced reduced loss of body weight, which is possibly the reason for the shorter weaning to service interval (Alexopoulos et al. 2004a). An improved balance of gut microbiota leading to improvement of body condition of sows and young animals may have been a possible explanation. For the piglets, an indirect effect of *Bacillus* products could have been the case, as fat and lactose content of dam's milk was higher, and thus more nutritional (Alexopoulos et al. 2001, 2004a). Advantages such as better feed intake and bigger average live weight and size and vitality of the litter had been recorded after administration of *Bacillus cereus* var. *toyoi* and *Enterococcus faecium* in gestating sows (Böhmer et al. 2006; Taras et al. 2005, 2006).

Treatment of sows during early pregnancy, as well as treatment of their litters with feed supplemented with *B. cereus* var. *toyoi*, had as a result increased intestinal IgA secretion both in sows and piglets, lower incidence of diarrhea, and reduced carriage of pathogenic *E. coli* strains in piglets, while the absolute numbers and distributions of immune cells in the piglets had been altered (Scharek et al. 2007a, b; Schierack et al. 2007).

Attempts to apply a probiotic on long-term basis (e.g., for two sequential reproduction cycles) had been made by Kritas et al. (2015) using a *Bacillus subtilis* C-3102 spore-based probiotic on sows and their litters. Significant benefits such as improved sow body condition during pregnancy, increased sow feed consumption, reduced sow weight loss during lactation, reduced sow weaning-estrus interval, and higher body weight of piglets at weaning had been observed in both reproductive cycles (Kritas et al. 2015). Additionally, higher piglet birth weight and number of weaned piglets, as well as reduced *Escherichia coli* and *Clostridium* spp. counts in piglet feces, had been observed in the second cycle litters of probiotic-treated sows (Kritas et al. 2015).

Studies to further explain positive results of probiotics in sows have been designed. *B. cereus* var. *toyoi* supplementation of sows and their piglets had a positive impact on the health status of the offspring after a challenge with *Salmonella*, likely due to an altered immune response marked by reduced frequencies of CD8+ $\gamma\delta$ T cells in the peripheral blood and the jejunal epithelium (Scharek-Tedin et al. 2013). In a recent study, sows were fed with *Enterococcus faecium* NCIMB 10415 during pregnancy and lactation, and evidence has been provided for the expression of mCD14 by the porcine mammary epithelium and an immunological effect of mCD14(+) milk cells on the piglets' intestinal immune system (Scharek-Tedin et al. 2015).

### 4.4.2 Nursery and Growing-Finishing Pigs

During weaning, piglets experience biological stress such as physiological, environmental, and social challenges (e.g., separation from their mother, moving to another building and pen with new penmates and new type of feed, establishing group hierarchy, etc.). These stressful events lead to intestinal and immune system dysfunctions that result in reduced health, growth, and feed intake of pigs, particularly during the first week after weaning (Campbell et al. 2013). Proper growth of pigs after weaning is greatly influenced by pathogens that often profit from the reduced immune response of the animal in each stage and cause clinical or subclinical disease. Postweaning diarrhea is the most common clinical manifestation during this period (Campbell et al. 2013). To control this specific problem, antibiotics have been used worldwide for many years. Antimicrobials, when used as growth promoters, are claimed to improve daily weight gain by 3–9% and feed utilization by 2–7%, with fewer scour problems (Visek 1978; Doyle 2001). For this reason they are customarily used in pigs even on high-health status farms. They appear to act by reducing the pathogenic bacteria and modifying the microbiota in the gut, providing more nutrient availability for the animal itself and less substrate for the bacterial organisms to use for their own growth (Visek 1978).

During growing and finishing period, pigs are becoming more resistant to adverse conditions and less prone to diseases. Although feed antibiotics become less indispensable at this age (except from therapeutic reasons), some farmers continue antibiotic use as growth promoters. Dritz et al. (2002) showed that antibiotics are justified for use only in nursery pigs but not in growers and finishers. Once more recent concerns regarding the potential transfer of antibiotic resistance to human pathogens had directed researchers to alternative solutions.

For many years, weaning (nursery) stage is the target age group for probiotics and prebiotics in the field practice, as pigs of this age are considered more vulnerable to physiological changes, as well as to pathogens. Yeasts are highly resistant to inactivation during their passage along the alimentary canal, and they have been reported to produce a variety of beneficial production responses in piglets (Jurgens et al. 1997; Mathew et al. 1998; Van Heugten et al. 2003). *Saccharomyces cerevisiae* ssp. *boulardii* supplementation to piglets was associated with better body growth, while yeast-supplemented pigs had a thinner intestinal adherent mucous layer, higher proliferation of epithelial cells, and more mucosal macrophages (Bontempo et al. 2006). Administration of *Bacillus cereus* var. *toyoi* or *Saccharomyces cerevisiae* spp. *boulardii* to weaned pigs for 3–4 weeks had increased the growth performance with concomitant increase of villus length in the small intestine and a decrease in the number of goblet cells in the large intestine (Baum et al. 2002). Mathew et al. (1998) had recorded increased body weight gain and feed intake of piglets after feeding *S. cerevisiae*, while administration of *S. cerevisiae*, even in concomitance with ETEC infections, had reduced pig illness and mortality (Trevisi et al. 2015). LAB, e.g., *Pediococcus acidilactici*, seem to positively influence weight and postweaning average daily gain of weaned piglets, and part of this effect can be attributed to improvement of physiological parameters (increased villi height and crypts depth, larger number of proliferating enterocytes) (Di Giancamillo et al. 2008). The benefits of intestinal IgA secretion and reduction of translocation of enterotoxigenic *E. coli* have also been observed with *S. boulardii* or *P. acidilactici* given to piglets (Lessard et al. 2009). Similar findings on modulation of IgA development, together with a decreased ileal prevalence of ETEC, have been reported with a strain of *Lactobacillus sobrius* (Konstantinov et al. 2008).

Wang et al. (2013) reported that weaned piglets supplemented with *L. fermentum* presented higher feed intakes and as a consequence grew faster than negative control piglets. However, feed conversion was unaffected by *L. fermentum* supplementation. Moreover, *Bifidobacterium animalis* subsp. *lactis* improved the growth performance in weaning piglets and the ratio of bifidobacteria to *E. coli* in the gut (Modesto et al. 2009). Administration of *Enterococcus faecium* significantly improved growth and feed conversion of weaning pigs, while at the same time, the number of lactobacilli had been increased and the coliform counts had been reduced in the ileum (Mallo et al. 2010).

The positive effect of *Bacillus* probiotics on the control of certain pathogens in animals has been shown in several studies, where they appear to control enteric diseases associated with enterotoxigenic *Escherichia coli* (ETEC) or other enteric pathogens, one of which is postweaning diarrhea syndrome (PWDS) in pigs (Kozasa 1986; Kyriakis et al. 1999; Bhandari et al. 2008). Probiotics can reduce incidence and severity of diarrhea and mortality and ETEC counts, while they can improve weight gain and feed conversion ratio (Kyriakis et al. 1999; Marubashi et al. 2012). *Bacillus pumilus* spore treatment decreased ileal *E. coli* counts in a manner similar to the medicated treatment (apramycin and pharmacological levels of zinc oxide) but without the adverse effects on growth performance, *Lactobacillus* counts, cecal short-chain fatty acid concentration, and possible liver toxicity experienced with that medicated treatment (Prieto et al. 2014).

Porcine edema disease caused by Shiga toxin 2e-producing *Escherichia coli* (STEC) is another important disease of weaned pigs. Oral administration of a *Bacillus subtilis* strain in weaned piglets had prevented experimentally induced edema disease through the suppression of the growth of STEC in the ileum (Tsukahara et al. 2013).

In some studies, probiotics seem to play a helpful role not only against bacterial but also against viral pathogens in pigs. Administration of *Bifidobacterium lactis* can reduce the severity of weanling diarrhea associated with rotavirus and *E. coli*, possibly via a mechanism of enhanced immune-mediated protection (Shu et al. 2001).

Probiotics are generally thought of as acting in the intestine and thus combating mainly enteric pathogens. Several studies (Cangemi de Gutierrez et al. 2001; Hori et al. 2001) show that probiotics may act also in other sites of the body (respiratory, urinary, etc.), presumably by other ways than those expected (e.g., inhibition, competition for nutrients) and, namely, by being involved in immunological mechanisms. After oral administration of LAB in mice, increased IgA cells at both the intestinal and bronchial levels have been observed (Perdigon et al. 1999). Furthermore, intranasal administration of *Lactobacillus casei* in mice before infection of upper respiratory tract with influenza virus resulted in a significant tenfold reduction of virus titers recovered from nasal cavities and a significant increased survival of mice compared to controls (69% vs. 15%). Cellular immunity as well as interleukin 12, interferon-$\gamma$, and TNF-$\alpha$ were increased (Hori et al. 2001). A clinical study in humans has shown that daily ingestion of probiotics for a period of 3 weeks reduced nasal colonization with pathogenic bacteria (*Staphylococcus aureus*, *S. pneumoniae*, beta-hemolytic streptococci) (Gluck and Gebbers 2003). An attempt to determine whether oronasal administration of *Lactobacillus casei* can assist vaccination efficacy against the most worldwide distributed swine pathogen, the porcine reproductive and respiratory syndrome (PRRS) virus, was performed in a controlled challenge trial (Kritas and Morrison 2007). Pigs that were challenged intranasally with a wild PRRSV strain gained significantly more weight when they had been vaccinated against PRRS or when they had received *L. casei* (Kritas and Morrison 2007). Positive effects of a fecal-prepared probiotics and of an *Enterococcus faecium* strain against other respiratory pathogens such as *Mycoplasma hyopneumoniae* and influenza virus, respectively, had been documented in pigs (Schachtschneider et al. 2013; Wang et al. 2014).

Pigs in many farms worldwide (e.g., more than 75% of nursery pigs in the USA) receive antibiotics in their feed as growth promoters (Dewey et al. 1999) because it was found that antibiotics provided an extra improvement of 4–8% in ADG and FCR in healthy weaned pigs (Doyle 2001). In a large-scale field study employing 21,000 pigs of a high-health nursery, a *Bacillus* probiotic was substituted for a low-dose antibiotic scheme during a 7-week period without causing any negative pig health or productivity consequences (Kritas and Morrison 2005). In other words, under pragmatic conditions, both antibiotic and probiotic pig groups behaved similarly with regard to their productivity but also the cost of production (Kritas and Morrison 2005). Thus, farmers may be able to maintain the performance of their nursery pigs while reserving antibiotics administered through feed for therapy or prevention of more serious health conditions.

As growing-finishing pigs have a mature GI tract, with high digestive enzyme activity, immune capacity, and disease resistance, the influence of probiotics in these pigs is relatively limited. Nevertheless, there are studies showing a benefit in body weight and carcass quality after probiotic administration. Supplementation of a *Bacillus*-based probiotic had improved weight gain and reduced mortality of growing-finishing pigs (Davis et al. 2008]. In-feed administration of *Bacillus toyoi* spores in nursery and growing and finishing pigs resulted in lower incidence of postweaning diarrhea and improvement of weight gain (>4.5%) and of feed conversion ratio (>6.6%), while more than 45% of the carcasses was classified in the top three categories of the EUROP scale (S, E, and U) compared to the control group (Kyriakis et al. 2003). Basically similar observations were made by Alexopoulos et al. (2004b) with a *Bacillus subtilis* and *Bacillus licheniformis* containing probiotic and by Kritas et al. (2013) with *Bacillus subtilis* C-3102. Dietary supplementation of probiotics (*Bacillus subtilis* and *Clostridium butyricum*) during finishing stage increased growth performance (ADG, FCR) throughout the experiment and exerted beneficial effects on meat characteristics (color scores, marbling scores, drip loss values, pH, LM area and firmness) and on apparent total tract digestibility of the pigs (Meng et al. 2010).

Prebiotics are known to influence the GI tract microbiota in pigs. MOS isolated from the *S. cerevisiae* cell wall had beneficial effect on the intestinal microbiota, as well as on animal growth (Shim et al. 2005). It has been suggested that they suppress the growth of *E. coli*, *S. typhimurium*, *Clostridium botulinum*, and *C. sporogenes* and conversely stimulate the growth of *B. longum*, *L. casei*, *L. acidophilus*, and *L. delbrückei*.

Protective effects have been attributed to combinations of probiotics with prebiotics, e.g., synbiotics. Maltodextrins and *L. paracasei* had increased the efficiency of piglets, by reducing pathogenic *E. coli* growth and adhesion in the digestive tract (Bomba et al. 2002, 2006). Feeding preparations containing *L. fermentum*, *L. brevis*, *L. salivarius*, or *E. faecium* with lactulose or lactitol had improved efficiency of pigs (Bomba et al. 2002; Piva et al. 2005). A mixture of fructooligosaccharides and *L. paracasei* has been demonstrated to have a stimulating effect on the growth of natural intestinal microorganisms, to decrease the numbers of undesirable microbiota including coliforms, *Clostridium*, and *Enterobacteriaceae* and to improve the morphology of intestinal villi of the pigs (Spencer et al. 1997; Nemcova et al. 1999; Breves et al. 2001; Bomba et al. 2002; Shim et al. 2005). Moreover, a higher effectiveness of synbiotics was shown when they were given to animals during the preweaning period (Shim et al. 2005).

## 4.5 The Effect of Probiotics and Prebiotics in Horses

There are very little data concerning the use of probiotics in horses, and these data are controversial mainly due to the small number of observations. Their administration is mainly related with the control of intestinal diseases. Horses that were supplemented with *Saccharomyces boulardii* presented shorter duration and less watery diarrhea (Desrochers et al. 2005). However, in another study, the administration of *S. boulardii* in horses affected with antimicrobial-associated enterocolitis did not confer significant differences with regard to the occurrence of fecal consistency or cessation of watery diarrhea (Boyle et al. 2013). The administration of a probiotic based on lactobacilli and bifidobacteria in diarrheic foals has a limited potential for therapeutic modification of the gastrointestinal microbiota or for reducing pathogen shedding (Schoster et al. 2014, 2015, 2016). Parraga et al. (1997) did not observe any effect of probiotic administration on the prevalence of fecal shedding of *Salmonella*, the prevalence of postoperative diarrhea, the length of antimicrobial therapy, and the length of hospitalizations during the postoperative period in horses with colic.

After intravenous administration of *Propionibacterium acnes* in mares with persistent endometritis, improved pregnancy and increased live foal rates had been documented (Rohrbach et al. 2007).

## 4.6 "Killed" Probiotics

Although probiotic microorganisms are considered safe, there is always skepticism on whether the administration of live organisms is not without risk, particularly in certain populations (von Wright 2005; Boyle et al. 2006). Therefore, it is of considerable interest to determine if the health benefits of probiotics can be attained without the risks associated with administration of a live organism (Adams 2010). In a review of Kataria et al. (2009), it is presented that heat-killed, ultraviolet-inactivated *Lactobacillus rhamnosus* GG, better known as LG, and even components of these agents may be just as effective and considerably safer for the host. These agents seem also to induce milder immunological reactions to the host. In animals, administration of *Saccharomyces cerevisiae* cell components to broilers improved growth performance, meat tenderness, and ileal villus development (Zhang et al. 2005). Nevertheless, there is a substantial argument on whether dead microbes or elements of their carcasses "may be a safer alternative."

## 4.7 Are the Results Consistent? (Modified from Kritas and Morrison 2003)

Although probiotics have been used for many years, systematic knowledge is rather shallow. Protocols for testing probiotics have varied substantially and do not allow for direct comparison. Each probiotic preparation is different, based on a single or a mixture of microorganisms, and not all microorganisms (and/or their metabolites) behave similarly against a certain pathogenic strain. Thus, it is expected, for instance, that the effect of *Bacillus licheniformis*, *Streptococcus faecium*, or *Lactobacillus casei* against the same strain of *E. coli* will not be the same. As the action of probiotics could be a dual one, e.g., live probiotic cells influence both the gastrointestinal microbiota and the immune response, while the components of dead cells exert an anti-inflammatory response in the gastrointestinal tract, the relative proportions of live and dead cells in a probiotic culture may vary (Adams 2010). Variable amounts of dead cells might contribute to the variation in response often seen with live probiotic cultures (Adams 2010).

Testing of probiotics has often been performed in nontarget species, and the target animal may carry different receptors involved in pathogenesis. Besides, the nature of the microbiota across different herds or flocks and even between animals or birds within the same group highly varies, and thereby there is concern that probiotics and prebiotics, which may be effective under limited conditions, may never be usefully deployed across the whole industry. Finally, the health status of a farm and factors that determine it (e.g., pig flow) may account for significant variability in results. The farm context in which the organism is used is likely to be critical; the use of probiotics is more likely to result in measurable economic gains in animals living in suboptimal conditions rather than in those reared in the highest welfare and husbandry conditions. It is known, for instance, that with higher productivity rates of the broilers, the effect of probiotics becomes smaller (Timmerman et al. 2006).

All these effects create an impression of inconsistency of results, and therefore some researchers remain skeptical about their reported benefits.

However, inconsistency of results is a characteristic of every new development. Our opinion is that probiotics are not a panacea and should be used after critical thinking. They can be introduced in cases similar to those described as successful in literature. If novel applications are sought, limited number of target animals should be used and the product should be tested. Failure does not necessarily mean that probiotics do not work. Their effect is dependent on several factors as previously mentioned. Trial-and-error thinking should be practiced at least in the first years of investigating their effect. Their potential to substitute for antibiotics in simple illnesses and to produce "ecological" meat products may also guide practitioners' decisions.

## References


Adams CA (2010) The probiotic paradox: live and dead cells are biological response modifiers. Nutr Res Rev 23:37–46

Aiello SE (1998) Chemotherapeutics introduction. In: Merck veterinary manual, 8th edn. Merck and Co/National Publishing, Whitehouse Station/Philadelphia, pp 1738–1745

Alexopoulos C, Karagiannidis A, Kritas SK et al (2001) Field evaluation of a bioregulator, containing live *Bacillus cereus* spores, on health status and performance of sows and their litters. J Vet Med A 48:137–145

Alexopoulos C, Georgoulakis IE, Tzivara A et al (2004a) Field evaluation of the efficacy of a probiotic containing *Bacillus licheniformis* and *Bacillus subtilis* spores, on the health status and performance of sows and their litters. J Anim Physiol Anim Nutr (Berl) 88:381–392

Alexopoulos C, Georgoulakis I, Tzivara A et al (2004b) Field evaluation of the effect of a probiotic-containing *Bacillus licheniformis* and *Bacillus subtilis* spores on the health status, performance, and carcass quality of grower and finisher pigs. J Vet Med A 51:306–312

Anderson DB, McCracken VJ, Aminov RI et al (1999) Gut microbiology and growth-promoting antibiotics in swine. Nutr Abstr Rev Ser B Livestock Feeds Feeding 70:101–188

Baum B, Liebler-Tenorio EM, Enss ML et al (2002) *Saccharomyces boulardii* and *Bacillus cereus* var *Toyoi* influence the morphology and the mucins of the intestine of pigs. Z Gastroenterol 40:277–284

Bhandari SK, Xu B, Nyachoti CM et al (2008) Evaluation of alternatives to antibiotics using an *Escherichia coli* K88+ model of piglet diarrhoea: effects on gut microbial ecology. J Anim Sci 86:836–847

Böhmer BM, Kramer W, Roth-Maier DA (2006) Dietary probiotic supplementation and resulting effects on performance, health status, and microbial characteristics of primiparous sows. J Anim Physiol Anim Nutr 90:309–315

Bomba A, Nemcova R, Gancarcikova S et al (2002) Improvement of the probiotic effect of micro-organisms by their combination with maltodextrins, fructo-oligosaccharides and polyunsaturated acids. Br J Nutr 88:95–99

Bomba A, Jonecova Z, Koscova J et al (2006) The improvement of probiotics efficacy by synergistically acting components of natural origin: a review. Biologia 61:729–734

Bontempo V, Di Giancamillo A, Savoini G et al (2006) Live yeast dietary supplementation acts upon intestinal morphofunctional aspects and growth in weanling piglets. Anim Feed Sci Technol 129:224–236

Boyle RJ, Robins-Browne RM, Tang MLK (2006) Probiotic use in clinical practice: what are the risks? Am J Clin Nutr 83:1256–1264

Boyle AG, Magdesian KG, Durando MM et al (2013) Saccharomyces boulardii viability and efficacy in horses with antimicrobial- induced diarrhea. Vet Rec 172:128

Breves G, Szentkuti L, Schroder B (2001) Effects of oligosaccharides on functional parameters of the intestinal tract of growing pigs. Dtsch Tierarztl Wochenschr 108(6):246–248

Campbell JM, Crenshaw JD, Polo J (2013) The biological stress of early weaned piglets. J Anim Sci Biotechnol 4(1):19

Cangemi de Gutierrez R, Santos V, Nader-Macias ME (2001) Protective effect of intranasally inoculated *Lactobacillus fermentum* against *Streptococcus pneumoniae* challenge on the mouse respiratory tract. FEMS Immunol Med Microbiol 31:187–195

Cheng G, Hao H, Xie S et al (2014) Antibiotic alternatives: the substitution of antibiotics in animal husbandry? Front Microbiol 5:217. https://doi.org/10.3389/fmicb201400217

Choct M (2009) Managing gut health through nutrition. Br Poultry Sci 50:9–15

Collado MC, Grzeskowiak C, Salminen S (2007) Probiotic strains and their combination inhibit in vitro adhesion of pathogens to pig intestinal mucosa. Curr Microbiol 55:260–265

Corcionivoschi N, Drinceanu D, Pop IM et al (2010) The effect of probiotics on animal health. Anim Sci Biotechnol 43(1):35–41

Davis ME, Parrott T, Brown DC et al (2008) Effect of a Bacillus-based direct-fed microbial feed supplement on growth performance and pen cleaning characteristics of growing-finishing pigs. J Anim Sci 86:1459–1467

Desrochers AM, Dolente BA, Roy MF et al (2005) Efficacy of *Saccharomyces boulardii* for treatment of horses with acute enterocolitis. J Am Vet Med Assoc 227:954–959

Dewey CE, Cox BD, Straw BE et al (1999) Use of antimicrobials in swine feeds in the United States. Swine Health Prod 7(1):19–25

Di Giancamillo A, Vitari F, Savoini G et al (2008) Effects of orally administered probiotic *Pediococcus acidilactici* on the small and large intestine of weaning piglets. A qualitative and quantitative micro-anatomical study. Histol Histopathol 23:651–664

Dibner JJ, Richards JD (2005) Antibiotic growth promoters in agriculture: history and mode of action. Poult Sci 84:634–643

Dominguez-Bello MG, Costello EK, Contreras M et al (2010) Delivery mode shapes the acquisition and structure of the initial microbiota across multiple body habitats in newborns. Proc Natl Acad Sci U S A 107:11971–11975

Doyle ME (2001) Alternatives to antibiotic use for growth promotion in animal husbandry. Food Research Institute Briefings, University of Winsconsin, Madison, April issue

Dritz SS, Tokach MD, Goodband RD et al (2002) Effects of administration of antimicrobials in feed on growth rate and feed efficiency of pigs in multisite production systems. J Am Vet Med Assoc 220:1690–1169

Duerkop BA, Vaishnava S, Hooper LV (2009) Immune responses to the microbiota at the intestinal mucosal surface. Immunity 31:368–376

Dufresne L, 1998: Alimentary tract disorders of growing pigs. In: Proceedings of 15th International Pig Veterinary Society Congress, Birmingham, England, 5–9 July 1998, vol 1, pp 71–77

English P, Smith W, McLean A (1984) Replacement policy. In: English P, Smith W, McLean A (eds) The sow: improving her efficiency. Farming Press, Ipswich, p 83

Fuller R (1989) Probiotics in man and animals. J Appl Bacteriol 66:365–378

Gareau MG, Wine E, Sherman PM (2009) Early life stress induces both acute and chronic colonic barrier dysfunction. NeoReviews 10:191–197

Giannakopoulos CG, Kyriakis SC, Saoulidis K et al (2001) The effect of salinomycin on health status and performance of sows and their litters: a dose titration study. J Vet Med A 48:257–265

Gill HS (2003) Probiotics to enhance anti-infective defences in the gastro-intestinal tract. Best Pract Res Clin Gastroenterol 17:755–773

Gluck U, Gebbers JO (2003) Ingested probiotics reduced nasal colonization with pathogenic bacteria (*Staphylococcus aureus*, *Streptococcus pneumoniae*, beta-hemolytic streptococci). Am J Clin Nutr 77:517–520

Hooper LV, Midtvedt T, Gordon JI (2002) How host-microbial interactions shape the nutrient environment of the mammalian intestine. Annu Rev Nutr 22:283–307

Hori T, Kiyoshima J, Shida K et al (2001) Effect of intranasal administration of *Lactobacillus casei* Shirota on influenza virus infection of upper respiratory tract in mice. Clin Diagn Lab Immunol 8:593–597

Jin LZ, Ho YW, Abdulla N et al (2000) Digestive and bacterial enzyme activities in broilers fed diets supplemented with *Lactobacillus* cultures. Poult Sci 79:886–891

Jurgens MH, Rikabi RA, Zimmerman DR (1997) The effect of dietary active dry yeast supplement on performance of sows during gestation-lactation and their pigs. J Anim Sci 75:593–597

Kantas DV, Vassilopoulos V, Kyriakis SC et al (1998) A dose titration study on the effect of virginiamycin on gilt/sow and piglet performance. J Vet Med A 45:525–533

Kataria J, Li N, Wynn JL et al (2009) Probiotic microbes: do they need to be alive to be beneficial? Nutr Rev 67(9):546–550

Kenny M, Smidt H, Mengheri E et al (2011) Probiotics – do they have a role in the pig industry? Animal 5(3):462–470

Koenig JE, Spor A, Scalfone N et al (2011) Succession of microbial consortia in the developing infant gut microbiome. Proc Natl Acad Sci U S A 108(Suppl 1):4578–4585

Konstantinov SR, Smidt H, Akkermans ADL et al (2008) Feeding of *Lactobacillus sobrius* reduces *Escherichia coli* F4 levels in the gut and promotes growth of infected piglets. FEMS Microbiol Ecol 66:599–607

Kovacs-Zomborszky M, Kreizinger F, Gombos S et al (1994) Data on the effects the probiotics "Lacto Sacc". Acta Vet Hung 42(1):3–14

Kozasa M (1986) Toyocerin (*Bacillus toyoi*) as growth promoter for animal feeding. Migrobiol Aliments Nutr 4:121–124

Kritas SK, Morrison RB (2003) A critical review of feeding probiotics to pigs. In: Proceedings of AD Leman swine conference, Saint Paul, 13–16 Sept, vol 30, pp 252–255

Kritas SK, Morrison RB (2005) Evaluation of probiotics as a substitute for antibiotics in a large pig nursery. Vet Rec 156:447–448

Kritas SK, Morrison RB (2007) Effect of orally administered *Lactobacillus casei* on Porcine Reproductive and Respiratory Syndrome (PRRS) virus vaccination in pigs. Vet Microbiol 119(2–4):248–255

Kritas SK, Marubashi T, Filioussis G et al (2013) Field evaluation of Calsporin, a probiotic based on viable spores of *Bacillus subtilis* C-3102: health, performance, and carcass quality of grower-finisher pigs. In: The sixth Asian Pig Veterinary Society Congress, Ho Chi Minh City, Vietnam, 23–25 Sept 2013

Kritas SK, Marubashi T, Filioussis G et al (2015) Reproductive performance of sows was improved by administration of a sporing bacillary probiotic (*Bacillus subtilis* C-3102). J Anim Sci 93(1):405–413

Kyriakis SC, Vassilopoulos V, Demade I et al (1992) The effect of virginiamycin on sow and litter performance. Anim Prod 55:431–436

Kyriakis SC, Sarris K, Kritas SK, Saoulidis K, Tsinas AC, Tsiloyiannis VK (1995) The effect of Salinomycin on the control of Clostridium perfringens type-A infection in growing pigs. J Vet Med B 42:355–359

Kyriakis SC, Tsiloyiannis VK, Lekkas S et al (1997) The efficacy of enrofloxacin in-feed medication, by applying different programmes for the control of post weaning diarrhoea syndrome of piglets. J Vet Med B 44:513–521

Kyriakis SC, Tsiloyiannis VK, Vlemmas J et al (1999) The effect of probiotic LSP 122 on the control of post-weaning diarrhoea syndrome of piglets. Res Vet Sci 67:223–228

Kyriakis SC, Georgoulakis I, Spais A et al (2003) Evaluation of Toyocerin, a probiotic containing *Bacillus toyoi* spores, on health status and productivity of weaned, growing and finishing pigs. Asian Aust J Anim Sci 16(9):1326–1331

Lessard M, Dupuis M, Gagnon N et al (2009) Administration of *Pediococcus acidilactici* or *Saccharomyces cerevisiae boulardii* modulates development of porcine mucosal immunity and reduces intestinal bacterial translocation after *Escherichia coli* challenge. J Anim Sci 87:922–934
Mallo JJ, Rioperez J, Honrubia P (2010) The addition of *Enterococcus faecium* to diet improves piglet's intestinal microbiota and performance. Livest Sci 26:243–256
Marubashi T, Gracia MI, Vilà B et al (2012) The efficacy of the probiotic feed additive Calsporin (*Bacillus subtilis* C-3102) in weaned piglets: combined analysis of four different studies. J Appl Anim Nutr 1(e2):1–5
Mathew AG, Chattin SE, Robbins CM et al (1998) Effects of a direct-fed yeast culture on enteric microbial populations, fermentations acids, and performance of weanling pigs. J Anim Sci 76:2138–2145
Mazmanian SK, Round JL, Kasper D (2008) A microbial symbiosis factor prevents inflammatory disease. Nature 453:620–625
Meng QW, Yan L, Ao X et al (2010) Influence of probiotics in different energy and nutrient density diets on growth performance, nutrient digestibility, meat quality, and blood characteristics in growing-finishing pigs. J Anim Sci 88(10):3320–3326
Modesto M, D'Aimmo MR, Stefanini I (2009) A novel strategy to select Bifidobacterium strains and prebiotics as natural growth promoters in newly weaned pigs. Livest Sci 122:248–258
Murali SE, Kavitha BTVV, JGI S (2010) Probiotics as potential therapies in human gastrointestinal health. Int J Adv Pharm Sci 1:96–110
Nemcova R, Bomba A, Gancarcikova S et al (1999) Study of the effect of *Lactobacillus paracasei* and fructooligosaccharides on the faecal microflora in weanling piglets. Berl Munch Tierarztl Wochenschr 112:225–228
O'Hara AM, Shanahan F (2006) The gut flora as a forgotten organ. EMBO Rep 7:688–693
Ouwehand AC, Salminen S, Isolauri E (2002) Probiotics: an overview of beneficial effects. Antonie Van Leeuwenhoek 82:279–289
Parraga ME, Spier SJ, Thurmond M et al (1997) clinical trial of probiotic administration for prevention of Salmonella shedding in the postoperative period in horses with colic. J Vet Intern Med 11:36–41
Perdigon G, Alvarez S, Medina M et al (1999) Influence of the oral administration of lactic acid bacteria on producing cells associated to bronchus. Int J Immunopathol Pharmacol 12:97–102
Perdigon G, Fuller R, Raya R (2001) Lactic acid bacteria and their effect on the immune system. Curr Issues Intest Microbiol 2(1):27–42
Piva A, Casadei G, Gatta PP et al (2005) Effect of lactitol, lactic acid bacteria, or their combinations (synbiotic) on intestinal proteolysis in vitro, and on feed efficiency in weaned pigs. Can J Anim Sci 85:345–353
Prieto ML, O'Sullivan L, Tan SP et al (2014) Evaluation of the efficacy and safety of a marine-derived Bacillus strain for use as an in-feed probiotic for newly weaned pigs. PLoS One 9(2):e88599
Robertson JL, Tournout J (1994) Probiotic concept applied to the sow-piglet couple. In: Chulalongkorn University, Faculty of Veterinary Science, Bangkok (ed) Proceedings of 13th IPVS Congress Chulalongkorn University, Faculty of Veterinary Science, Bangkok, Thailand, p 263
Rohrbach BW, Sheerin PC, Cantrell CK et al (2007) Effect of adjunctive treatment with intravenously administered *Propionibacterium acnes* on reproductive performance in mares with persistent endometritis. J Am Vet Med Assoc 231(1):107–113
Salminen S, Isolauri E, Salminen E (1996) Clinical uses of probiotics for stabilizing the gut mucosal barrier: successful strains and future challenges. Antonie Van Leeuwenhoek 70:347–358
Salzman NH, Ghosh D, Huttner KM et al (2003) Protection against enteric salmonellosis in transgenic mice expressing a human intestinal defensin. Nature 422:522–526
Schachtschneider KM, Yeoman CJ, Isaacson RE et al (2013) Modulation of systemic immune responses through commensal gastrointestinal microbiota. PLoS One 8(1):e53969. https://doi.org/10.1371/journalpone0053969. Epub 2013 Jan 11

Scharek L, Altherr BJ, Tölke C et al (2007a) Influence of the probiotic *Bacillus cereus* var *toyoi* on the intestinal immunity of piglets. Vet Immunol Immunopathol 120(3–4):136–147

Scharek L, Guth J, Filter M et al (2007b) Impact of the probiotic bacteria *Enterococcus faecium* NCIMB 10415 (SF68) and *Bacillus cereus* var *toyoi* NCIMB 40112 on the development of serum IgG and faecal IgA of sows and their piglets. Arch Anim Nutr 61(4):223–234

Scharek-Tedin L, Pieper R, Vahjen W (2013) *Bacillus cereus* var *Toyoi* modulates the immune reaction and reduces the occurrence of diarrhea in piglets challenged with *Salmonella Typhimurium* DT104. J Anim Sci 91(12):5696–5704

Scharek-Tedin L, Kreuzer-Redmer S, Twardziok SO (2015) Probiotic treatment decreases the number of CD14-expressing cells in porcine milk which correlates with several intestinal immune parameters in the piglets. Front Immunol 6:108

Schierack P, Wieler LH, Taras D (2007) Bacillus cereus var toyoi enhanced systemic immune response in piglets. Vet Immunol Immunopathol 118(1–2):1–11

Schoster A, Weese JS, Guardabassi L (2014) Probiotic use in horses - what is the evidence for their clinical efficacy? J Vet Intern Med 28(6):1640–1652

Schoster A, Staempfli HR, Abrahams M et al (2015) Effect of a probiotic on prevention of diarrhea and Clostridium difficile and Clostridium perfringens shedding in foals. J Vet Intern Med 29(3):925–931

Schoster A, Guardabassi L, Staempfli HR et al (2016) The longitudinal effect of a multistrain probiotic on the intestinal bacterial microbiota of neonatal foals. Equine Vet J 48(6): 689–696

Shim SB, Verstegen WA, Kim IH et al (2005) Effects of feeding antibiotic-free creep feed supplemented with oligofructose, probiotics or synbiotics to suckling piglets increases the preweaning weight gain and composition of intestinal microbiota. Arch Anim Nutr 59:419–427

Shu Q, Qu F, Gill HS (2001) Probiotic treatment using *Bifidobacterium lactis* HN019 reduces weanling diarrhea associated with rotavirus and *Escherichia coli* infection in a piglet model. J Pediatr Gastroenterol Nutr 33(2):171–177

Si JM, Yu YC, Fan YJ et al (2004) Intestinal microecology and quality of life in irritable bowel syndrome patients. World J Gastroenterol 10:1802–1805

Siggers RH, Siggers J, Boye M et al (2008) Early administration of probiotics alters bacterial colonization and limits diet-induced gut dysfunction and severity of necrotizing enterocolitis in preterm pigs. J Nutr 138(8):1437–1444

Snel J, Harmsen HJM, Van der Wielen PWJJ et al (2002) Dietary strategies to influence the gastrointestinal microflora of young animals, and its potential to improve intestinal health. In: Blok MC, Vahl HA, De Lange L, Van de Braak AE, Hemke G, Hessing M (eds) Nutrition and health of the gastrointestinal tract. Wageningen Academic, The Netherlands, pp 37–69

Spencer KJ, Touchette KJ, Liu H et al (1997) Effect of spray-dried plasma and fructooligosaccharide on nursery performance and small intestinal morphology of weaned pigs. J Anim Sci 75(Suppl 1):199

Taras D, Vahjen W, Macha M et al (2005) Response of performance characteristics and fecal consistency to long-lasting supplementation with the probiotic strain Bacillus cereus var toyoi to sows and piglets. Arch Anim Nutr 59:405–417

Taras D, Vahjen W, Macha M et al (2006) Performance, diarrhea incidence, and occurrence of *Escherichia coli* virulence genes during long-term administration of a probiotic *Enterococcus faecium* strain to sows and piglets. J Anim Sci 84:608–617

Timmerman HM, Koning CJM, Mulder L et al (2004) Monostrain, multistrain and multispecies probiotics – a comparison of functionality and efficacy. Int J Food Microbiol 96:219–233

Timmerman HM, Veldman A, van den Elsen E et al (2006) Mortality and growth performance of broilers given drinking water supplemented with chicken-specific probiotics. Poult Sci 85(8):1383–1388

Tizard IR (2013) Veterinary immunology, 9th edn. Elsevier, Amsterdam, pp 204–257

Trevisi P, Colombo M, Priori D et al (2015) Comparison of three patterns of feed supplementation with live *Saccharomyces cerevisiae* yeast on postweaning diarrhea, health status, and blood

metabolic profile of susceptible weaning pigs orally challenged with *Escherichia coli* F4ac. J Anim Sci 93(5):2225–2233
Tsukahara T, Tsuruta T, Nakanishi N et al (2013) The preventive effect of *Bacillus subtilis* strain DB9011 against experimental infection with enterotoxemic *Escherichia coli* in weaning piglets. Anim Sci J 84(4):316–321
Van Heugten E, Funderburke DW, Dorton KL (2003) Growth performance, nutrient digestibility, and fecal microflora in weanling pigs fed live yeast. J Anim Sci 81:1004–1012
Visek WJ (1978) The mode of growth promotion by antibiotics. J Anim Sci 46:1447–1469
Vondruskova H, Slamova R, Trckova M et al (2010) Alternatives to antibiotic growth promoters in prevention of diarrhoea in weaned piglets: a review. Vet Med 55(5):199–224
Walsh MC, Gardiner GE, Hart OM et al (2008) Predominance of a bacteriocin-producing *Lactobacillus salivarius* component of a five-strain probiotic in the porcine ileum and effects on host immune phenotype. FEMS Microbiol Ecol 64:317–327
Wang A, Yu H, Gao X et al (2009) Influence of *Lactobacillus fermentum* I5007 on the intestinal and systemic immune responses of healthy and *E coli* challenged piglets. Antonie Van Leeuwenhoek 96:89–98
Wang AN, Cai CJ, Zeng XF et al (2013) Dietary supplementation with *Lactobacillus fermentum* I5007 improves the anti-oxidative activity of weanling piglets challenged with diquat. J Appl Microbiol 114:1582–1591
Wang Z, Burwinkel M, Chai W et al (2014) Dietary *Enterococcus faecium* NCIMB 10415 and zinc oxide stimulate immune reactions to trivalent influenza vaccination in pigs but do not affect virological response upon challenge infection. PLoS One 9(1):e87007
Whittemore C (1998) Optimisation of feed supply to growing pigs and breeding sows. In: Whittemore C (ed) The science and practice of pig production. Blackwell Sciences, Oxford, p 421
Williams BA, Verstegen MWA, Tamminga S (2001) Fermentation in the large intestine of single-stomached animals and its relationship to animal health. Nutr Res Rev 14:207–227
von Wright A (2005) Regulating the safety of probiotics – the European approach. Curr Pharm Des 11:17–23
Yasui H, Mike A, Ohwaki M (1989) Immunogenicity of *Bifidobacterium breve* and change in antibody production in peyer's patches after oral administration. J Dairy Sci 72:30–35
Yu HF, Wang AN, Li XJ et al (2008) Effect of viable *Lactobacillus fermentum* on the growth performance, nutrient digestibility and immunity of weaned pigs. J Anim Feed Sci 17:61–69
Zhang AW, Lee BD, Lee SK et al (2005) Effects of yeast (*Saccharomyces cerevisiae*) cell components on growth performance, meat quality, and ileal mucosa development of broiler chicks. Poult Sci 84(7):1015–1021
Zoetendal EG, Collier CT, Koike S et al (2004) Molecular ecological analysis of the gastrointestinal microbiota: a review. J Nutr 134:465–472

# 5 Probiotics and Prebiotics for the Health of Poultry

Rebin Aswad Mirza

## Abbreviations

| | |
|---|---|
| AGP | Antibiotic growth promoters |
| DFM | Dried feed microbial |
| FOS | Fructooligosaccharides |
| GIT | Gastrointestinal tract |
| Hct | Haematocrit |
| IBDV | Infection bursa disease virus |
| Ig | Immunoglobulin |
| LAB | Lactic acid bacteria |
| PCV | Pocket cell volume |
| SCFA | Short-chain fatty acid |
| WHC | Water holding capacity |

## 5.1 Introduction

The poultry industry has become an important economic activity in many countries, thanks to the developments in several areas such as nutrition, genetics and management strategies to maximize the efficiency of growth performance and meat production. The mortality of chickens due to intestinal pathogens such as *Escherichia coli*, *Salmonella*, *Campylobacter* and *Clostridium perfringens* continues to cause problems, especially with high stocking densities associated with intensive production systems. Prevention and control of diseases have led during recent decades to a substantial increase in the use of veterinary medicines.

Campylobacteriosis, salmonellosis and infections related to *Escherichia coli* are the most prevalent zoonotic diseases in humans globally. The contaminated poultry meat by *Campylobacter jejuni* can be considered one of the most important sources of enteric infections in humans. Meat can be contaminated with various types of

R.A. Mirza
Department of Animal Resource, College of Agriculture,
Salahaddin University – Erbil, Kurdistan, Iraq
e-mail: rebinaswed@yahoo.com



D. Di Gioia, B. Biavati (eds.), *Probiotics and Prebiotics in Animal Health and Food Safety*, https://doi.org/10.1007/978-3-319-71950-4_5

food-borne pathogenic bacteria, which can cause a variety of health complications in human being, particularly enteric illnesses (Akbar and Anal 2011). Reducing the proportion of *Campylobacter*-infected poultry flocks or reducing the number of *Campylobacter* in live poultry will consequently lower the risk to consumers considerably (Westrell et al. 2009). Eggs can be contaminated with *Salmonella*, either present as shell contamination due to contact with faeces after laying (migration of the organism through the shell is possible) or as egg content contamination due to colonization of the hen's oviduct. *Salmonella enterica*, in particular, is known to be closely associated with eggs (Fan et al. 2014). The use of probiotics, which can help to improve the natural defence of animals against pathogenic bacteria, is an effective alternative and approach to antibiotic administration for livestock to reduce bacterial contamination.

For the past four decades, antibiotics have been used as additives in poultry feed to enhance the growth performance and protect birds from the harmful influence of pathogenic enteric microorganisms. Antibiotic feed additives were banned by the European Union in 2006 due to concerns over the rise of widespread antibiotic resistance in human pathogens. Consequently, poultry producers are seeking for alternatives to maintain efficient poultry production.

Probiotics, prebiotics and synbiotics can be used as an attempt to reduce the chances of infection in poultry. There are various definitions of probiotics, according to FAO/WHO (FAO/WHO 2002), probiotics are defined as mono or mixed cultures of 'live microorganisms which when administered in adequate amounts confer a health benefit on the host'. Prebiotics are defined as 'non-digestible feed ingredients that beneficially affect the host by selectively stimulating the growth and/or activity of one or a limited number of bacteria in the colon' (Gibson and Roberfroid 1995). The combination of a probiotic and prebiotic is called synbiotic which is referred to both beneficial microorganisms and their substrates. Synbiotics may have synergistic effects on the intestinal tract of animals. The prebiotic present in the synbiotic product may promote the growth of the co-administered probiotic or may promote the growth of other beneficial organisms in the gut.

A number of probiotic products are commercially available for poultry production, such as Bactocell® (*Pediococcus acidilactici* CNCM MA 18/5M), BioPlus 2B® (*Bacillus licheniformis* DSM 5749 and *Bacillus subtilis* DSM 5750), Cylactin® LBC® (*Enterococcus faecium* NCIMB 10415), *Lactobacillus acidophilus* D2/CSL® (*Lactobacillus acidophilus* CECT 4529), Microferm® (*Enterococcus faecium* DSM 5464), Oralin® (*Enterococcus faecium* NCIMB 10415), Protexin® (multistrain probiotic) (Lan et al. 2003; Ayasan et al. 2006; Gunal et al. 2006), *Saccharomyces cerevisiae* (Zhang et al. 2005) and Thepax® (Yousefi and Karkoodi 2007). Prebiotics such as mannanoligosaccharides (Flemming et al. 2004), fructooligosaccharides (Verdonk and Leeuwen 2004) and inulin (Roberfroid 2007; Sofia and Gibson 2007; Rehman et al. 2008) enhance the growth of intestinal bacteria and may affect the intestinal histology.

## 5.2 Probiotics and Prebiotics in Poultry Diet

In the short lifespan of broiler chickens, any delay in microbial colonization of the intestinal tract can leave the bird's intestine open to infections. The idea of using probiotics in poultry farms is due to the bird's behaviour in nature. In the natural environment, the mother is always responsible for feeding their hatching chicks with a feed, which has been stored in her crop. This feed is fermented in the mother's crop, mixed with beneficial microbes and then passed to the hatching chicks' beak favouring the intestinal and crop colonization of newborns. The colonization is also provided through eating of faeces. This vertical transmission allows protection of hatching chicks from pathogenic microbes (Fuller 2001). However, commercially reared chickens are hatched in sterile incubators. The young chickens lack contact with the natural environment, so colonization of the intestinal tract is often a more prolonged process taking around 21 days for broilers to develop a balanced intestinal microbiota (Barnes 1979; Amit-Romach et al. 2004). This period represents about 50% of a broiler's lifespan, and it has been found that the later intestinal colonization occurs, the most vulnerable the intestinal ecosystem is to colonization by pathogenic microorganisms. After the first 21 days of life, other challenges such as stress, feed changes, antibiotic interventions and diseases can also upset the gastrointestinal microbiota and can lead to poor weight gain or considerable loss of stock (Gasson et al. 2004). Moreover, acid production in the young chick is limited at first but gradually increases. HCl gastric secretion has a deep impact on microbiota selection (Rynsburger and Classen 2007). Therefore, an early supplementation of probiotics and prebiotics at hatching could be important and useful in avian species.

## 5.3 Anatomy and Histology of the Bird Gastrointestinal Tract

The digestion of food begins in the beak of the chicken. In the beak, saliva is mixed with the food so that it can be easily swallowed. The swallowed food then moves to a storage organ called the crop and then through the true stomach of the bird called the proventriculus. Here, the food is further mixed with more enzymes (such as pepsin) that assist with the breakdown of food and digest proteins to amino acids. The food then moves to a grinding organ called the gizzard. Grit and gravel, which have been picked up by the bird, help to grind or crush the food particles in this organ. The food then passes through the duodenal loop and into the small intestine, where absorption of food particles primarily occurs. Undigested particles then pass through two pouches called ceca, where the water is absorbed from the food. The remaining undigested food particles pass through the colon and rectum and are excreted through the cloaca.

The small intestine is differentiated into three main regions, namely, the duodenum, the jejunum and the ileum. The small intestine is considered as the most important part of the GI tract, because the majority of the enzymatic digestion occurs and the food mass will remain in this part of GI tract for more than 8 h. The small intestine is also the most important centre for the presence of microorganisms inside the digestive tract, together with the two caeca. The small intestine is histologically composed of four layers from inside to outside: mucosa, submucosa, muscularis and serosa (Fig. 5.1). The inner lining of the intestines (mucosa) is composed of fingerlike form called villi. The role of these protrusions is to increase the surface area exposed to the absorption, and an increase of the length of villi leads to a higher efficiency of the digestion process and absorption and also favours protection

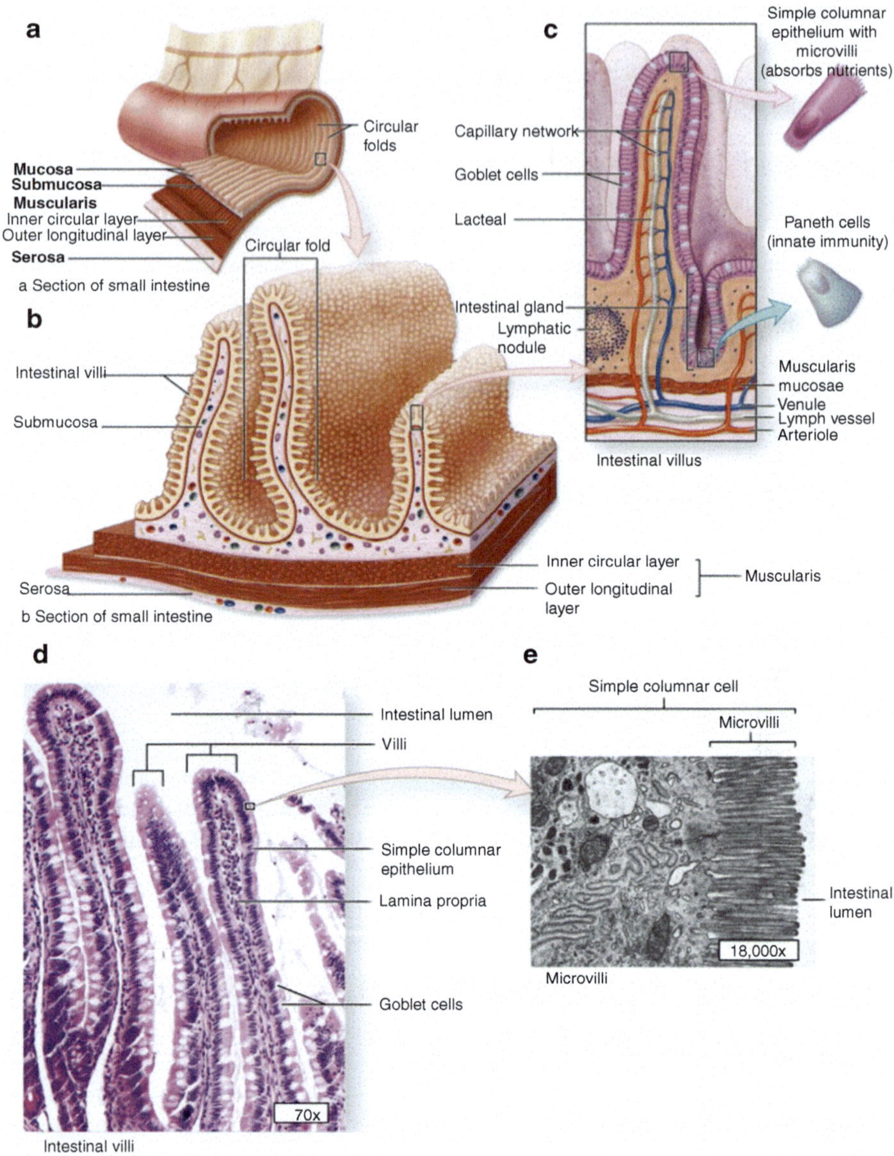


**Fig. 5.1** Histological structure of the small intestine consisting of four layers: mucosa, submucosa, muscularis and serosa (Mescher 2013)

against many kinds of pathogens. Submucosa is a layer of dense irregular connective tissue that supports the mucosa. Muscularis is composed of several thin layers of smooth muscle fibres, keeping the mucosal surface and underlying glands in a constant state of gentle agitation to expel contents of glandular crypts and enhance contact between epithelium and the contents of the lumen. The serosa consists of a thin layer of loose connective tissue covered by mesothelium. There are many columnar epithelial cells called enterocytes on the walls of villi, and all the enterocytes contain a large number of microvilli which are brush border-like (Fig. 5.1).

## 5.4 Ecology of Microbiota in the Chicken Gastrointestinal Tract

The microbiota of the digestive tract can be divided into two groups. The first, harmful bacteria, may be involved in the induction of infection, intestinal putrefaction and toxin production. The second group, commensal bacteria, may be involved in beneficial activities for the host, such as vitamin production, stimulation of the immune system and suppression of pathogenic bacteria. But there are also some microorganisms that are considered opportunistic pathogens taking advantage of an unbalanced microbiota (Jeurissen et al. 2002). Microorganisms inhibiting the GIT can be also classified as autochthonous bacteria (established microbiota), which colonize the gut through environmental exposure and normal feeding (Gusils et al. 1999), and allochthonous bacteria (transitory microbiota) introduced as dietary supplements into the GI tract through the feed or drinking water as direct-fed microbial (DFM) or probiotics (Patterson and Burkholder 2003). Scientific data indicate that allochthonous bacteria introduced via probiotic supplements may prevent infection and colonization of the GI tract by opportunistic pathogens (Fuller 1989; Abdel-Raheem et al. 2012).

The GI tract consists of a diverse community of bacteria. The development of this community begins at hatching, thanks to the bacteria present in the environment, the feed, and the people handling the chicks post-hatch. These inoculation routes can affect gut microbiota development. Microbes are spread throughout the entire length of the GI tract, where they show locative variation in community composition biogeographically (Fig. 5.2).

Chicken gut microbiota has been studied using various approaches. The earliest studies were based on culture-dependent methods (Barnes et al. 1972). These methods can be biased and inaccurate as most bacteria are unable to be cultured due to unknown growth requirements (Zhu et al. 2002; Wei et al. 2013). Previous reports highlighted that only up to 60% of caeca gut microbiota were culturable (Barnes et al. 1972). More advanced techniques were introduced in the early 2000s, in which molecular fingerprinting methods such as denaturing gradient gel electrophoresis (DGGE) (Gong et al. 2008), temporal temperature gradient gel electrophoresis (TTGE) (Zhu et al. 2002) and terminal-restriction fragment length polymorphism (T-RFLP) (Gong et al. 2002) were used to study gut microbiota of chicken. In recent years, microbiota analyses increasingly rely in high-throughput next-generation sequencing (HT-NGS) technologies, which provide large-scale analysis with unprecedented depths and coverages (Mohd Shaufi et al. 2015).

In the chicken caeca, taxonomic richness and diversity typically increase from day of hatch to market age of commercial broilers at 6 weeks (Fig. 5.3). Mohd Shaufi et al. (2015) studied the microbiota of ilea and caeca of broiler chickens using

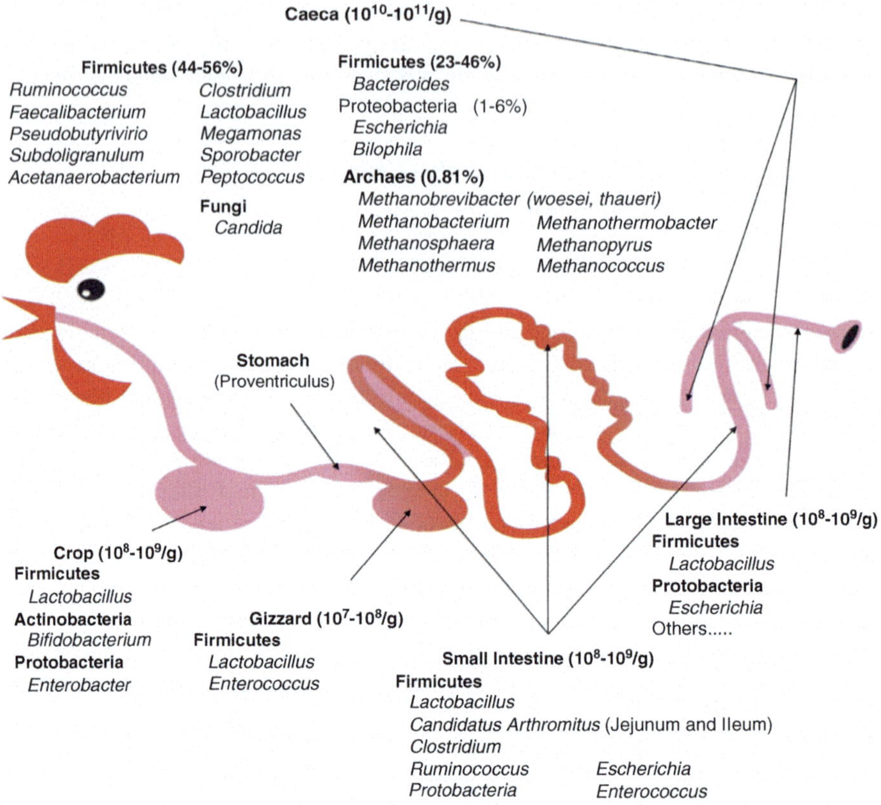


**Fig. 5.2** Major types of surveyed bacteria along the gastrointestinal tract of chicken (adapted from Yeoman et al. 2012)

high-throughput next-generation sequencing in 7-, 14-, 21- and 42-day-old birds. They found that *Firmicutes* was the most abundant phylum (49–85%) in both ilea and caeca of chicken at all ages. In the caecum, *Firmicutes* increased from 69% at day 7–76% at day 14 and decreased to 49% at day 42, while in the ileum, it increased slowly from 67 to 85% as the chicken aged. In ilea, *Proteobacteria* was the second most abundant phylum (5–32%), except at day 14 in which *Bacteroidetes* (22%) was more dominant. The presence of *Proteobacteria* was not obvious in caeca where it only can be detected at days 7 (5%) and 21 (3%). *Bacteroidetes* (18–21%) was consistently found as the second most abundant group at each time point in the caeca.

Apajalahti et al. (2004) used molecular techniques based on 16S rDNA analyses to investigate and identify individual species. Using these techniques, they have found that only 10% of the GI microorganisms belong to previously known species and 35% are unknown species but belong to known genera, while 55% are unknown bacteria. The study reported 640 different species belonging to 140 genera. Most of these microorganisms are anaerobes with many species able to hydrolyse polysaccharides to monosaccharides, which are primarily fermented to short-chain volatile fatty acids, hydrogen and carbon dioxide.

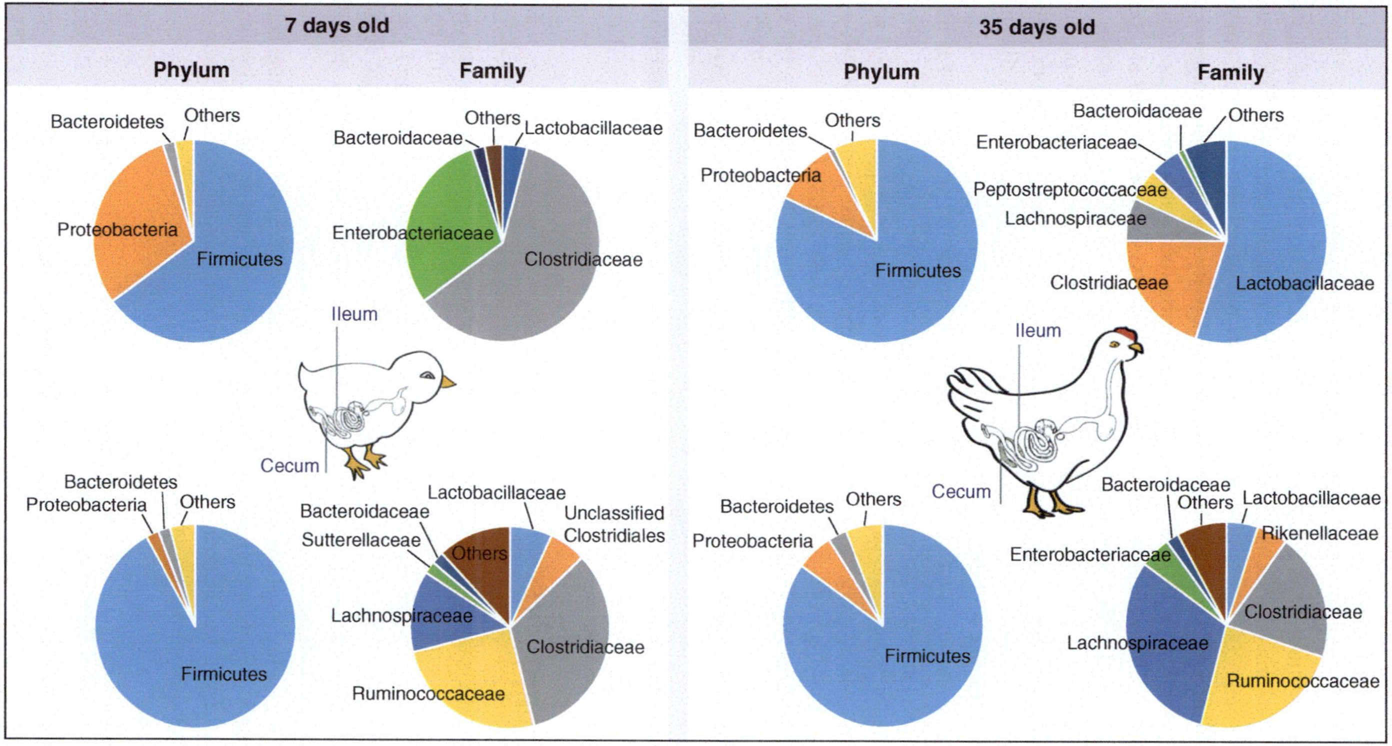


**Fig. 5.3** The chicken gut microbiome. The graphs provide an overview of the relative abundance of dominant bacterial phyla and families of the broiler chicken ileal (top level) and caecal (bottom level) microbiota in two different ages, 7 and 35 days old (Pourabedin and Zhao 2015)

## 5.5 Action of Probiotics in the Gastrointestinal Tract of Poultry

The effect of probiotic feed additives in poultry mainly consists of (1) maintaining normal intestinal microbiota by competitive exclusion and antagonism (Kizerwetter-Swida and Binek 2009), (2) altering metabolism by increasing digestive enzyme activity and decreasing bacterial harmful enzyme activity (e.g. β-glucuronidase) and ammonia production (Yoon et al. 2004), (3) improving feed consumption and digestion (Awad et al. 2006) and (4) stimulating the immune system (Brisbin et al. 2008).

Since 1991, Rolfe described competitive exclusion, bacterial antagonism and immune modulation as the major mechanisms involved in beneficial probiotic effects. Enhancements of colonization resistance and/or direct inhibitory effects against pathogens are the major mechanisms involved in probiotic trials that have reduced the incidence and duration of diseases.

The gastrointestinal tract is the largest immune organ in the body and is negatively affected by stress. Commercial poultry production will ultimately always have multiple stressors such as dietary changes, catching, transport and feed withdrawal. Stress will effectively and rapidly alter the intestinal population allowing opportunistic pathogens to adhere to the gastrointestinal tract. *Lactobacillus* and *Bifidobacterium* species are examples of beneficial bacteria that populate the GIT, and their populations decrease when birds become stressed (Hong et al. 2005). A probiotic administration could be able to repair or repopulate deficiencies within the intestinal microbiota, stimulating the immune system against pathogenic infection and producing antimicrobial compounds such as volatile fatty acids (Patterson and Burkholder 2003; Ahmad 2006; Callaway et al. 2008).

### 5.5.1 Competitive Exclusion

The concept of competitive exclusion implies that cultures of selected, beneficial microorganisms, supplemented to the feed, compete with potentially harmful bacteria in terms of adhesion sites and organic substrates, mainly carbon and energy sources (Schneitz 2005). Probiotics may colonize and multiply in the gut, thereby blocking receptor sites and preventing the attachment of other bacteria including harmful species such as enteropathogenic *Salmonella* or *E. coli*.

Competitive exclusion of pathogens is thought to be one of the most important beneficial mechanisms of probiotic bacteria (Rolfe 2000), and it is based on bacteria-to-bacteria interaction mediated by competition for available nutrients and mucosal adhesion sites (Figs. 5.4 and 5.5) (Patterson and Burkholder 2003).

### 5.5.2 Bacterial Antagonism

The presence of acid-productive microorganisms in the GI tract has important antagonistic activity against pathogenic bacteria. Bacteria are able to compete with other microorganisms producing and secreting some substances with bacteriostatic

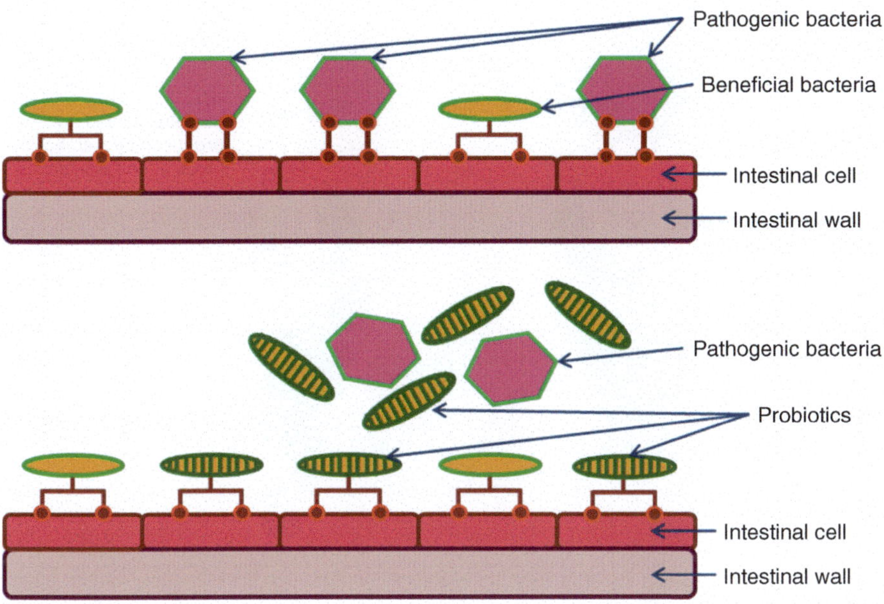


**Fig. 5.4** The beneficial bacteria when added to diet of poultry compete for binding sites on the intestinal epithelium

and bactericidal effect (Edens 2003). Bacteria are able to produce antibacterial substances, which have been shown to have an inhibitory activity against poultry pathogens both Gram-positive and Gram-negative bacteria (Jin et al. 1998). Moreover, the production of hydrogen peroxide by some strains of LAB has a strong bactericidal effect on different pathogens (Jin et al. 1996).

Different strains of probiotic microorganisms have demonstrated the capability of producing substances with bactericidal or bacteriostatic properties such as bacteriocins, hydrogen peroxide and diverse organic acids. These substances have a detrimental impact on harmful bacteria (Ewing and Cole 1994).

Bacteriocins are biologically active proteins, naturally produced by diverse microbes in different environments (Willey and Van Der Donk 2007). Despite significant structural and characteristic differences, bacteriocins display potent antimicrobial activities against a wide range of viruses, bacteria and fungi and have been recognized as a novel class of antimicrobials to control food-borne pathogens (Zasloff 2002). Bacteriocins are produced by most of the genera of lactic acid bacteria. Many bacteria, including intestinal commensals, could synthesize at least one bacteriocin. Therefore, intestinal bacteriocin-producing bacteria may achieve a competitive advantage and function as innate barrier against pathogens in the host.

### 5.5.3 Immune Modulation

Another important mechanism of probiotic action is the stimulation of the immune system. Newborn chicks have a sterile digestive system, and before their organism

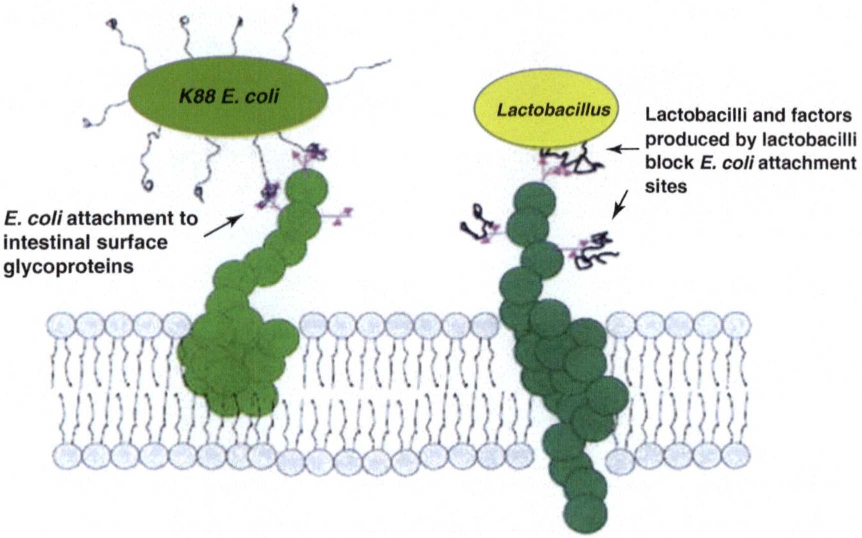


**Fig. 5.5** Antagonistic activity of lactobacilli against *E. coli* through the secretion of some adherents which prevents *E. coli* adhesion to the intestinal receptors (Patterson and Burkholder 2003)

will be able to produce its own antibodies, microorganisms from the environment begin to colonize the digestive system. The development and activation of the humoral and cellular gut-associated immune system is largely affected by the development of the gut microbiota (Ouwehand et al. 1999). Regular use of probiotics has a striking effect on the stimulation of the immune system through enhancing the production of natural interferons/cytokines; increasing macrophage, lymphocyte and natural killer (NK) cell activity; increasing immunoglobulin (IgG, IgM and IgA); and stimulating the production of γ-interferon (Koenen et al. 2004; Haghighi et al. 2006; Yang et al. 2009; Alkhalf et al. 2010). Probiotics increase the number of lymphocytes and lymphoid cells in lamina propria and intra-epithelial lymphocytes (IEL) in the small intestine and are found to inhibit the growth of infective microorganisms.

According to Lan et al. (2005), microbial communities can support the animal's defence against invading pathogens by stimulating gastrointestinal immune response. Recent scientific investigations have supported the important role of probiotics as a part of a healthy diet for both human and animals and may be a way to provide a safe, cost-effective and natural approach that sets up a barrier against microbial infection. Thereby, this strategy can result in health maintenance and disease prevention (Parves et al. 2006). Consumption of LAB may have favourable effects on the immune system.

## 5.6 Action of Prebiotics in the Gastrointestinal Tract of Poultry

Prebiotics are non-digestible food ingredients that beneficially affect the host by selectively stimulating the growth and/or activity of one or a limited number of bacteria in the colon. For a dietary substrate to be classified as a prebiotic, at least three criteria are required: (1) the substrate must not be hydrolysed or absorbed in the stomach or small intestine, (2) it must be selective for beneficial commensal bacteria in the large intestine, and (3) fermentation of the substrate should induce beneficial luminal/systemic effects within the host (Gibson et al. 2004). Research on the influences of prebiotics on the activity of the microbiota of broilers is limited and the influences are dissimilar, depending on the type of prebiotics. The ability of a probiotic LAB strain to survive in the GI tract may be promoted by co-administration of probiotic substances that stimulate the growth of the strain in the lumen (Salminen et al. 1998).

Feeding prebiotics from chicory (fructans) to broilers may improve weight gain, feed conversion and carcass weight. Feeding chicory fructans may also have systemic effects like a decrease in serum cholesterol levels and deposit of fat tissue (Yusrizal and Chen 2003). The selective interaction between prebiotics and the intestinal microbiota results in increased intestinal colonization resistance. Kleessen et al. (2003) evidenced a decrease of total aerobes, *Enterobacteriaceae* and *C. perfringens* in caecum as well as reduced levels of endotoxins in the blood through the administration of high-fructan Jerusalem artichoke compared with control birds. Therefore, Jerusalem artichokes stimulate growth of broiler chickens and protect them against endotoxins and potential caecum pathogens.

Mannanoligosaccharides (MOS) are efficacious prebiotic compounds that act by binding and removing pathogens from the intestinal tract and stimulating the immune system (Patterson and Burkholder 2003). Pathogenic bacteria mainly adhere to the intestinal cells of the host with type 1 fimbriae, and this attachment enables the bacteria to cause disease in the host (Fig. 5.6). Type 1 fimbriae are adhesion organelles expressed by many Gram-negative bacteria that enhance the bacteria's ability to attach to the host and cause diseases (Connell et al. 1996). Mannose is the main component of MOS and has the property of binding the type 1 fimbriae.

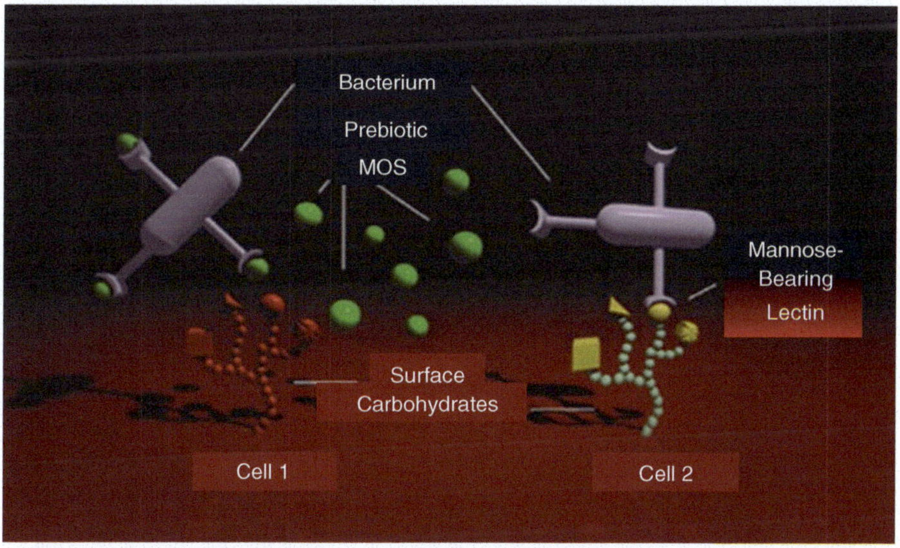


**Fig. 5.6** Blocking bacterial attachment and thus inhibiting host colonization by MOS as prebiotic

Therefore, mannose promotes the transit of undesirable bacteria through the intestine without colonization (Newman 1994). *Salmonella typhimurium* colonization of the intestine was decreased when 2.5% mannose was applied in the drinking water of broilers (Griggs and Jacob 2005).

Another prebiotic widely used in poultry diet is inulin. Inulin derived from some kind of plants such as Jerusalem artichoke, chicory, garlic, onion, asparagus, leek, banana and dandelion (Van Loo et al. 1995). Jerusalem artichoke and chicory are natural sources of inulin (Kaur and Gupta 2002; Stolzenburg 2005). Chicory is the most commonly used plant for the extraction of inulin-type fructans (De Leenheer 2007). Supplementation of chicory inulin was found to positively affect the performance in monogastric animals (chicken, pig, rabbit and rat), especially in young animals (Rehman et al. 2007, 2008; Liu 2008; Rebole et al. 2010; Awad et al. 2011).

## 5.7 Effects of Probiotics and Prebiotics on Poultry

### 5.7.1 Growth Performance

There are several reviews discussing the effect of probiotics, prebiotics and synbiotics on poultry performance. A growing body of scientific research supports the role of probiotics, prebiotics and synbiotics as effective alternatives to the use of AGP in animal nutrition (Patterson and Burkholder 2003; Pelicano et al. 2004). The selection of probiotic microorganisms, their administration and usage instructions (like preparation, dosages and condition of animals) are very crucial for getting the desired health and productive results (Huang et al. 2004). LAB may enhance digestion by increasing surface area and length of intestinal villi (Banasaz et al. 2002). The gut microbiota affects the digestion, absorption and metabolism of dietary carbohydrates, protein, lipids and minerals and the synthesis of vitamins (Jin et al. 1997). Most of the volatile fatty acids formed by intestinal bacteria are absorbed and metabolized by the host, contributing to host energy requirements. Maintaining a balanced gut health is a key aspect to ensure bird performance and health. If an imbalance in gut microbiota occurs, nutrient digestion and absorption may be affected which, in turn, may affect bird health and performance. Moreover, also bird management and the environment can significantly affect the balance of the gut microbiota.

Kalavathy et al. (2008) showed that the average body weight of a 42-day-old Hubbard broiler supplemented with probiotics (*Lactobacillus* stains, 1 g/kg feed) was significantly ($P < 0.05$) higher than control group. Mountzouris et al. (2010) observed that basal diet containing a probiotic concentration $10^8$ CFU/kg significantly increased body weight of broilers in comparison with control group. Dizaji et al. (2013) showed that the addition of prebiotics (1 kg of Active MOS/ton), probiotics (150, 100 and 50 gm of Protexin/ton of the starter, grower and final diets, respectively) and synbiotics (1 kg of Amax4x/ton) in Ross 308 broiler feed significantly improved average body weight of broiler chickens at 42 days old. The higher performance was recorded for synbiotic group compared with other groups. Mookiah et al. (2014) showed that the use of prebiotic isomalto-oligosaccharides (IMO), 11 probiotic *Lactobacillus* strains and combination of both (synbiotic) in poultry feed significantly improved weight gain of broiler chickens and feed conversion ratio compared with control group. Addition of probiotic and prebiotic to the poultry diets has shown beneficial effects on growth performance of poultry as listed in Table 5.1.

**Table 5.1** Effects of probiotics, prebiotics and synbiotics on growth and performance in poultry production

| Type of supplements used | Administration | General effect of performance | Reference |
|---|---|---|---|
| Bio-MOS | Feed 2 g/kg | Improved the growth performance of birds | Hooge (2004) |
| MOS[a] | 500 g/ton | Improved daily weight gain, feed intake and feed conversion ratio | Flemming et al. (2004) |
| Probiotic and prebiotic (MOS) | 1 kg/ton from 1–42 days separately | Improved feed conversion ratio | Pelicano et al. (2004) |
| Fermecto® (*Aspergillus mycelium*) | Feed | Weight of breast and thigh to body weight significantly increased | Piray et al. (2007) |
| *Lactobacillus*—All-Lac XCL 5x™ (challenged with *Salmonella enteritidis*) | Spray-mixing 5 g/400 mL/2000 chicks in distilled water | No significant effect on body weight, weight gain, feed intake, feed conversion ratio and Liveability | Riberio et al. (2007) |
| LAB (FM-B11) | Drinking water $10^9$ cfu LAB/mL | No significant effect on body weight | Rodriguez et al. (2007) |
| Bactocell® (*Pediococcus acidilactici*) | Feed 1.5 kg/ton/42 days | Increased body weight significantly | Rowghani et al. (2007) |
| LAB (FM-B11) + lactose | Probiotic in drinking water and lactose in feed | Increase body weight significantly | Rodriguez et al. (2007) |
| Synbiotic (Biomin/IMBO)[b] | Feed (1 kg/ton of the starter diets and 0.5 kg/ton of the grower diets) | Increased the growth performance and improved intestinal morphology and nutrient absorption | Awad et al. (2008) |
| Prebiotic (FOS) | Feed | Improved broiler's weight gain about 5–8% and feed conversion ratio about 2–6% | Yang et al. (2009) |

[a]Mannanoligosaccharides
[b]A combination of *Enterococcus faecium* and prebiotic derived from chicory

### 5.7.2 Gut Microbiota

The chicken GI tract harbours a very complex microbiota, with over 600 different bacterial species from more than 100 bacterial genera (Torok et al. 2011). Microbial populations in the gastrointestinal tracts of poultry have an important impact on the normal digestive processes and in maintaining animal health. Well-characterized probiotic strains have been selected to evaluate modulation of the avian gut microbiota and protection against a variety of pathogens; in particular, there has been a recent increase in the investigation of the effect of feeding *Lactobacillus* spp. to broilers. Studies have focused on strains previously selected in vitro for adhesion properties and antimicrobial activity (Patterson and Burkholder 2003).

Probiotics have been demonstrated to improve microbial balance in the gastrointestinal tract through bacterial antagonisms, competitive exclusion and immune stimulation (Rolfe 1991; Brisbin et al. 2008). Prebiotics may control or manipulate microbial composition and/or activity, thereby maintaining a beneficial microbiota that suppresses the growth of pathogens (Gibson et al. 2004). The combination of probiotics and prebiotic may improve the survival rate of probiotics during their passage through the digestive tract, thus contributing to the stabilization and/or enhancement of the probiotic effect.

The importance of prebiotics as enhancer of the growth and performance of probiotic bacteria has been documented in chicken (Rebole et al. 2010). There are many studies that reported probiotic and prebiotic inhibition of harmful bacteria via occupying cell wall spaces inside the intestinal mucosa. Mountzouris et al. (2007) demonstrated that the probiotic product Biomin Poultry 5 Star composed of probiotic bacteria isolated from the crop (*Lactobacillus reuteri*), jejunum (*Enterococcus faecium*), ileum (*Bifidobacterium animalis*) and caecum (*Pediococcus acidilactici* and *Lactobacillus salivarius*) of healthy adult chickens has a potential impact on pathogen inhibition and modulation on intestinal microbiota. They found that the population of bacteria belonging to *Bifidobacterium* spp., *Lactobacillus* spp. and Gram-positive cocci significantly ($P \leq 0.05$) increased in the probiotic treatment compared to the control and antibiotic treatments.

Selective *Lactobacillus plantarum* or *Bifidobacterium longum* strains, combined with galacto-oligosaccharides (GOS), fructooligosaccharides (FOS) or xylo-oligosaccharides (XOS), demonstrated a modulation activity of the gut microbiota with the increase of beneficial microbial groups and the reduction of the load of pathogens such as *C. jejuni* (Santini et al. 2010; Baffioni et al. 2012). Studies showed that synbiotic supplementation was more successful in reducing *C. jejuni* and *Campylobacter* spp. when administered lifelong starting from the first day of life with respect to a shorter administration period (Baffioni et al. 2017). Wali and Beal (2011) showed that *Lactobacillus salivarius* (NCIMB 41606) isolated from the chicken gut significantly reduced the growth of *Salmonella typhimurium* in a chicken-simulated digestive system in vitro. Chaveerach et al. (2004) showed that *Lactobacillus* strains isolated from the chicken gut had antibacterial activity against most strains of *Campylobacter jejuni*.

In addition, Taheri et al. (2009) tested 62 LAB strains isolated from healthy chickens for their antagonistic activity against several pathogens. They were able to inhibit the growth of *E. coli* O78:K80, *Salmonella enterica* and *Salmonella typhimurium*, while the halo zones of inhibition varied among strains. Kizerwetter-Swida and Binek (2009) found that lactobacilli have higher antibacterial effects against the Gram-positive pathogenic bacteria (*Staphylococcus aureus* and *Clostridium perfringens*) with respect to *E. coli* and *Salmonella*.

Prebiotics in the intestinal tract causes the removal of pathogenic bacteria that might attach to the surface of the intestinal epithelial cells (Newman 1994). Oyofo et al. (1989) showed that dietary prebiotic were successful in inhibiting the intestinal colonization of *Salmonella typhimurium*. Studies on the effects of inulin found that foods containing Jerusalem artichoke inulin at the level of 5 g/day significantly increased bifidobacteria (Ramnani et al. 2010).

### 5.7.3 Gut Histology

Probiotics and prebiotics also promote changes in the intestinal environment through reducing the pH, increasing short-chain fatty acid concentration, supplying enzymes that aid digestion and increasing enzyme activity in the GI tract. Pelicano et al. (2005) showed that administration of a probiotic and a prebiotic to poultry increased the villus height leading to an increased intestinal surface area and therefore to an increased digestion and absorption of nutrients. One of the roles of probiotics, prebiotics and synbiotics is the ability to change the morphology of the digestive tract, increasing villi length and crypt depth (Pelicano et al. 2005). An increase in villi length refers to high digestion and absorption efficiency with the presence of good microbial balance and healthy status.

Samli et al. (2007) reported that the administration of a probiotic supplement containing *Enterococcus faecium* to broiler diet increased the ileal villus height. Songsak et al. (2008) reported that when different levels 0, 0.08, 0.8 and 8.1% of cassava yeast probiotic were added to the diet of broiler chicks, there was a significant increase in ileum villus height at 42 days compared to the control. Samanya and Yamauchi (2002) demonstrated that birds administered with *Bacillus subtilis* var. natto for 28 days showed a higher villus extension than the control group. Santin et al. (2001) recorded that broilers fed with *Saccharomyces cerevisiae* had higher villus height than that of the control group during the first 7th day. Zhang et al. (2005) showed that when 0.5% of *Saccharomyces cerevisiae* yeast was added to the diet of male broiler chicks, there was a significant increase in the villus height in the ileum at 21 days compared to the control.

Pelicano et al. (2005) observed an improvement in histological indexes of the intestinal mucosa with the use of probiotics and prebiotics at 21 days of age. Xu et al. (2003) also reported that broilers fed with fructooligosaccharides (4 g/kg) had higher villi in the jejunum and ileum than the control group. Rehman et al. (2007) demonstrated that supplementation of dietary inulin increased the jejunal villus length and crypt depth in broilers after 35 days.

Awad et al. (2008) reported that the addition of the synbiotic Biomin IMBO (a combination of *Enterococcus faecium* and chicory inulin) with diet increased the villus height/crypt depth ratio and villus height in ileum. However, the ileal crypt depth was decreased compared to the control. The intestinal mucosal architecture can reveal useful information on the intestinal function. Pelicano et al. (2007) reported that there was a significant increase in intestinal villus height of broiler chicks at 42 days when synbiotic was used compared to the control. Hassanpour et al. (2013) indicated that 0.1% synbiotic (Biomin IMBO) significantly increased villus height.

### 5.7.4 Short-Chain Fatty Acid in the Gastrointestinal Tract

Fermentation of prebiotics by commensal bacteria results in the production of SCFAs. Short-chain fatty acids are the main energy source for colonocytes, particularly butyric acid, which is the preferred energy substrate of caecal/colonic epithelium. On the other hand, bacterial fermentation in the caeca leads to the formation of short-chain fatty acids, which are necessary metabolites for intestinal epithelial cells, but they also decrease luminal pH and create a less favourable environment for pathogenic species in the GI tract (Topping and Clifton 2001).

Butyrate is a major source of energy for enterocytes and colonocytes (Chapman et al. 1995) and has a fundamental role in maintaining a healthy GI tract. Lawhon et al. (2002) reported that butyrate and propionate were more efficient compared to other types of SCFAs in inhibiting *Salmonella typhimurium*, whereas other researchers observed that acetic acid was more effective (Van der Wielen et al. 2000).

### 5.7.5 Meat Quality

In recent years, the high growth rate and improvements in meat quality and properties of carcasses have been beneficial to the poultry industry, especially in broiler production. Currently, an important research area is the use of probiotics, prebiotics and synbiotics as feed additives as an alternative to antibiotics. There are many reports concerning the effect of using probiotics, prebiotics and synbiotics on feed performance (Abdel-Raheem et al. 2012; Gunal et al. 2006; Satbir and Sharma 1999), but carcass and meat quality of broilers have not been studied. Broiler chickens have a rapid growth rate and have been genetically selected for high live body weight. Generally, probiotics are used to correct dysfunctions in the gastrointestinal tract caused by stress factors.

Zhou et al. (2010) studied the effect of *Bacillus coagulans* ZjU0616 supplemented as probiotic in different concentrations (0 Control, $1.0 \times 10^6$ cfu $g^{-1}$ (T1), $2.0 \times 10^6$ cfu $g^{-1}$ (T2) and $5.0 \times 10^6$ cfu $g^{-1}$ (T3)) on the breast chemical composition and meat quality of Guangxi Yellow chickens. There were no significant differences in breast chemical composition (moisture %, crude protein %, crude fat % and crude ash %) among treatments. On the other hand, tenderness significantly decreased in T2 and T3 compared to the control group.

Mahajan et al. (2000) reported that the supplementation of the probiotic Lacto-Sacc composed of *Lactobacillus acidophilus* and *Streptococcus faecium* to broiler diet resulted in meat with a higher ($P < 0.001$) percentage of moisture, proteins, ashes and WHC and a lower fat percentage at the end of the 6-week feeding trial. Endo and Nakano (1999) reported that the use of probiotics (*Bacillus*, *Lactobacillus*, *Streptococcus*, *Clostridium*, *Saccharomyces* and *Candida* spp.) improved the characteristics of carcass and meat quality in male broilers.

Colour is an important quality attribute that influences consumer acceptance of many food products, including poultry meat. Consumers often reject products in which the colour varies from the expected normal appearance. Consequently, colour is often used to determine the economic value of food (Qiao et al. 2001). Broiler quality improvement may depend on the selected feed ingredient. Appearance is the major criterion for purchase, selection and initial evaluation of meat quality (Fletcher 2002). Other quality attributes, such as tenderness, cooking loss and shelf-life are important for the consumer after purchasing the product (Jeremiah 1982; Husak et al. 2008). Variations in colour of broiler breast meat fillets were significant correlated with muscle pH. Breast meat may appear darker because of high muscle pH (Karaoglu et al. 2004).

### 5.7.6 Haematological Parameters

Haematological parameters of animals are used as indexes of their health status. Haematocrit (Hct) value is also used as an indicator of animal health and represents the percentage of packed blood cells to plasma volume.

Al-Kassie et al. (2008) reported that when the probiotic *Aspergillus niger* and the prebiotic *Taraxacum officinale* were added to the Arbor Acres broiler diet at a rate of 10 g/kg, there was a significant increase in Hb concentration and only prebiotic significantly increased PCV% at 42 days old compared to the control.

The significant increases of Hct in the chicks fed on probiotics and prebiotics may be due to the acidic condition of the GI tract caused by additives supplementation which resulted in better iron salt absorption from the small intestine. This may also increase vitamin production by beneficial bacteria which may positively affect blood-forming processes (Kander 2004).

Physiological and pathological stresses in avian species affect neuroendocrine system (glucocorticoids, catecholamines, epinephrine, norepinephrine, prolactin and growth hormones) and reduce lymphocyte production (Marketon and Glaser 2008). When birds are stressed, glucocorticoid hormones are secreted, and the stress level increases further (Dhabhar et al. 1996). Stress could cause an increased stimulation of the adrenal gland to produce hormones. This causes an increase in heterophils/lymphocytes ratio (H/L *ratio*) (Gross and Siegel 1983). Thus, H/L ratio could be used as an indicator for the health of animals, and any increase of H/L ratio refers to an increase in stress case (James and Stanley 1989). Paryad and Mahmoudi (2008) reported that when different levels 0, 0.5, 1.5 and 2% of *Saccharomyces cerevisiae* were added to the diet of broiler chicks, there was a significant decrease in H/L ratio at 42 days old. While, Sarinee et al. (2008) assumed that when the probiotic was added to the drinking water of male Cobb broiler chicks, there was no significant effect in the H/L ratio at 28 and 42 days old compared to the control group. Al-Kassie et al. (2008) found a significant decrease in H/L ratio of broiler fed on the diet supplemented with 10 g/kg of prebiotic *Taraxacum officinale* at 42 days old compared to the control. Heterophil granules contain antimicrobial substances that can be released through degranulation to kill phagocytized bacteria (He et al. 2005). Lymphocytes are a type of white blood cells (WBCs) which form part of the body's immune system and help the body fight the infections. Lymphocytes attack foreign bodies by either producing antibodies or swallowing pathogens.

### 5.7.7 Cholesterol Content

Cholesterol is a critical fatty substance necessary for the proper function of every cell in the body. Cholesterol is a structural component of cell membrane and plasma lipoproteins, and it is important in the synthesis of steroid hormones and bile acids. Mostly synthesized in the liver, some amount of cholesterol is also introduced with the diet, especially if rich in saturated fats. Paryad and Mahmoudi (2008) reported a significant decrease in serum cholesterol after administration of different levels 0, 0.5, 1.5 and 2% of probiotic supplement to the broiler diet at 42 days of age. Mansoub (2010) found that when the diet of male Ross 308 broiler chicks was supplemented with 1% *Lactobacillus casei*, there was a significant decrease in serum cholesterol at 42 days old compared to the control group.

The significant reduction in serum cholesterol of broiler chickens fed on probiotic-supplemented diet could be attributed to a reduced absorption and/or synthesis of cholesterol in the gastrointestinal tract (Mohan et al. 1996). Furthermore, some probiotic bacteria may interfere with cholesterol absorption by deconjugating bile salts (Li et al. 2007; Liong and Shah 2006). Abdulrahim et al. (1996) demonstrated that a *Lactobacillus acidophilus* strain reduced the cholesterol in the blood by deconjugating bile salts in the intestine, thereby preventing them from acting as precursors in cholesterol synthesis. Several *Lactobacillus* strains have a high bile salt hydrolytic activity, which is responsible for deconjugation of bile salts (Surono 2003).

The effects of probiotics and prebiotics on serum cholesterol concentrations are inconsistent among studies. Some studies have shown that probiotics and prebiotics exhibit lipid-lowering properties. This capability might be related to changes in the composition of the intestinal microbiota, which ferments prebiotics to produce short-chain fatty acids in the gut, thereby causing a decrease in the systemic levels of blood lipids and cholesterol.

Another explanation for these contrasting results might be ascribed to the different dosages applied, the duration of the administration period as well as to the different probiotic strains and type of prebiotic used (Angel et al. 2005; O'Dea et al. 2006; Patterson and Burkholder 2003).

The lower cholesterol concentration in groups fed on probiotic and synbiotic may be due to microorganisms present in the probiotic product that have the ability of use cholesterol for their metabolism, lowering the cholesterol absorption in the gastrointestinal tract (Nelson and Gilliland 1984; Mohan et al. 1995). In addition, some probiotic microorganisms have shown the inhibition of hydroxymethylglutaryl-coenzyme A, which is involved in the cholesterol synthesis (Fukashima and Nakano 1995). Studies underlined that prebiotics can have hypocholesterolemic effect through binding bile acids, thereby reducing lipid absorption in intestine with the of increasing cholesterol elimination and hepatic synthesis of new bile acid (Zhang et al. 2003).

Recent researches have revealed that probiotics affect gene expression of carrier proteins, which are responsible for cholesterol absorption. The protein called Niemann-Pick C1-like 1 (NPC1L1), which is abundantly expressed on the surface of enterocytes, plays a key role on the absorption of cholesterol from the intestine. Reduction or inhibition of the expression levels of this protein leads to a decrease in plasma cholesterol levels. The probiotic *Lactobacillus acidophilus* ATCC 4356 reduces NPCIL-1 gene expression and inhibits the cellular uptake of micellar cholesterol in Caco-2 cells.

### 5.7.8 Immune Modulation

Probiotics maintain the proper balance of useful microbial populations in the intestine of birds, which is important for an efficient feed conversion, growth, productivity and stimulation of birds' immune mechanisms to counteract pathogens. Prebiotic supplementation has been shown to increase the production of IgA in laying hens (Kim et al. 2009). Immunoglobulin A inhibits the attachment and penetration of bacteria in the lumen, increases the production of mucus (McKay and Perdue 1993) and prevents inflammation that could cause epithelial tissue damage (Russell et al. 1989).

The establishment of a normal microbiota constitutes a key component of gut health, through colonization resistance mechanisms, and has implications for proper development of the gut and full maturation of the mucosal immune system (Oakley et al. 2014). The communication between the microbiota and the immune system is principally mediated by interaction between microbes and pattern recognition receptors (PRRs) expressed by the intestinal epithelium and various local antigen-presenting cells, resulting in activation or modulation of both innate and adaptive immune responses (Sommer and Backhed 2013). The composition of the GI microbiota is known to affect many host functions including nutrient utilization, gut epithelium feeding and the development and activity of the gut immune system (Hill et al. 2010). The interaction between the immune system of the gut and commensal microbiota in chickens starts immediately after hatching and leads to a low level of inflammation characterized by an increased cytokine and chemokine expression as well as a number of immune-associated proteins (Crhanova et al. 2011). As a result, there is an infiltration of heterophils and lymphocytes into the lamina propria or the gut epithelium and normalization of the gut immune system (Mwangi et al. 2010).

The stimulation of cell-mediated immunity would constantly help fight against viral infections and thus can reduce the flock mortality occurring due to immunosuppressive diseases. Koenen et al. (2004) reported that different strains of *Lactobacillus* spp. have modulating effects on immune system of layer- and meat-type chickens. Furthermore, Nayebpor et al. (2007) reported that DFM enhanced the humoral immune response in broiler chickens. Antibodies such as immunoglobulin A (IgA) are produced by plasma cells of the immune system and are involved in protecting the body from potentially harmful bacteria. Probiotic bacteria have been shown to alter host immune responses to infection by stimulating secretory IgA production (Fukushima et al. 1998).

The gut is often referred to as the largest immune organ as it harbours more lymphocytes than any other organs, and its size and the amount of surface area in contact with the autochthonous and allochthonous bacteria are also significant factors. The intestinal epithelium enterocytes provide a barrier against pathogens and enable the immune system to detect potential pathogens in the lumen (Dhama et al. 2011).

The bursa of Fabricius is an organ of the immune system and is responsible for maturation of B lymphocytes (Alloui et al. 2005). The size of the bursa is an indication of the immune functions, and the relative weight of bursa to live body weight was recorded to compare the results between different treatments. Withers et al. (2005) have observed that there are two distinct types of follicle in the recovering bursa, large follicles with a cortex and medulla and small follicles without these structural compartments (Fig. 5.7). Birds with only small follicles did not produce detectable antibodies against IBDV or subsequently administered antigen. The presence of the larger follicles was correlated to ability to produce Ig responses. In contrast, the small follicles were not able to support the complete programme of bursal B-cell development.

Elrayeh and Yildiz (2012) reported that in their study supplementation of 0.7% inulin in the diet of broilers did not affect the weight of bursa of Fabricius compared to the control. Dizaji et al. (2013) reported that addition of prebiotic to broilers' diet did not show any significant effect on BF weight compared to the control group.

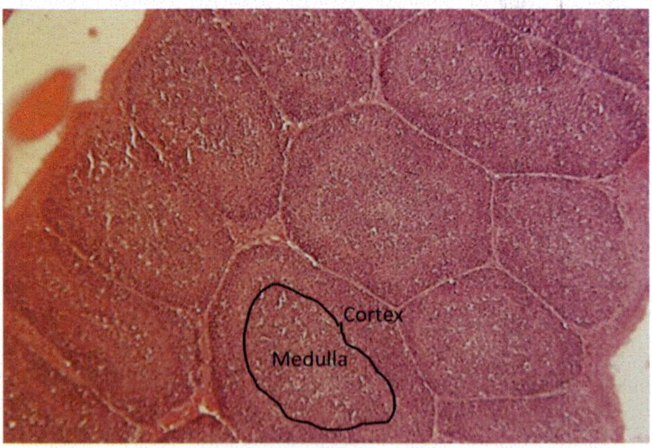


**Fig. 5.7** Structure of follicle of bursa of Fabricius

**Conclusions**

Probiotic and prebiotic increased the performance of broiler chickens. The higher production performance observed in broilers fed with probiotic and prebiotic may be due to suppression of pathogenic bacteria, reduction in pH value and increase of SCFA in the intestine. These beneficial mechanisms lead to an increased intestinal absorption surface and villus height, thus improving animal overall performance.

## References


Abdel-Raheem SM, Sherief MS, Khaled MA (2012) The effects of prebiotic, probiotic and synbiotic supplementation on intestinal microbial ecology and histomorphology of broiler chickens. Int J Agro Vet Med Sci 6(4):277–289

Abdulrahim SM, Haddadin MSY, Hashlamoun EAR, Robinson RK (1996) The influence of *Lactobacillus acidophilus* and bacitracin on layer performance of chickens and cholesterol content of plasma and egg yolk. Br Poult Sci 37:341–346

Ahmad I (2006) Effect of probiotics on broilers performance. Int J Poult Sci 5:593–597

Akbar A, Anal AK (2011) Food safety concerns and food-borne pathogens, Salmonella, Escherichia coli and Campylobacter. FUUAST J Biol 1(1):5–17

Al-Kassie GAM, Al-Jumaa YMF, Jameel YJ (2008) Effect of probiotic (*Aspergillus niger*) and prebiotic (*Taraxacum officinale*) on blood picture and biochemical properties of broiler chicks. Int J Poult Sci 7(12):1182–1184

Alkhalf A, Alhaj M, Al-Homidan I (2010) Influence of probiotic supplementation on immune response of broiler chicks. Egyptian Poultry Science 30:271–280

Alloui MN, Sellaouil S, Djaaba S (2005) Morphometrical and anatomopathological survey of the bursa of fabricius in broiler chickens. Int Soc Anim Hygiene 2:52–55

Amit-Romach E, Sklan D, Uni Z (2004) Microflora ecology of the chicken intestine using 16S ribosomal DNA primers. Poult Sci 83:1093–1098

Angel R, Dalloul RA, Doerr J (2005) Performance of broiler chickens fed diets supplemented with a direct-fed microbial. Poult Sci 84:1222–1231

Apajalahti J, Kettunen A, Graham H (2004) Characteristics of the gastrointestinal microbial communities, with special reference to the chicken. Worlds Poult Sci J 60:223–263

Awad WA, Bohm J, Razzazi-Fazeli E, Ghareeb K, Zentek J (2006) Effect of addition of a probiotic microorganism to broiler diets contaminated with deoxynivalenol on performance and histological alterations of intestinal villi of broiler chickens. Poult Sci 85:974–979

Awad WA, Ghareeb K, Bohm J (2008) Intestinal structure and function of broiler chickens on diets supplemented with a synbiotic containing *Enterococcus faecium* and oligosaccharides. Int J Mol Sci 9:2205–2216

Awad WA, And GK, Bohm J (2011) Evaluation of the chicory inulin efficacy on ameliorating the intestinal morphology and modulating the intestinal electrophysiological properties in broiler chickens. J Anim Physiol Anim Nutr 95:65–72

Ayasan T, Ozcan BD, Baylan M, Conogullari S (2006) The effects of dietary inclusion of probiotic protexin on egg yield parameters of Japanese quails (Coturnix Coturnix Japonica). Int J Poult Sci 5:776–779

Baffioni L, Gaggia F, Di Gioia D, Santini C, Mogna L, Biavati B (2012) A Bifidobacterium-based synbiotic product to reduce the transmission of *C. jejuni* along the poultry food chain. Int J Food Microbiol 157:156–161

Baffioni L, Gaggia F, Garofolo G, Di Serafino G, Buglione E, Di Giannatale E, Di Gioia D (2017) Evidence of *Campylobacter jejuni* reduction in broilers with early synbiotic administration. Int J Food Microbiol 251:41–47

Banasaz M, Norin E, Holma R, Midtvedt T (2002) Increased enterocyte production in gnotobiotic rats mono-associated with *Lactobacillus rhamnosus* GG. Appl Environ Microbiol 68:3031–3034

Barnes EM, Mead GC, Barnum DA, Harry EG (1972) The intestinal Flora of the chicken in the period 2 to 6 weeks of age, with particular reference to the anaerobic bacteria. Br Poult Sci 13:311–326

Barnes EM (1979) The intestinal microflora of poultry and game bird during life and after storage. Soc Appl Bacteriol 46(3):407–419

Brisbin JT, Zhou H, Gong J, Sabour P, Akbari MR, Haghighi HR, Yu H, Clarke A, Sarson AJ, Sharif S (2008) Gene expression profiling of chicken lymphoid cells after treatment with *Lactobacillus acidophilus* cellular components. Clin Vaccine Immunol 32:563–574

Callaway T, Edrington T, Anderson R, Harvey R, Genovese K, Kennedy C (2008) Probiotics, prebiotics, and competitive exclusion for prophylaxis against bacterial disease. Anim Health Res Rev 9(2):217–225

Chapman MA, Grahn MF, Hutton M, Williams NS (1995) Butyrate metabolism in the terminal ileal mucosa of patients with ulcerative colitis. Br J Surg 82:36–38

Chaveerach P, Lipman LJA, van Knapen F (2004) Antagonistic activities of several bacteria on in vitro growth of 10 strains of campylobacter jejuni/coli. Int J Food Microbiol 90:43–50

Connell I, Agace W, Klemm P, Schembri M, Mărild S, Svanborg C (1996) Type 1 fimbrial expression enhances Escherichia coli virulence for the urinary tract. Proc Natl Acad Sci 93:9827–9832

Crhanova M, Hradecka H, Faldynova M, Matulova M, Havlickova H, Sisak F, Rychlik I (2011) Immune response of chicken gut to natural colonization by gut micro-flora and to Salmonella enterica serovar enteritidis infection. Infect Immun 79(7):2755–2763

De Leenheer L (2007) Production and use of inulin: industrial reality with a promising future. In: Van Bekkum H, Röper H, Voragen F (eds) Carbohydrates as organic raw materials III. V.C.H. Publishers, New York, pp 67–92

Dhabhar FS, Miller AH, Stein M, Mcewen BS, Spencer RL (1996) Stress induced changes in blood leukocyte distribution—role of adrenal steroid hormones. J Immunol 157:1638–1644

Dhama K, Vinay V, Sawant PM, Ruchi T, Vaid RK, Chauhan RS (2011) Applications of probiotics in poultry: enhancing immunity and beneficial effects on production performances and health—a review. J Immunol Immunopathol 13(1):1–19

Dizaji BR, Afshin Z, Amirhossein G, Saman F, Orhan R (2013) Influences of different growth promoters on intestinal morphology of broiler chickens. Eur J Exp Biol 3(2):32–37

Edens F (2003) An alternative for antibiotic use in poultry' probiotics. Revista Brasileira de Ciência Avícola 5:75–97

Elrayeh AS, Yildiz G (2012) Effects of inulin and β-glucan supplementation in broiler diets on growth performance, serum cholesterol, intestinal length, and immune system. Turk J Vet Anim Sci 36(4):388–394

Endo T, Nakano M (1999) Influence of a probiotic on productivity, meat components lipid metabolism, caecal flora and metabolites and raising environment in broiler production. Anim Sci J 70:207–218

Ewing WN, Cole DJA (1994) The living gut: an introduction to micro-organisms in nutrition. Context Publications, Dungannon, UK

Fan S, Zheng J, Duan Z, Yang N, Xu G (2014) The influences of SE infection on layers' production performance, egg quality and blood biochemical indicators. J Anim Sci Biotechnol 5(4):1–6

FAO/WHO (2002) Joint FAO/WHO (Food and Agriculture Organization/World Health Organization) working group report on drafting guidelines for the evaluation of probiotics in food, London, ON, Canada, pp 1–11

Flemming JS, Freitas JRS, Fontoura P, Montanhini NR, Arruda JS (2004) Use of mannanoligosaccharides in broiler feeding. Braz J Poult Sci 6(3):159–161

Fletcher DL (2002) Poultry meat quality. Worlds Poult Sci J 78:1323–1327. 58(2):131–145

Fukashima M, Nakano M (1995) The effect of probiotic on faecal and liver lipid classes in rats. Br J Nutr 73:701–710

Fukushima Y, Kawata Y, Hara H, Terada A, Mitsuoka T (1998) Effect of probiotic formula on intestinal immunoglobulin A production in healthy children. Int J Food Microbiol 42:39–44
Fuller R (1989) Probiotics in man and animals. J Appl Bacteriol 66:365–378
Fuller R (2001) The chicken gut microflora and probiotic supplements. J Poult Sci 38:189–196
Gasson MJ, Narbad A, Woodward MJ, Ragione R l (2004) In vivo characterization of *Lactobacillus johnsonii* FI9785 for use as a defined competitive exclusion agent against bacterial pathogens in poultry. Lett Appl Microbiol 38(3):197–205
Gibson GR, Roberfroid MB (1995) Dietary modulation of the human colonic microbiota: introducing the concept of prebiotics. J Nutr 125:1401–1412
Gibson GHR, Probert M, Van Loo J, Robert RA, Roberfroid MB (2004) Dietary modulation of the human colonic microbiota: updating the concept of prebiotics. Nutr Res Rev 17:259–275
Griggs JP, Jacob JP (2005) Alternatives to antibiotics for organic poultry production. J Appl Poult Res 14:750–756
Gong J, Forster RJ, Yu H, Chambers JR, Wheatcroft R, Sabour PM, Chen S (2002) Molecular analysis of bacterial populations in the ileum of broiler chickens and comparison with bacteria in the cecum. FEMS Microbiol Ecol 41:171–179
Gong J, Yu H, Liu T, Gill JJ, Chambers JR, Wheatcroft R, Sabour PM (2008) Effects of zinc bacitracin, bird age and access to range on bacterial microbiota in the ileum and caeca of broiler chickens. J Appl Microbiol 104:1372–1382
Gross WB, Siegel HS (1983) Evaluation of heterophile/lymphocyte ratio as a measure of stress in chickens. Avian Dis 27:972–979
Gunal M, Yayli G, Kaya O, Karahan N, Sulak O (2006) The effects of antibiotic growth promoter, probiotic or organic acid supplementation on performance, intestinal microflora and tissue of broilers. Int J Poult Sci 5:149–155
Gusils C, Perez AC, Gonzalez S, Oliver G (1999) *Lactobacilli* isolated from chicken intestines: potential use as probiotics. J Food Prot 63(3):252–256
Haghighi HR, Gong J, Gyles CL, Hayes MA, Zhou H, Sanei B, Chambers JR, Sharif S (2006) Probiotics stimulate production of natural antibodies in chickens. Clin Vaccine Immunol 13(9):975–980
Hassanpour H, ZamaniMoghaddam AK, Khosravi M, Mayahi M (2013) Effects of synbiotic on the intestinal morphology and humoral immune response in broiler chickens. Livest Sci 153:116–122
He H, Lowry VK, Ferro PM, Kogut MH (2005) CpG-oligodeoxynucleotide stimulated chicken heterophil degranulation is serum cofactor and cell surface receptor dependent. Dev Comp Immunol 29:255–264
Hill DA, Hoffmann C, Abt MC, Du Y, Kobuley D, Kirn TJ (2010) Metagenomic analyses reveal antibiotic-induced temporal and spatial changes in intestinal microbiota with associated alterations in immune cell homeostasis. Mucosal Immunol 3(2):148–158
Hong HA, Duc LH, Cutting SM (2005) The use of bacterial spore formers as probiotics. FEMS Microbiol Rev 29:813–835
Hooge D (2004) Meta-analysis of broiler chicken pen trials evaluating dietary mannan oligosaccharide. Int J Poult Sci 3:163–174
Huang MK, Choi YJ, Houde R, Lee JW, Lee B, Zhao X (2004) Effects of Lactobacilli and an acidophilic fungus on the production performance and immune responses in broiler chickens. Poult Sci 83(5):788–795
Husak RL, Sebranek JG, Bregendahl K (2008) A survey of commercially available broilers marketed as organic, free-range, and conventional broilers for cooked meat yields, meat composition, and relative value. Poult Sci 87:2367–2376
James M, Stanley EC (1989) Multiple concurrent stressors in chicks: 3. Effects on plasma corticosterone and the heterophil: lymphocyte ratio. Poult Sci 68:522–527
Jeremiah LE (1982) A review of factors influencing consumption selection and acceptability of meat purchases. J Consum Stud Home Econ 6:137–154
Jeurissen SH, Lewis F, van der Klis JD, Mroz Z, Rebel JM, ter Huurne AA (2002) Parameters and techniques to determine intestinal health of poultry as constituted by immunity, integrity, and functionality. Curr Issues Intest Microbiol 3:1–14

Jin LZ, Ho YW, Abdullah N, Jalaudin S (1996) Influence of dried Bacillus Subtilis and lactobacilli cultures on intestinal microflora and performance in broilers. Asian-Australas J Anim Sci 9:99–107

Jin LZ, Ho YW, Abdullah N, Jalaludin S (1997) Probiotics in poultry: modes of action. Worlds Poult Sci J 53:351–368

Jin LZ, Ho YW, Abdullah N, Ali MA, Jalaludin S (1998) Effects of adherent *Lactobacillus* cultures on growth, weight of organs and intestinal microflora and volatile fatty acids in broilers. Anim Feed Sci Technol 70(3):197–209

Kalavathy R, Abdullah N, Jalaludin S, Wong CMVL, Ho YW (2008) Effects of *Lactobacillus* cultures and oxytetracycline on the growth performance and serum lipids of chickens. Int J Poult Sci 7:385–389

Kander M (2004) Effect of Bifidobacterium sp. on the health state of piglets, determined on the basis of hematological and biochemical indices. Electron J Pol Agric Univ 2(7)

Karaoglu M, Aksu MI, Esenbuga N, Kaya M, Macit M, Durdag H (2004) Effect of dietary probiotic on the pH and colour characteristics of carcasses, breast fillets and drumsticks of broilers. Anim Sci 78:253–259

Kaur N, Gupta AK (2002) Applications of inulin and oligofructose in health and nutrition. J Biosci 27(7):703–714

Kim CH, Shin KS, Woo KC, Paik IK (2009) Effect of dietary oligosaccharides on the performance, intestinal microflora and serum immunoglobulin contents in laying hens. Korean J Poult Sci 36:125–131

Kizerwetter-Swida M, Binek M (2009) Protective effect of potentially probiotic Lactobacillus strain on infection with pathogenic bacteria in chickens. Pol J Vet Sci 12:15–20

Kleessen B, Elsayed NAAE, Loehren U, Schroedl W, Krueger M (2003) Jerusalem artichokes stimulate growth of broilers chickens and protect them against endotoxins and potential cecal pathogens. J Food Prot 66:2171–2175

Koenen ME, Kramer J, Van Der Hulst R, Heres L, Jeurissen SHM, Boersma WJE (2004) Immunomodulation by probiotic lactobacilli in layer- and meat-type chickens. Br Poult Sci 45:355–366

Lan PT, Binh TL, Benno Y (2003) Impact of two probiotics lactobacillus strains feeding on fecal lactobacilli and weight gains in chickens. J Gen Appl Microbiol 49:29–36

Lan Y, Verstegen MWA, Tamminga S, Williams BA (2005) The role of the commensal gut microbial community in broiler chickens. Worlds Poult Sci J 61(01):95–104

Lawhon SD, Maurer R, Suyemoto M, Altier C (2002) Intestinal short-chain fatty acids alter *Salmonella typhimurium* invasion gene expression and virulence through BarA/SirA. Mol Microbiol 46:1451–1464

Li XJ, Piao XS, Kim SW, Liu P, Wang L, Shen YB, Jung SC, Lee HS (2007) Effects of chitooligosaccharide supplementation on performance, nutrient digestibility, and serum composition in broiler chickens. Poult Sci 86:1107–1114

Liong MT, Shah NP (2006) Effects of a *Lactobacillus casei* synbiotic on serum lipoprotein, intestinal microflora and organic acids in rats. J Dairy Sci 89:1390–1399

Liu H (2008) Influence of chicory feeding on performance and gut development in broilers. M.Sc. thesis, Department of Animal Nutrition and Management, Swedish University of Agricultural Sciences

Mahajan P, Sahoo J, Panda PC (2000) Effect of probiotic (Lacto-Sacc) feeding and season on poultry meat quality. Indian J Poult Sci 35(3):297–301

Mansoub NH (2010) Effect of probiotic bacteria utilization on serum cholesterol and triglycerides contents and performance of broiler chickens. Global Vet 5(3):184–186

Marketon JIW, Glaser R (2008) Stress hormone and immune function. Cell Immunol 252:16–26

McKay DM, Perdue MH (1993) Intestinal epithelial function: the case for immune physiological regulation. Dig Dis Sci 38:1377–1387

Mescher AL (2013) Junqueira's basic histology: text and atlas, 13th edn. McGraw-Hill, New York, NY

Mohan B, Kadirvel R, Bhaskaran M, Natarajan A (1995) Effect of probiotic supplementation on serum/yolk cholesterol and on egg shell thickness in layers. Br Poult Sci 36:799–803

Mohan B, Kadirvel R, Natarajan A, Bhaskaran M (1996) Effect of probiotic supplementation on growth, nitrogen utilisation and serum cholesterol in broilers. Br Poult Sci 37:395–401

Mohd Shaufi MA, Sieo CC, Chong CW, Gan HM, Ho YW (2015) Deciphering chicken gut microbial dynamics based on high-throughput 16S rRNA metagenomics analyses. Gut Pathogens 7:4

Mookiah S, Chin CS, Kalavathy R, Norhani A, Yin WH (2014) Effects of dietary prebiotics, probiotic and synbiotics on performance, caecal bacterial populations and caecal fermentation concentrations of broiler chickens. J Sci Food Agr 94(2):341–348

Mountzouris KC, Tsirtsikos P, Kalamara E, Nitsch S, Schatzmayr G, Fegeros K (2007) Evaluation of the efficacy of a probiotic containing *Lactobacillus*, *Bifidobacterium*, *Enterococcus*, and *Pediococcus* strains in promoting broiler performance and modulating cecal microflora composition and metabolic activities. Poult Sci 86:309–317

Mountzouris KC, Tsitrsikos P, Palamidi I, Arvaniti A, Mohnl M, Schatzmayr G, Fegeros K (2010) Effects of probiotic inclusion levels in broiler nutrition on growth performance, nutrient digestibility, plasma immunoglobulins, and cecal microflora composition. Poult Sci 89:58–67

Mwangi WN, Beal RK, Powers C, Wu X, Humphrey T, Watson M, Bailey M, Friedman A, Smith AL (2010) Regional and global changes in TCRalphabeta T cell repertoires in the gut are dependent upon the complexity of the enteric microflora. Dev Comp Immunol 34(4):406–417

Nelson CR, Gilliland SE (1984) Cholesterol uptake by *lactobacillus acidophilus*. J Dairy Sci 67:50

Newman K (1994) Manna-oligosaccharides: natural polymers with significant impact on the gastrointestinal microflora and the immune system. In: Lyons TP, Jacques KA (eds) Biotechnology in the feed industry, Proceedings of Alltech's 10th annual symposium. Nottingham University Press, Nottingham, UK, pp 167–174

Nayebpor M, Farhomand P, Hashemi A (2007) Effects of different level of direct feed microbial (Primalac) on growth performance and humoral response in broiler chickens. J Anim Vet Adv 6(11):1308–1313

Oakley BB, Lillehoj HS, Kogut MH, Kim WK, Maurer JJ, Pedroso A, Margie DL, Stephen RC, Timothy JJ, Nelson AC (2014) The chicken gastrointestinal microbiome. FEMS Microbiol Lett 360(2):100–112

O'Dea EE, Fasenko GM, Allison GE, Korver DR, Tannock GW, Guan LL (2006) Investigating the effects of commercial probiotics on broiler chick quality and production efficiency. Poult Sci 85(10):1855–1863

Ouwehand AC, Kirjavainen PV, Shortt C, Salminen S (1999) Probiotics: mechanisms and established effects. Int Dairy J 9:43–52

Oyofo BA, DeLoach JR, Corrier DE, Norman JO, Ziprin RL, Molenhauer HH (1989) Prevention of *Salmonella typhimurium* colonization of broilers with D-mannose. Poult Sci 68:1357–1360

Parves S, Malik KA, Kang AHS, Kim HY (2006) Probiotic and their fermented food products are beneficial for health. J Appl Microbiol 100(6):1171–1185

Paryad A, Mahmoudi M (2008) Effect of different levels of supplemental yeast (Saccharomyces cerevisiae) on performance, blood constituents and carcass characteristics of broiler chicks. Afr J Agric Res 3(12):835–842

Patterson JA, Burkholder KM (2003) Application of prebiotics and probiotics in poultry production. Poult Sci 82:627–631

Pelicano ERL, Souza PA, Souza HB, Leonel FR, Zeola NMBL, Boiago MM (2004) Productive traits of broiler chickens fed diets containing different growth promoters. Braz J Poult Sci 6(3):177–182

Pelicano ERL, Souza PA, Souza HBA, Figueiredo DF, Boiago MM, Carvalho SR, Bordon VF (2005) Intestinal mucosa development in broiler chickens fed natural growth promoters. Braz J Poult Sci 7(4):221–229

Pelicano ERL, Souza PA, Souza HBA, Figueiredo DF, Amaral CMC (2007) Morphometry and ultra-structure of the intestinal mucosa of broilers fed different additives. Braz J Poult Sci 9(3):173–180

Piray AH, Kermanshani H, Tahmasbi AM, Bahrampour J (2007) Effects of cecal cultures and aspergillus meal prebiotic (Fermacto) on growth performance and organ weights of broiler chickens. Int J Poult Sci 6(5):340–344

Pourabedin M, Zhao X (2015) Prebiotics and gut microbiota in chickens. FEMS Microbiol Lett 362(15):1–8

Qiao M, Fletcher DL, Smith DP, Northcutt JK (2001) The effect of broiler breast meat color on pH, moisture, water holding capacity, and emulsification capacity. Poult Sci 80:676–680

Ramnani P, Gaudier E, Bingham M, Bruggen P, Touhy KM, Gibson GR (2010) Prebiotic effect of fruit and vegetables shots containing Jerusalem artichoke inulin: a human intervention study. Br J Nutr 104(02):233–240

Rebole A, Ortiz LT, Rodriguez ML, Alzueta C, Trevino J, Velasco S (2010) Effects of inulin and enzyme complex, individually or in combination, on growth performance, intestinal microflora, cecal fermentation characteristics, and jejunal histomorphology in broiler chickens fed a wheat- and barley-based diet. Poult Sci 89:276–286

Rehman H, Rosenkranz C, Böhm J, Zentek J (2007) Dietary inulin affects the morphology but not the sodium-dependent glucose and glutamine transport in the jejunum of broilers. Poult Sci 86:118–122

Rehman H, Hellweg P, Taras D, Zentek J (2008) Effects of dietary inulin on the intestinal short-chain fatty acids and microbial ecology in broiler chickens as revealed by denaturing gradient gel electrophoresis. Poult Sci 87:783–789

Riberio AML, Vogt LK, Canal CW, Cardoso MRI, Labres RV, Streck AF, Bessa MC (2007) Effects of prebiotics and probiotics on the colonization and immune response of broiler chickens challenged with *Salmonella* Enteritidis. Braz J Poult Sci 9(3):193–200

Roberfroid MB (2007) Inulin type fructans: functional food ingredients. J Nutr 137:2493S–2502S

Rodriguez AT, Higgins SE, Vicente JLS, Wolfenden AD (2007) Effect of Lactose as a prebiotic on turkey body weight under commercial conditions. Poult Sci Assoc 16:635–641

Rolfe RD (1991) Population dynamics of the intestinal tract. In: Blankenship LC (ed) Colonization control of human bacterial Enteropathogens in poultry. Academic, San Diego, CA, pp 59–75

Rolfe RD (2000) The role of probiotic cultures in the control of gastrointestinal health. J Nutr 130:3965–4025

Rowghani E, Arab M, Akbarian A (2007) Effect of probiotic and other feed additive on performance and immune response of broiler chicks. Int J Poult Sci 6(4):261–265

Russell MW, Reinholdt J, Kilian M (1989) Anti-inflammatory activity of human IgA antibodies and their Fab α fragments: inhibition of IgG-mediated complement activation. Eur J Immunol 19:2243–2249

Rynsburger JM, Classen HL (2007) Effect of age on intestinal pH of broiler chickens. Paper presented at the international poultry scientific forum, January 23–25, 2007, Atlanta, GA

Salminen S, Bonley C, Bourron-Ruanlt M, Cummings JH, Franck A, Gibson GR, Isolauri E, Moreau MC, Roberfroid M, Rowland I (1998) Functional food science and gastrointestinal physiology and function. Br J Nutr 80:147S–171S

Samanya M, Yamauchi K (2002) Histological alterations of intestinal villi in chickens fed dried *Bacillus subtilis* var. natto. Comp Biochem Physiol 133:95–104

Samli HE, Senkoylu N, Koc F, Kanter M, Agma A (2007) Effects of *Enterococcus faecium* and dried whey on broiler performance, gut histomorphology and microbiota. Arch Anim Nutr 61:42–49

Santin E, Maiorka A, Macari M, Grecco M, Sanchez JC, Okada TM, Myasaka AM (2001) Performance and intestinal mucosa development of broiler chickens fed diets containing *Saccharomyces cerevisiae* cell wall. J Appl Poult Res 10:236–244

Santini C, Baffioni L, Gaggia F, Granata M, Gasbarri R, Di Gioia D, Biavati B (2010) Characterization of probiotic strain: an application as feed additives in poultry against *Campylobacter jejuni*. Int J Food Microbiol 141:S98–S108

Sarinee K, Thongsong BT, Vivat C (2008) Blood haematological-cholesterol profile and antibody titer response of broilers with added probiotic containing both bacteria and yeast or an antibiotic in drinking water. Thail J Vet Med 38(4):45–56

Satbir S, Sharma VP (1999) Performance of broiler chicks under different energy and probiotic levels during summer season. Indian J Poult Sci 34(1):34–37

Stolzenburg K (2005) Jerusalem artichokes—raw material for inulin and fructose production. Zuckerindustrie 130:193–200

Schneitz C (2005) Competitive exclusion in poultry: 30 years of research. Food Control 16:657–667

Sofia K, Gibson GR (2007) Prebiotic capacity of inulin-type fructans. J Nutr 137:2503S–2506S

Sommer F, Backhed F (2013) The gut microbiota—masters of host development and physiology. Nat Rev Microbiol 11(4):227–238

Songsak C, Chinrasri O, Somchan T, Ngamluam S, Soychuta S (2008) Effect of dietary inclusion of cassava yeast as probiotic source on growth performance, small intestine (ileum) morphology and carcass characteristic in broilers. Int J Poult Sci 7:246–250

Surono IS (2003) In vitro probiotic properties of indigenous Dadih lactic acid bacteria. Asian Australas J Anim Sci 16:726–731

Taheri HR, Moravej H, Tabandeh F, Zaghari M, Shivazad M (2009) Screening of lactic acid bacteria toward their selection as a source of chicken probiotic. Poult Sci 88:1586–1593

Topping DL, Clifton M (2001) Short-chain fatty acids and human colonic function: roles of resistant starch and nonstarch polysaccharides. Physiol Rev 81:1031–1064

Torok VA, Hughes RJ, Mikkelsen LL, Perez-Maldonado R, Katherine B, Ron Mac A, Nigel JP, Kathy OK (2011) Identification and characterization of potential performance-related gut microbiotas in broiler chickens across various feeding trials. Appl Environ Microbiol 77:5868–5878

Van der Wielen PW, Biesterveld S, Notermans S, Hofstra H, Urlings BA, van Knapen F (2000) Role of volatile fatty acids in development of the cecal microflora in broiler chickens during growth. Appl Environ Microbiol 66:2536–2540

Van Loo J, Coussement P, Deleenheer L, Hoebregs H, Smits G (1995) On the presence of inulin and oligofructose as natural ingredients in the western diet. Crit Rev Food Science and Nutrition 35:525–552

Verdonk JMAJ, Van Leeuwen P (2004) The application of inulin-type fructans in diets for veal calves and broilers. In: Inulin and oligofructose: FeelGood factors for health and well-being, 4th ORAFTI research conference, Paris, 12–13 Feb, pp 50–51

Wali NA, Beal J (2011) Survival of *Lactobacillus salivarius* (MCIMB 41606) in an in vitro model of the chicken digestive process and its effect on the survival of *salmonella typhimurium* NALR SALL344. Int J Probiotics Prebiotics 6(3):93–97

Wei S, Morrison M, Yu Z (2013) Bacterial census of poultry intestinal microbiome. Poult Sci 92:671–683

Westrell T, Ciampa N, Boelaert F, Helwigh B, Korsgaard H, Chriel M, Ammon A, Makela P (2009) Zoonotic infections in Europe in 2007: a summary of the EFSA. Eur Secur 14:1–3

Willey JM, Van Der Donk WA (2007) Lantibiotics: peptides of diverse structure and function. Annu Rev Microbiol 61:477–501

Withers DR, Young JR, Davison TF (2005) Infectious bursal disease virus-induced immunosuppression in the chick is associated with the presence of undifferentiated follicles in the recovering bursa. Viral Immunol 18:127–137

Xu ZR, Hu CH, Xia MS, Zhan XA, Wang MQ (2003) Effects of dietary fructooligosaccharide on digestive enzyme activities, intestinal microflora and morphology of male broilers. Poult Sci 82:1030–1036

Yang Y, Iji PA, Choct M (2009) Dietary modulation of gut microflora in broiler chickens: a review of the role of six kinds of alternatives to in-feed antibiotics. World's Poult Sci J 65:97–114

Yeoman CJ, Chia N, Jeraldo P, Sipos M, Goldenfeld ND, White BA (2012) The microbiome of the chicken gastrointestinal tract. Anim Health Res Rev 13(1):89–99

Yoon CN, Park CS, Han JH, Nam SK, Kwon YM (2004) Effect of feeding multiple probiotics on performance and fecal noxious gas emission in broiler chicks. Korean J Poult Sci 3:229–235

Yousefi M, Karkoodi K (2007) Effect of probiotic thepax and Saccharomyces cerevisiae supplementation on performance and egg quality of laying hens. Int J Poult Sci 6:52–54

Yusrizal, Chen TC (2003) Effect of adding chicory fructans in feed on broiler growth performance, serum cholesterol and intestinal length. Int J Poult Sci 2(3):214–219

Zasloff M (2002) Antimicrobial peptides in health and disease. N Engl J Med 347:1199–1200

Zhang WF, Li DF, Lu WQ, Yi GF (2003) Effects of isomalto-oligosaccharides on broiler performance and intestinal microflora. Poult Sci 82:657–663

Zhang AW, Lee BL, Lee KW, An GH, Song KB, Lee CH (2005) Effect of yeast (*Saccharomyces cerevisiae*) cell components on growth performance, meat quality and ileal mucosa development of broiler chicks. Poult Sci 84:1015–1021

Zhu XY, Zhong T, Pandya Y, Joerger RD (2002) 16S rRNA-based analysis of microbiota from the cecum of broiler chickens. Appl Environ Microbiol 68:124–137

Zhou X, Wang Y, Gu Q, Li W (2010) Effect of dietary probiotic, *Bacillus coagulans*, on growth performance, chemical composition, and meat quality of Guangxi yellow chicken. Poult Sci 89:588–593

# 6 Probiotics and Prebiotics for the Health of Cattle

Laureano Sebastián Frizzo, Marcelo Lisandro Signorini, and Marcelo Raúl Rosmini

## 6.1 Intensive Production Farms in Cattle Breeding and Health Problems

Throughout the years the livestock production systems have evolved modifying the natural resistance of animals against the diseases. These systems are characterized by new feeding methods (especially using unnatural feeds such as milk replacers), the intensive farming which limits the maternal contact and uses artificial habitat conditions, the use of animals with better growth parameters, and the use of antimicrobials substances. All these conditions increase the stress on animals, and digestive disorders become more frequent as a result of imbalance in the intestinal microbiota and generate a reduction in the natural resistance against contamination or pathogen colonization (James et al. 1984; Savage 1987; Fuller 1992; Mulder et al. 1997).

L.S. Frizzo
Laboratory of Food Analysis, Institute of Veterinary Science (ICiVet Litoral), National University of the Littoral, National Council of Scientific and Technical Research (UNL/CONICET), Esperanza, Province of Santa Fe, Argentina

Department of Public Health, Faculty of Veterinary Science, National University of the Littoral, Esperanza, Province of Santa Fe, Argentina
e-mail: lfrizzo@fcv.unl.edu.ar

M.L. Signorini
National Council of Scientific and Technical Research (CONICET), National Institute of Agricultural Technology (INTA) EEA Rafaela, Rafaela, Province of Santa Fe, Argentina

Department of Public Health, Faculty of Veterinary Science, National University of the Littoral, Esperanza, Province of Santa Fe, Argentina

M.R. Rosmini (✉)
Department of Public Health, Faculty of Veterinary Science, National University of the Littoral, Esperanza, Province of Santa Fe, Argentina
e-mail: mrosmini@unl.edu.ar



D. Di Gioia, B. Biavati (eds.), *Probiotics and Prebiotics in Animal Health and Food Safety*, https://doi.org/10.1007/978-3-319-71950-4_6

Intestinal disorders in farm animals are caused by imbalances in the intestinal microbiota. These disorders are frequent in intensive farming systems due to confinement of large numbers of animals in small areas, stress, and competition for feed, the reduced maternal contact, nonnatural feeding, and transport. Under these conditions the intestinal microbiota colonization is altered and facilitates the activity of pathogenic microorganisms (Rosmini et al. 2004). This intestinal imbalance in favor of pathogens brings about consequences on the animal health, reducing the profitability of farms and favoring disease transmission to humans by direct contact with ill animals or their feces or through the food chain (Rosmini et al. 2004; Frizzo et al. 2012, 2013; Signorini et al. 2012).

To prevent and control intestinal infections, a current practice (especially in intensive rearing systems) is to use antibiotics, a strategy which may increase the emergence and spread of antibiotic-resistant bacteria in meat and dairy products (McEwen and Fedorka-Cray 2002). The growing concern about the spread of microorganisms resistant to antibiotics in humans has determined the ban on the use of antibiotics as growth promoters in livestock (Schwarz et al. 2001). The food consumption with antibiotic residues causes many problems in public health, highlighting allergies and gastrointestinal disorders due to the alteration of the intestinal microbiota (Vassalo et al. 1997; Saarela et al. 2000). Another consequence of the antibiotic therapy is the negative impact on the beneficial microorganisms and the gastrointestinal imbalance (Parker 1990; Salminen et al. 1998).

Therefore, feed companies and researchers have been looking for alternative products and strategies that help to maintain intestinal health of the animal to prevent or reduce the prevalence of pathogens in the food chain. The use of probiotics has increased as an alternative therapy that prevents the use of antibiotics and, thus, reduces the emergence and spread of antibiotic-resistant bacteria and residual antibiotics in animal foods (Santini et al. 2010).

Therefore, the use of antibiotics has been geared more to preventing animal health problems during primary production than as preslaughter strategy to reduce the spread of foodborne pathogens and to reduce the impact on public health. Although antibiotics have enabled a significant increase in food production, the problems which have been generated have been important, and there is strong pressure from consumers and regulatory agencies to prohibit their use.

In this sense, many authors have evaluated the effect of probiotic supplementation on growth performance and protection against pathogens in chickens (Pascual et al. 1999; Santini et al. 2010), pigs (Casey et al. 2007; Mallo et al. 2010; Ross et al. 2010), and calves (Abe et al. 1995; Abdala et al. 2001; Timmerman et al. 2005; Adams et al. 2008; Frizzo et al. 2010a) with promising results.

The knowledge that the use of probiotics may replace antibiotic therapies with less aggressive methods has resulted in a new vision in the pharmaceutical industry. This technology requires the isolation of probiotics from specific ecosystems such as a herd or particular geographic region, the selection and characterization of the bacteria responsible for probiotic action, the production of probiotic on an industrial scale, and its reintroduction into the animals through its diet. In many cases, the use of non-selected probiotics has generated low or no beneficial effect on growth performance (Fuller 1989). That situation may be a consequence of using probiotics which were isolated from other regions or even from other animal species.

Raising calves can be done intensively with minimal maternal contact or extensively in the presence of their mother during lactation, depending on the type of cattle production. Rearing calves for meat production is performed, at least in countries like Argentina, extensively. In this system, the calf remains with its mother from birth to weaning season (approximately 6 months after birth).

On the other hand, the most important concern in the dairy farms is to produce more quantities of milk and with good quality. So, among other measures, the artificial breeding of calves is essential. This system has as main purpose accelerated the transformation process from monogastric to polygastric digestive system and thus leveraged the cheaper feed.

In the healthy animal, each portion of the intestine is colonized by a typical microbiota, which adapts and develops into a beneficial symbiosis with the host (Kurzak et al. 1998). The animals' digestive tract raised in natural production systems (extensive production systems) is colonized spontaneously and naturally. The colonizer microbiota comes from the environment that surrounds it, being essential in the contact with the mother during the first days of life, since this would supply the microbiota. When the animals are reared in intensive production systems, the probability to acquiring the natural microbiota is greatly reduced, and the intestine is more vulnerable to be colonized by pathogens. The effects of microbiota and their metabolic activities require special attention in the context of animal production in which the efficiency of animal growth is the primary goal (Collier et al. 2003). The probiotic supplementation is a useful alternative to improve the intestinal colonization by beneficial microorganisms during the first days of life.

## 6.2 Intestinal Microbiota Role

The gastrointestinal tract of calves is sterile at birth, and intestinal microorganisms are introduced from fecal, vaginal, and environmental microbiota. The impact of the intestinal microbiota is critical to host nutritional status and is of particular interest in farm animals that are reared in intensive systems (Rosmini et al. 2004) because the balance of the intestinal ecosystem can be altered by farming systems. This can be due to separation from their mothers, feeding with milk replacers and elimination of the benefits of cows' milk, inadequate colostrum intake, stressful situations, and use of antibiotics. Such practices may cause morbidity and mortality of young calves which can be related to economic losses.

The use of autochthonous microorganisms with probiotic activities provides an efficient alternative for treating and preventing some animal diseases (Rosmini et al. 2004). Under normal conditions, probiotic administration would not be necessary because animals acquire the protective intestinal microorganisms directly from maternal and environmental sources. Nevertheless, intensive rearing conditions oblige farmers to wean calves early (thus limiting the contact between calves and their mothers), feed them nonnatural feed (e.g., replacers), and introduce them to highly stressful environments. All of these conditions make the animals more susceptible to colonization by pathogenic microorganisms (Frizzo et al. 2012).

The indigenous intestinal microbiota of the calf is a complex microbial community that plays an important role in nutrition and health. This ecosystem is under the influence of the healthy status of the host, and this, in turn, is influenced by its environment. The gastrointestinal health may thus be defined as the ability to maintain a balance within the constantly changing ecosystem of the gastrointestinal tract (Melin et al. 1997). In intensive rearing, especially in early-weaned calves, the possibility of acquiring a natural autochthonous microbiota is strongly diminished, and, as a result, pathogenic microorganisms are much more likely to colonize the intestine (Rosmini et al. 2004).

*Salmonella* spp. and *Escherichia coli* are the most common bacterial etiologic agents of calf diarrhea during the first weeks of life (Rodríguez Armesto et al. 1996). Increased isolation frequency of *Salmonella* spp. indicates that the modern cattle breeding is favorable for development of this pathogen, especially when there are deficiencies in hygienic practices during rearing. The use of probiotic bacteria as a supplement in farm animal feeds, especially in intensive cattle production systems, is based on properties of the bacteria that improve feed nutrient conversion, and as the ability of these microorganisms to act against pathogenic bacteria (Frizzo et al. 2005). At the same time, probiotic microorganisms contribute to the safety of raw materials to be used in food consumed by humans. In Argentina, there are some commercial products intended for animal feed that are marketed as beneficial supplements due to their probiotic properties. However, no probiotic inoculum isolated from the indigenous microbiota of animals belonging to national livestock farms is found in the market or has been described in the literature. The ability of probiotic microorganisms to inhibit or counteract the negative effects of pathogens in live animals is a property that has been widely studied in laboratory animals (Maia et al. 2001; Moura et al. 2001; Frizzo et al. 2005) but not farm animals. Experimental models of intestinal disease could be used to evaluate the ability of an experimental probiotic inoculum to prevent the translocation of a pathogen to internal organs and the production of lesions in calves infected with *Salmonella*. The animals supplemented with probiotics and lactose could have advantages in their response against intestinal pathogen. We previously observed that some lactic acid bacteria (LAB) are capable of colonizing the intestinal tract of mice without affecting feed intake and protecting the animals against Salmonella Dublin DSPV 595T (Frizzo et al. 2005, 2006). In addition, LAB can colonize the gastrointestinal tract of calves without translocating to other internal organs (Frizzo et al. 2010a) and improve the growth of dairy calves exposed to nutritional stress such as diets with high lactose contents (Frizzo et al. 2010b, 2011a).

## 6.3 Probiotic and Prebiotic Concepts

Lactic acid bacteria (LAB) are natural components of the normal intestinal microbiota in both humans and animals (Schneider et al. 2004) and have been used to control the effects of pathogens such as *Salmonella* spp. (Gill et al. 2001) and

*Escherichia coli* (Shu and Gill 2002). These two pathogens are the most frequent bacterial etiologic agents in calf scours during the first week of life (Barrington et al. 2002; Millemann 2009).

Probiotics are live microorganisms which, when administered in adequate amounts, confer a health benefit on the host (Hill et al. 2014). Administration of a probiotic inoculum of bovine origin may favor establishment of a stable and balanced intestinal microbiota which would improve calf health (Abe et al. 1995). To produce a beneficial effect, administration of the inoculum must be continuous because the inoculated strains leave the intestinal tract. The strain is selected taking into account its benefits for the host through in vitro and in vivo studies of its probiotic properties. The technological features of the strain must be also evaluated, because they should demonstrate possible production of these cultures and their stability and survival during storage (Dunne et al. 2001). The viability and number of microorganisms inoculated is vital because the suggested minimum level (SML) of bacteria to produce beneficial effects is $10^6$ cfu/mL (Vinderola et al. 2000).

The genera *Lactobacillus* and *Streptococcus* are the most commonly used microorganisms as probiotics in animal production (Abu-Tarboush et al. 1996; Stephens et al. 2010). Many authors reported beneficial effects of probiotic preparations on animal growth (Frizzo et al. 2011b; Signorini et al. 2012). Young calves supplemented with probiotics showed a similar incidence of diarrhea and observed the same or better growth performance parameters (daily weight gain, feed conversion rate, etc.) compared with animals supplemented with antibiotics (Morrill et al. 1995; Timmerman et al. 2005). The trend toward natural rearing of animals without exposure to chemicals, pesticides, and herbicides is a good reason to increase the interest in probiotics in livestock farms (Reid and Friendship 2002).

The probiotic effect is mediated by three mechanisms: competition for specific niches on the intestinal mucosa, the competition for nutrients, and the production of bactericidal or bacteriostatic compounds (Fuller 1992). The inhibitory effects of LAB on undesirable microorganisms may also be due to the production of organic acids (e.g., lactic, acetic, and propionic acid) which reduce the intestinal pH and also by the production of hydrogen peroxide (Nousiainen and Setälä 1998). The production of specific antibacterial compounds such as bacteriocins (nisin and pediocin) has been mentioned among factors of the beneficial probiotic microorganisms (Klaenhammer 1988; Schillinger and Lucke 1989; Daeschel 1993). The efficiency in the bacteriocin production of some indigenous microorganisms and the purification, characterization, and overproduction of these proteins by genetic engineering methods have been explored both at laboratory and industrial scale (Remiger et al. 1999; Ross et al. 1999). Compared to most of the antibiotics, the bacteriocins are relatively specific and affect a limited number of bacterial species. This specificity may be particularly advantageous for applications aimed at a single strain or species without disturbing other microbial populations (Diez-Gonzalez 2007). The use of bacteria which produce bacteriocins as food security strategy in primary production is considered one of the best interventions to reduce the gastrointestinal colonization of farm animals by pathogens which produce foodborne diseases (Callaway et al. 2003). These bacteriocin-producing bacteria can be easily administered to animals in fresh or dry form mixed with feed crops or by drinking water. The LAB can be incorporated in the diet continuously or sporadically depending on the ability of each probiotic strain to colonize the gastrointestinal tract. The administration of bacteriocin-producing bacteria may have a direct effect on the reduction of foodborne pathogens, and additionally, the colonization of the intestinal tract with these beneficial bacteria might prevent the reintroduction of pathogenic bacteria (Brashears et al. 2003; Diez-Gonzalez 2007). The beneficial effect of probiotic microorganisms in both animal health and production of food for human consumption is known. One factor clearly associated with this effect is the production of substances with inhibitory capacity against pathogenic bacteria by probiotic microorganisms.

The prebiotics are carbohydrates or other organic compounds which are indigestible by animal enzymes and they are not hydrolyzed by gut acids or absorbed in the upper gastrointestinal tract but they are digested by beneficial microbes (Walker and Duffy 1998). It has been found that prebiotic populations increase bifidobacteria and lactobacilli in the intestine of young pigs (Smiricky-Tjardes et al. 2003). To be considered a prebiotic, a compound must conform to the following guidelines: (a) it must be resistant to digestion in the upper gastrointestinal tract (remain unaltered through hydrolytic enzymatic digestion), (b) it must selectively stimulate one or a limited number of beneficial bacteria integrating intestinal microbiota, and (c) it must benefit host health by improving colonic microbiota composition (Morrison et al. 2010). Two popular oligosaccharides used in domestic livestock are mannan-oligosaccharides (MOS) and fructo-oligosaccharides (FOS) (Verdonk et al. 2005; Hill et al. 2008; Morrison et al. 2010). An additional benefit of prebiotic treatment is that some bacterial species that have a competitive advantage can produce antimicrobial substances (e.g., bacteriocins, colicins) which may directly inhibit pathogenic bacteria. A further consideration related to the use of probiotics in ruminants is that probiotics must be able to prevent degradation by ruminal microbiota and thus require specific strategies adapted to allow sufficient amounts reach the intestine of ruminants. The combined use of probiotics with prebiotics (known as symbiotic) may produce a synergistic effect in reducing populations of foodborne pathogens in food animals before slaughter (Callaway et al. 2003). The use of prebiotics has been increased as an alternative therapy that prevents the use of antibiotics and, thus, reduces the emergence and spread of antibiotic-resistant bacteria and residual antibiotics in dairy foods, meat, and milk (Hill et al. 2008). Inversely, a consensus has not been reached as to whether prebiotics may be effective in improving animals' growth performance or reducing the prevalence of gastrointestinal diseases in young calves.

## 6.4 Probiotic and Prebiotic Strategy for Intensive Production Farms in Cattle Rearing

Several pathogens, alone or most often in combination with other pathogens, are etiologic agents of diarrhea in young calves. Most of these agents are predominantly transmitted by the fecal-oral route from the feces of infected animals to the mouths of susceptible animals (Barrington et al. 2002). It is very important to reduce the prevalence of gastrointestinal infections in young calves because when animals are sick at this stage, their subsequent growth is delayed, thus affecting their productivity (Rosmini et al. 2004). The incidence of intestinal disease is especially high in intensive rearing systems, where exposure to pathogens is increased due to the confinement of large numbers of animals in small spaces. Lactic acid bacteria (LAB) are natural components of the normal intestinal microbiota in both humans and animals (Schneider et al. 2004) and have been used to control the deleterious effects of gastrointestinal pathogens.

A meta-analysis was conducted (Signorini et al. 2012) with the aim to assess the effect of probiotics on diarrhea incidence and the intestinal microbial balance. LAB supplementation has been shown to exert a protective effect and to reduce the incidence of diarrhea (Relative Risk, RR = 0.437). However, this positive effect was not observed in all the conditions. Considering the feed consumed by calves, the protective effect of the probiotics against diarrhea was observed only in trials that used whole milk (Fig. 6.1). In the same way, the probiotic effect was observed in those trials which used multistrain inocula (Fig. 6.2).

Additionally, the animals fed with probiotics improved the consistency of feces (low level of fecal consistency) in comparison with animals without probiotic supplementation, but this difference was not statistically significant (SMD = −0.4904, 95% CI −1.011 to 0.035) (Fig. 6.3). The relative risk probability of significant effects (probiotic positive effect) was 0.731 for fecal consistency. Many of the problems that affect the growth performance of young calves are related to low digestion and reduced absorption of nutrients due to colonization of pathogenic bacteria. However, nutritional diarrhea often precedes and predisposes the calf diarrhea syndrome caused by pathogenic microorganisms. In these cases, the use

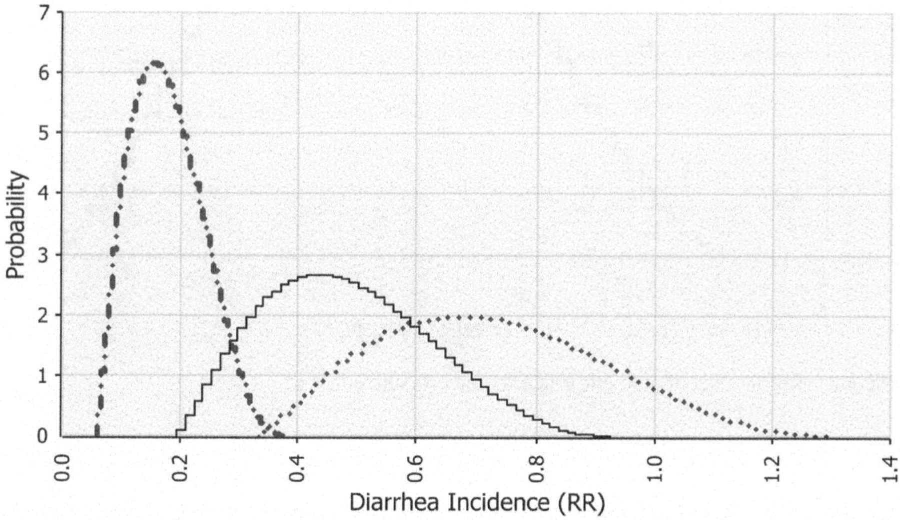


**Fig. 6.1** Global effect (solid line) and subanalysis comparison of probiotic efficacy on diarrhea incidence considering the feed: whole milk (cut line) or milk replacer (dash line)

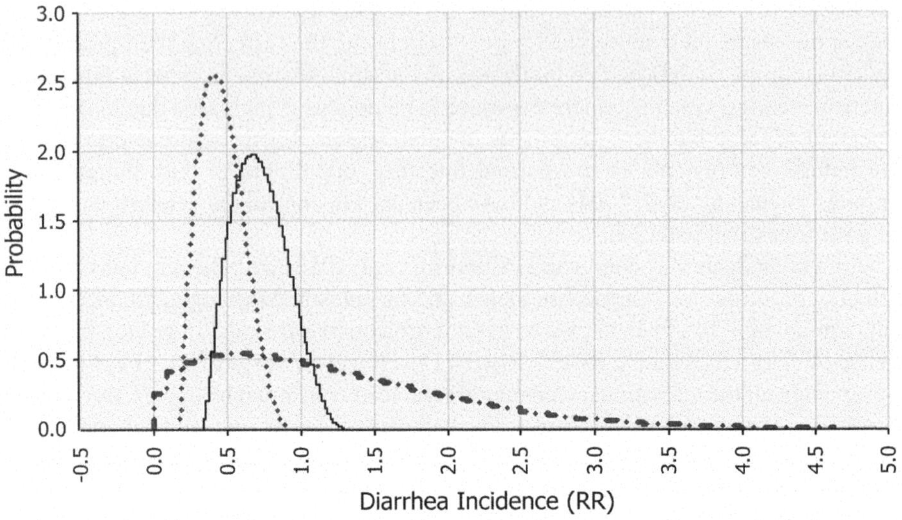


**Fig. 6.2** Global effect (solid line) and subanalysis comparison of probiotic efficacy on diarrhea incidence considering the characteristic of the inocula: monostrain inocula (cut line) or multistrain inocula (dash line)

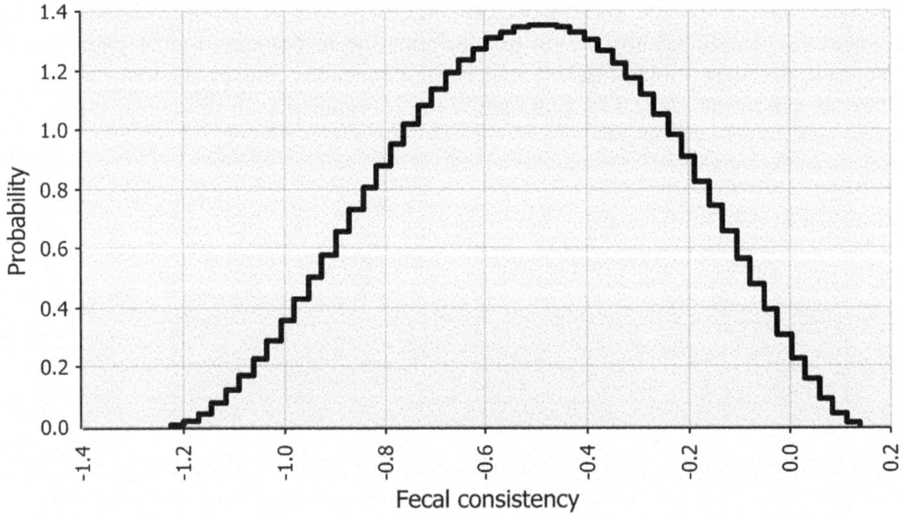


**Fig. 6.3** Global effect of probiotic efficacy on fecal consistency

of probiotics aims to prevent the diarrhea (Signorini et al. 2012). Additionally, the probiotic microorganism performance may vary from one animal to another of the same species, and for that reason, some authors (Gardiner et al. 2004; Timmerman et al. 2004) have recommended the administration of an inoculum formed by a mixture of different strains.

Lactic acid bacteria inoculum was used to evaluate a level protection capacity in calves with or without lactose supplements against *Salmonella* Dublin infection by evaluating histopathological lesions and pathogen translocation (Frizzo et al. 2012). In probiotic group animals, lesions observed during the entire series of necropsies were less severe. Microscopic lesions observed in the probiotic group after 80 h of *Salmonella* inoculation were similar to those found in the control group at 32 and 56 h post-infection. Paratyphoid nodules appeared later in groups treated with the LAB inoculum than in the control animals. This should have correlated with the lesions found in the jejunum and ileum, areas into which the nodes drain. Probiotic treatment was unable to delay the arrival of the pathogen to target organs. However, it was evident that the inoculum altered the response of the animals to pathogen attack because the severity of *Salmonella* infection was reduced and milder microscopic lesions developed in the group treated with lactose and lactic acid bacteria (Frizzo et al. 2012). *Salmonella* infection model used in those studies was established taking into account this information and the fact that inoculation of $2 \times 10^{10}$ cfu/animal produced a regular and homogeneous model of salmonellosis but demanded a high-level performance from the probiotic to counteract *Salmonella* deleterious effects. Bacterial translocation is a useful indicator to evaluate probiotic safety level (Locascio et al. 2001), because it is the first step in bacterial pathogenesis of many bacterial deleterious indigenous strains (Berg 1995). The ability to translocate could indicate possible probiotic infectivity (Zhou et al. 2000). Despite the high concentration of Salmonella administered to calves, the LAB inoculum of bovine origin was not capable of translocating to the internal organs in the extreme situations of intestinal imbalance generated by the pathogen (Frizzo et al. 2011b).

Growth performance of young calves is strongly related to the type of feed which they consume, the rearing system, and the intestinal microbiota balance. Probiotics may prevent intestinal microbial imbalances which are common in intensive rearing systems to reduce the incidence of disease. However, growth performance parameters are more sensitive than the health status parameters to assess the beneficial effect of probiotics applied to the calves' diet. A possible explanation might be the incidence of subclinical gastrointestinal diseases that can be detected only by a reduction in growth performance (Frizzo et al. 2010b). A meta-analysis was conducted (Frizzo et al. 2011c) with the aim to assess effects of probiotic supplementation on growth performance (e.g., body weight gain (BWG), feed efficiency) in young calves. In general, probiotic administration showed a beneficial effect on body weight gain (approximately 228 g/day) and on feed efficiency (814 g less feed consumed/kg of body weight gain) (Fig. 6.4). Different mechanisms of action of probiotics have been described (Fuller 1989; Blum et al. 1999): probiotics compete for nutrients and produce antibacterial compounds (e.g., organic acids, hydrogen peroxide, bacteriocins) in the intestinal lumen allowing them to occupy specific niches of the intestinal mucosa and activate the innate immune system of calves. The involvement of each of these mechanisms is directly related to the type of probiotic strain and feed consumed by the calves. The probiotic effect is more evident during the first few weeks of life, and this was especially

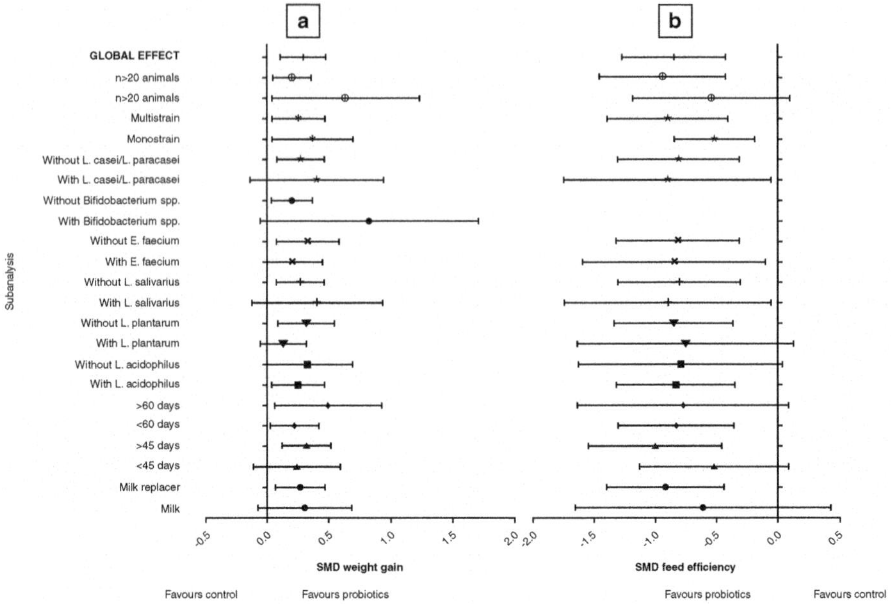


**Fig. 6.4** Summary of measures for subanalysis comparison of probiotic efficacy on weight gain (**a**) and feed efficiency (**b**) depending on type of feed, experiment time, bacterial species used, type of inoculum, and number of animals

clear in feed efficiency. Timmerman et al. (2005) report a clear increase in body weight gain in 1-week-old veal calves supplemented with probiotics but limited beneficial effects during the first 2 weeks of life.

Probiotic function may be related to an improvement in feed efficiency, especially in diets containing a high proportion of dry matter as grain and forage (Frizzo et al. 2010b), which has positive effect on ruminal development. An improvement in growth during this stage has a large impact on performance in subsequent rearing. Use of milk replacer and feed concentrates during the first few weeks of life may predispose calves to nutritional diarrheas and increase animal stress. The probiotic efficacy was feed-related because the positive effect only occurred when calves were fed milk replacer (Frizzo et al. 2011c). This improvement in performance produced by probiotics could help to improve production and economic indices of farms.

The meta-analysis study allowed to identify some characteristics of the experimental designs that have favored the expression of probiotic effect in animals. These aspects would allow establishing guidelines that could be adopted to standardize the experimental designs of trials conducted to assess the probiotic effect in calves in the future, which should be added to the basic premises reported by other authors for the use of probiotics in general (Fuller 1989, 2006; FAO/WHO 2001). Some of these rules may be (1) control the use of milk replacer during rearing (not ad libitum) of calves, trying to encourage early intake of starters; (2) there are more chances of finding beneficial effects on health indicators, designing experimental models that induce nutritional diarrhea (e.g., introduce some stressful substance to the diet such as lactose); (3) emphasize the assessment of growth performance during the first weeks of life, designing trials no longer than 6–9 weeks and because during this period there is the highest incidence of diarrhea; (4) trials longer than 9 weeks could be cumbersome and unnecessary to assess the probiotic effect but can be very useful to verify if the effect is maintained in adult cattle (this may be necessary in trials designed to analyze the capacity of a probiotic to control the spread of a foodborne pathogen prior to slaughter); (5) studies designed with more than 20 animals have more opportunities to find probiotic effects on feed efficiency than trials with lower number of animals; and (6) growth performance can be analyzed using both monostrain and multistrain inoculum. The application of these rules in the experimental designs can reduce the number of animals required and can also maintain a suitable model to measure the effects related to growth performance in young calves.

Special emphasis will be in the following paragraphs to highlight the importance of the viability of probiotics used. Satisfactory and sufficient amount of viable microorganisms upon administration to the animal is strictly necessary for the probiotic microorganisms to express their positive effects. These requirements lead to emphasize both the production of large amounts of biomass and in maintaining their viability.

Regarding the industrial production of probiotics, it is important to select the growth media considering the following factors: cost and ability to produce a large number of cells with high activity and allow the separation of microorganisms from the growth matrix. Bacterial growth requires appropriate culture media that provide the nutrients required by the microorganism for its development. There is an important variety of alternative raw materials that are commercially available and can be used as nutrients for large-scale fermentations, such as agricultural and industrial by-products and waste materials such as waste from sugarcane (Apás et al. 2008) or cheese whey (Rodrigues et al. 2006), which when they are enriched with protein hydrolysates can promote bacterial growth (Kwon et al. 2000; Fitzpatrick and O'Keeffe 2001; Drago et al. 2006; Soto et al. 2006).

Regarding the probiotic administration, the microorganisms should be administered periodically due to microorganisms which tend to leave the gastrointestinal tract if they are not frequently consumed (Abe et al. 1995). For that reason the probiotics for humans are included into feeds, and thus the consumption thereof is facilitated since they do not have to be administered as a dietary supplement but are ingested together with feed daily. This also has the advantage that some feeds are used as carriers, such as milk or fermented cheese which are stored under refrigeration, preserving the viability of the probiotic until the consumption.

In the case of probiotics for animals, there are no feeds with probiotics, but pharmaceutical formulations for individual administration: pasta, pills, capsules, powders, and granules. These products are suitable for the transport of probiotics to the farms, although in many cases there are no scientific supports to ensure a dose of probiotic strains capable of exerting their effects (Soto et al. 2009). For that reason, in most of the scientific studies, the probiotics are produced in the laboratory and preserved by freezing (Pérez Guerra et al. 2007; Adams et al. 2008) or refrigeration (Abu-Tarboush et al. 1996: Timmerman et al. 2005; Casey et al. 2007) with good results in terms of maintaining the viability of the bacteria, but in general, with little relevance on the farm. Based on these results, different researchers have evaluated the use of probiotic on calves and chickens by direct oral administration (Adams et al. 2008; Santini et al. 2010). Few authors have studied other methods of administration that can be brought to the farm, such as the study conducted by Pascual et al. (1999), who incorporated the probiotic in the feed during processing with the aim to bring it directly to the farm in the feed.

As described above, the conditions of preservation and transport of probiotics are closely related to the methodology of administration to animals on the farm. Therefore, it is essential to know what the real possibilities of administration in the farms are. One possibility is to incorporate the inoculum in the feed or drinking water at the point of consumption, or the other option is to incorporate the inoculum during the feed preparation which is the probiotic carrier. The latter option has the advantage that the probiotic does not need an extra handling, simplifying the operator's work at the farm, and prevents dosage errors. However, the disadvantage is that the probiotic has to adapt to the feed matrix. This requires that if the feed is powder, such as milk replacer for calves and pigs, probiotics should be dried by the freeze- or spray-drying. Another method used to protect the microorganisms against the environment hostilities is the encapsulation. Entrapment of bacteria in macrocapsules with feed pelleting size may be a solution to the addition of probiotics and at the same time protect the microorganisms against the adverse effects of the environment during storage, while the probiotics can be protected during passage through the gastrointestinal tract. Macrocapsule formation has been developed with the addition of polymers such as starch and alginate (Soto et al. 2011). The capsule has shown probiotic protection at room temperature. Other methods have also been developed for capsule formation from biomass suspended with whey (Weinbreck et al. 2010) and gelatin (Saxelin et al. 1995).

As it is important to induce the probiotic effect at farm level, it is necessary to develop an economical growth media, to ensure the viability of the inoculum during storage and industrialization stages, and to enable administration to animals in the field conditions. This approach will encourage the development of methodologies to generate inoculant probiotics for farm animals that may be added as a supplement in feeding, which allows the transfer of this technology to the industrial scale, allowing trade of this product and facilitating administration of the appropriate dose to the animals.

Encapsulation is currently being implemented to maintain the viability of probiotics. This consists of retaining the microorganisms within a porous gel matrix or within a semipermeable membrane containing a liquid core (Dembezynski and Jankowski 2002). Coating increases survival of the cells by protecting them from the adverse effects of the surrounding environment (Doleyres and Lacroix 2005) and protects bacteria from damage by subsequent processes such as drying of the microcapsules for storage at ambient temperatures (Champagne and Gardner 2001).

Microencapsulation provides advantages such as higher resistance to simulated gastric and intestinal conditions (Lian et al. 2003), biomass protection against possible contaminants (especially bacteriophages), and a decrease in production costs because of separation techniques such as centrifugation and filtration which are not necessary to concentrate the bacteria in the culture medium (Dembezynski and Jankowski 2002). Microcapsules also offer protection against oxygen for strict anaerobes.

When this immobilization technology is applied to bacteria added to food for humans, the disadvantage of producing a pearl size small enough to be imperceptible to the palate must be overcome. However, animal probiotic formulations, such as capsules, pills, and granules (O'Mahony et al. 2009; Soto et al. 2009), have the most appropriate size for the animal to be inoculated. Production of macrocapsules of a size similar to that of the feed starter pellet may allow bacterial preservation and may facilitate its administration to calves with feed.

Methods of freeze- or spray-drying of probiotics added to food are not the best option for viability preservation because direct contact of microorganisms with the product diminishes its bacterial counts. Encapsulation is an alternative to solve this problem (Muthukumarasamy et al. 2006) and has the additional advantage of the protection provided by the capsule to gastric conditions (Picot and Lacroix 2004). Due to the macrocapsules which have less surface contact than microcapsules per unit weight, they would protect bacteria against gastric conditions more efficiently. This is related to the physical barrier and the increased distance of the bacteria with the external environment (Lee and Heo 2000; Muthukumarasamy et al. 2006).

Environmental sensitivity of some probiotic strains frequently limits their practical use in non-refrigerated food and pharmaceutical supplements. In this sense, encapsulation may improve the viability of the strains during storage (Crittenden et al. 2006). Soto et al. (2011) coated and dried the capsules which kept their SML for 21 days at 18 °C and obtained a final product with a size similar to a feed starter pellet. This would allow it to be mixed homogeneously with the feed, thereby maintaining the bacterial viability necessary to exercise their probiotic effect during that period. Another possibility is to maintain probiotic capsules separate from the feed starter under refrigeration. In this way, probiotics may have an expiration time of at least 2 months and may be combined with the starter at the time it is being fed to the calves. This option also allows addition of capsules to milk or milk replacers for those calves that do not eat starter (Soto et al. 2011).

Prebiotic strategy for intensive production farms is an interesting tool that can be used to replace antibiotics and improve growth performance and health status of the animals. Oligosaccharides have been proposed as a means to manipulate the bacterial microbiota of the intestinal tract of animals, potentially reducing the incidence of disease. Furthermore, prebiotics have also been proposed as a strategy to reduce bacterial foodborne pathogens in food animals before slaughter (Abu-Tarboush et al. 1996), although the combined use of probiotics with prebiotics (known as symbiotic) would be more advisable because their synergistic effect could have a major impact. The potential role of prebiotics in improving the health and performance of calves is of increasing interest because of the public concern about the use of antimicrobials in cattle production (Terré et al. 2007). In the colon, prebiotics are readily fermented by the intestinal microflora. This may result in changes in the population of beneficial microorganisms while repressing the number of potential harmful bacteria. In addition, the production of volatile fatty acids by bacteria-fermenting prebiotics in animals may improve energy efficiency and alter intestinal morphology (Morrison et al. 2010).

A meta-analysis was carried out to assess the effect of prebiotic supplementation on growth performance (body weight gain (BWG) and feed efficiency (FE)) and on health status (fecal consistency index (FCI) and days with diarrhea (DD)) of young calves. Prebiotic supplementation did not show an increase in the BWG (standardized mean differences (SMD) = 0.0410, 95% confidence interval (CI) −0.1359 to 0.2180) and was not able to improve FE (SMD = −0.0656, 95% CI −0.3417 to 0.2105). Prebiotic supplementation was not able to improve FCI (SMD = −0.1403, 95% CI −0.3473 to 0.0668) and DD (SMD = 0.0630, 95% CI −0.3495 to 0.4755). Pooled estimates found that prebiotics added to milk replacer or whole milk were not successful in increasing BWG, FE, FCI, or DD in young calves.

Meta-analysis demonstrated that the prebiotics were unable to improve the daily gain and feed efficiency during rearing of young calves. Considering the fact that prebiotics have not been able to improve the growth performance, its use as a supplement for the production of food animals could be questioned. In intensive systems, improvement in these indicators is crucial to facilitate the implementation and use of probiotics in a mass and extended way over time. Another daunting issue is directly related to the characteristics of prebiotic substances, which, although not rare in nature, require purification systems to administer it to animals in appropriate concentrations.

Use of prebiotics can reduce the adhesion of certain bacterial species to the intestinal epithelium and can either prevent this imbalance in the intestinal tract, and stop the occurrence of diarrhea cases (Callaway et al. 2003), or reduce its prevalence in calves (Morrison et al. 2010). Although in this meta-analysis nonbeneficial effects of prebiotics on diarrhea and fecal consistency index were found, the results are inconclusive, and more trials, with specific experimental designs, should be conducted. Hill et al. (2008) suggested that prebiotics have potential only as preventatives to diarrhea when they are first administered to the healthy animal, and specifically they stated that there are no studies showing successful treatment of diarrhea using prebiotics.

No effects of prebiotics on health have been found in the meta-analysis. However, it is important to emphasize that only a limited number of studies have been conducted to evaluate the prebiotic effects on health. It is striking that, since the effect of prebiotics is produced directly on the benefit of indigenous microbial populations, these studies did not assess jointly health and the intestinal microbiota. The major advantage of prebiotics is a proliferative effect on native beneficial intestinal bacteria. Certain gut microflora have been shown to have positive effects on the whole body, including improved weight gain and immune function and decreased presence of pathogens (Morrison et al. 2010). The main factors that can explain the observed differences among the meta-analysis studies are related with the health status, the level of stress suffered by calves, and the exposure to intestinal pathogens during the rearing. The beneficial effect, due to lactic acid bacteria supplementation (e.g., growth performance, health, fecal microflora), can be detected more easily in farms that present high morbidity and mortality rates caused mainly by intestinal pathogens. A similar situation could occur with the supplementation with prebiotics. To improve the detection of prebiotic effects on the calves' health status, trials using experimental models with pathogens should be conducted. Trials should be designed with the aim to assess the prebiotic effects as a prophylactic tool to protect the young calves against the colonization of the digestive tract, stimulating development of the immune system and counteracting negative effects of such disease.

The wide variety of experimental designs found in the meta-analysis studies with prebiotics applied to calves acts as a source of heterogeneity in the meta-analysis and may reduce the consistency of the conclusions. Thus, meta-analysis has allowed identifying some characteristics of the experimental designs that have favored the expression of prebiotic effect in animals. These aspects would allow establishing guidelines that could be adopted to standardize the experimental designs of trials conducted to assess the prebiotic effect in calves in the future, which should be added to the basic premises reported by other authors for the use of prebiotics in general (Verdonk et al. 2005; Frizzo et al. 2010b). Some of these rules may be:

1. To combine the administration of prebiotics with probiotics with the aim to determine whether the incorporation of prebiotics according to the level of synergy achieved is appropriate. In these trials the suggested guidelines for trials with probiotics (Nargeskhani et al. 2010; Hill et al. 2008) should be taken into account.
2. The trials conducted to analyze the effect on health should include studies on the evolution of intestinal microbiota (beneficial and pathogens microorganisms).
3. Trials should incorporate intestinal health indicators (parameters related to the immune system and local response in the intestine).
4. There are more chances of finding beneficial effects on health indicators, designing experimental models that induce nutritional diarrhea.
5. Trials designed as experimental disease model can be useful to assess the prebiotic's effectiveness against specific pathogens.

The application of these rules in the experimental designs can reduce the number of animals required and can also maintain a suitable model to measure the effects related to growth performance and health in young calves.

## 6.5 Vision for the Future

The ban on the use of antibiotics in animal diet is extremely necessary to prevent resistance problems that may have serious consequences for public health. The increase in world population and the reduction of agricultural production areas will require the intensification of production systems. Therefore, it will be essential to have tools as alternatives to the antibiotics used in the human medicine to help manage animal and zoonotic pathogens and thus improve animal and public health. Probiotics and prebiotics can be used as effective intervention measures in primary production becoming crucial elements to reduce the spread of antibiotic-resistant bacteria.

## References

Abe F, Ishibashi N, Shimamura S (1995) Effect of administration of Bifidobacteria and lactic acid bacteria to newborn calves and piglets. J Dairy Sci 78:2838–2846

Abdala AA, Zimmerman G, Calvinho LF et al (2001) Evaluación de la eficacia de un probiótico incorporado a un sustituto lácteo y a leche entera, en la crianza de terneros. Rev Med Vet 83:196–198

Abu-Tarboush HM, Al-Saiady MY, Keir El-Din AH (1996) Evaluation of diet containing Lactobacilli on performance, fecal coliform, and Lactobacilli of young milk calves. Anim Feed Sci Technol 57:39–49

Adams MC, Luoa J, Rayward D et al (2008) Selection of a novel direct-fed microbial to enhance weight gain in intensively reared calves. Anim Feed Sci Technol 145:41–52

Apás AL, Arena ME, Draksler D et al (2008) Utilization of sugarcane industrial residues as animal food and probiotic medium. J Biosci Bioeng 106:363–367

Barrington GM, Gay JM, Evermann JF (2002) Biosecurity for neonatal gastrointestinal diseases. Vet Clin Food Anim Pract 18:7–34

Berg RD (1995) Bacterial translocation from the gastrointestinal tract. Trends Microbiol 3:149–154

Blum S, Álvarez S, Haller D et al (1999) Intestinal microbiota and the interaction with immunocompetent cells. Anton Leeuw 76:199–205

Brashears MM, Galyean ML, Loneragan GH et al (2003) Prevalence of *Escherichia coli* O157:H7 and performance by beef feedlot cattle given *Lactobacillus* direct-fed microbials. J Food Protect 66:748–754

Callaway TR, Anderson RC, Edrington TS et al (2003) Preslaughter intervention strategies to reduce food-borne pathogens in food animals. J Anim Sci 81(Suppl. 2):E17–E23

Casey PG, Gardiner GE, Casey G et al (2007) A five-strain probiotic combination reduces pathogen shedding and alleviates disease signs in pigs challenged with *Salmonella enterica* Serovar Typhimurium. Appl Environ Microbiol 73:1858–1863

Champagne CP, Gardner NJ (2001) The effect of protective ingredients on the survival of immobilized cells of *Streptococcus thermophilus* to air and freeze-drying. Electron J Biotechnol 3:146–152

Collier CT, Smiricky-Tjardes MR, Albin DM et al (2003) Molecular ecological analysis of porcine ileal microbiota responses to antimicrobial growth promoters. J Anim Sci 81:3035–3045

Crittenden R, Weerakkody R, Sanguansri L et al (2006) Synbiotic microcapsules that enhance microbial viability during nonrefrigerated storage and gastrointestinal transit. Appl Environ Microbiol 72:2280–2282

Daeschel MA (1993) Applications and interactions of bacteriocin from lactic acid bacteria in food and beverages. In: Hoover DG, Stenson LR (eds) Bacteriocins of lactic acid bacteria. Academic, Nueva York, pp 63–70

Dembezynski R, Jankowski T (2002) Growth characteristics and acidifying activity of *Lactobacillus rhamnosus* in alginate/starch liquid-core capsules. Enzyme Microb Technol 31:111–115

Diez-Gonzalez F (2007) Applications of bacteriocins in livestock. Curr Issues Intest Microbiol 8:15–24

Doleyres Y, Lacroix C (2005) Technologies with free and immobilised cells for probiotic bifidobacteria production and protection. Int Dairy J 15:973–988

Drago S, Soto LP, Frizzo LS et al (2006) Efecto de plasma bovino hidrolizado sobre el crecimiento de *Lactobacillus salivarius* DSPV 315T. II Simposio Internacional de Bacterias Lácticas. Primer encuentro Red BAL Argentina, San Miguel de Tucumán, 11 al 13 de octubre. p 67

Dunne C, O'Mahony L, Murphy L et al (2001) *In vitro* selection criteria for probiotic bacteria of human origin: correlation with in vivo findings. Am J Clin Nutr 73(suppl):386S–392S

FAO/WHO (2001) Health and nutritional properties of probiotics in food including powder milk with live lactic acid bacteria. Report of a Joint FAO/WHO Expert Consultation. Food and agriculture organization of the United Nations and World Health Organization. Córdoba, Argentina

Fitzpatrick JJ, O'Keeffe U (2001) Influence of whey protein hydrolysate addition to whey permeate batch fermentations for producing lactic acid. Process Biochem 37:183–186

Frizzo LS, Peralta C, Zbrun V et al (2005) Respuesta de ratones inoculados con bacterias lácticas de origen bovino a un desafío con *Salmonella dublin*. FAVE Secc Cienc Vet 4:41–53

Frizzo LS, Soto LP, Bertozzi E et al (2006) Evaluación *in vitro* de las capacidades probióticas microbianas orientadas al diseño de inóculos probióticos multiespecie para ser utilizados en la crianza de terneros. FAVE Secc Cienc Vet 5:69–81

Frizzo LS, Bertozzi E, Soto LP et al (2010a) Studies on translocation, scute oral toxicity and intestinal colonization of potentially probiotic lactic acid bacteria administered during calf rearing. Livest Sci 128:28–35

Frizzo LS, Soto LP, Zbrun MV et al (2010b) Lactic acid bacteria to improve growth performance in young calves fed milk replacer and spray-dried whey powder. Anim Feed Sci Technol 157:159–167

Frizzo LS, Soto LP, Zbrun MV et al (2011a) Effect of lactic acid bacteria and lactose on growth performance and intestinal microbial balance of artificially reared calves. Livest Sci 140:246–252

Frizzo LS, Soto LP, Bertozzi E et al (2011b) Intestinal populations of Lactobacilli and coliforms after *in vivo Salmonella dublin* challenge and their relationship with microbial translocation in calves supplemented with lactic acid bacteria and lactose. Anim Feed Sci Technol 170:12–20

Frizzo LS, Zbrun MV, Soto LP et al (2011c) Effect of probiotics on the growth performance in young calves: a meta-analysis of randomized controlled trials. Anim Feed Sci Technol 169:147–156

Frizzo LS, Zbrun MV, Soto LP et al (2012) Pathogen translocation and histopathological lesions in an experimental model of *Salmonella* Dublin infection in calves receiving lactic acid bacteria and lactose supplements. J Vet Sci 13(3):261–270

Frizzo LS, Astesana DM, Soto LP et al (2013) La seguridad en la cadena agroalimentaria de la carne: problemáticas, estrategias y posibles soluciones pre-faena. Revista Fave Ciencias Agrarias 12:7–26

Fuller R (1989) Probiotics in man and animals. J Appl Bacteriol 66:365–378

Fuller R (1992) History and development of probiotics. In: Fuller R (ed) Probiotics: the scientific basis. Chapman & Hall, London, pp 1–8

Fuller R (2006) Reasons for the apparent variation in the probiotic response. Biologia 61:751–754

Gardiner GE, Casey PG, Casey G et al (2004) Relative ability of orally administered *Lactobacillus murinus* to predominate and persist in the porcine gastrointestinal tract. Appl Environ Microbiol 70:1895–1906

Gill HS, Shu Q, Lin H et al (2001) Protection against translocating *Salmonella typhimurium* infection in mice by feeding the immuno-enhancing probiotic *Lactobacillus rhamnosus* strain HN001. Med Microbiol Immunol 190:97–104

Hill TM, Bateman HG II, Aldrich JM et al (2008) Oligosaccharides for dairy calves. Prof Anim Sci 24:460–464

Hill C, Guarner F, Reid G et al (2014) Expert consensus document: the International Scientific Association for Probiotics and Prebiotics consensus statement on the scope and appropriate use of the term probiotic. Nat Rev Gastroenterol Hepatol 11:506–514

James RE, McGilliard ML, Hartman DA (1984) Calf mortality in Virginia dairy herd improvements herds. J Dairy Sci 67:908–918

Kurzak P, Ehrmann MA, Vogel R (1998) Diversity of lactic acid bacteria associated with ducks. Syst Appl Microbiol 21:588–592

Klaenhammer TR (1988) Bacteriocins of lactic acid bacteria. Biochimie 70:337–349

Kwon S, Lee PC, Lee EG et al (2000) Production of lactic acid by *Lactobacillus rhamnosus* with vitamin-supplemented soybean hydrolizate. Enzym Microb Technol 26:209–215

Lee K, Heo T (2000) Survival of *Bifidobacterium longum* immobilized in calcium alginate beads in simulated gastric juices and bile salt solution. Appl Environ Microbiol 66:869–873

Lian W, Hsiao H, Chou C (2003) Viability of microencapsulated bifidobacteria in simulated gastric juice and bile solution. J Food Microbiol 86:293–301
Locascio M, de Ruiz Holgado AP, Perdigón G et al (2001) Enteric bifidobacteria: isolation from human infants and challenge studies in mice. Can J Microbiol 47:1048–1052
Maia OB, Duarte R, Silva AM et al (2001) Evaluation of the components of a commercial probiotic in gnotobiotic mice experimentally challenged with *Salmonella enterica* subsp. *enterica* ser. Typhimurium. Vet Microbiol 79:183–189
Mallo JJ, Rioperez J, Honrubia P (2010) The addition of *Enterococcus faecium* to diet improves piglet's intestinal microbiota and performance. Livest Sci 133:176–178
McEwen SA, Fedorka-Cray JF (2002) Antimicrobial use and resistance in animals. Clin Infect Dis 34(Suppl 3):S93–S106
Melin L, Jensen-Waern M, Johannisson A et al (1997) Development of selected faecal microfloras and of phagocytic and killing capacity of neutrophils in young pigs. Vet Microbiol 54:287–300
Millemann Y (2009) Diagnosis of neonatal calf diarrhea. Revue Méd Vét 160(8-9):404–409
Morrill JL, Morrill ZJM, Feyerherm AM et al (1995) Plasma proteins and a probiotic as ingredients in milk replacer. J Dairy Sci 78:902–907
Morrison SJ, Dawson S, Carson AF (2010) The effects of mannan oligosaccharide and *Streptococcus faecium* addition to milk replacer on calf health and performance. Livest Sci 131:292–296
Moura LN, Neumann E, Vieira LQ et al (2001) Protection by *Lactobacillus acidophilus* UFV-H2B20 against experimental oral infection with *Salmonella enterica* subsp. *enterica* ser. Typhimurium in gnotobiotic and conventional mice. Braz J Microbiol 32:66–69
Mulder RW, Havenaar R, Huis in 't Veld JH (1997) Intervention strategies: the use of probiotics and competitive exclusion microbiotas against contamination with pathogens in pigs and poultry. In: Fuller R (ed) Probiotics: 2. Application and practical aspects. Chapman & Hall, London, pp 187–207
Muthukumarasamy P, Wojtas PA, Holley RA (2006) Stability of *Lactobacillus reuteri* in different types of microcapsules. J Food Sci 71:20–24
Nargeskhani A, Dabiri N, Esmaeilkhanian S et al (2010) Effects of mannanoligosaccharide-β Glucan or antibiotics on health and performance of dairy calves. Anim Nutr Feed Technol 10:29–36
Nousiainen J, Setälä J (1998) Lactic acid bacteria as animal probiotics. In: Salminem S, Von Wright A (eds) Lactic acid bacteria: microbiology and functional aspects. Marcel Dekker, New York, pp 437–473
O'Mahony D, Barry Murphy K, MacSharry J et al (2009) Portrait of a canine probiotic Bifidobacterium-from gut to gut. Vet Microbiol 139:106–112
Parker D (1990) Manipulation of the functional activity of gut by dietary and other means (antibiotics/probiotics). J Nutr 60:639–648
Pascual M, Hugas M, Badiola JI et al (1999) *Lactobacillus salivarius* CTC2197 prevents *Salmonella enteritidis* colonization in chickens. Appl Environ Microbiol 65:4981–4986
Pérez Guerra N, Fajardo Bernárdez P, Méndez J et al (2007) Production of four potentially probiotic lactic acid bacteria and their evaluation as feed additives for weaned piglets. Anim Feed Sci Tech 134:89–107
Picot A, Lacroix C (2004) Encapsulation of bifidobacteria in whey protein-based microcapsules and survival in simulated gastrointestinal conditions and in yoghurt. Int Dairy J 14:505–515
Reid G, Friendship R (2002) Alternatives to antibiotic use: probiotics for the gut. Anim Biotechnol 13:97–112
Remiger A, Eijsink VGH, Ehrmann MA et al (1999) Purification and partial amino acid sequence of plantaricin 1.25 α and 1.25 β two bacteriocins produced by *Lactobacillus plantarum* TMW 1.25. J Appl Microbiol 86:1053–1058
Rodríguez Armesto R, Peralta C, Ochoteco M et al (1996) Salmonelosis septicémica en terneros lactantes: nueva presentación para una vieja enfermedad. Primera parte. Revista Therios 25:251–260
Rodrigues LR, Teixeira JA, Oliveira R (2006) Low-cost fermentative medium for biosurfactant production by probiotic bacteria. Biochem Eng J 32:135–142

Rosmini MR, Sequeira GJ, Guerrero-Legarreta I et al (2004) Producción de probióticos para animales de abasto: importancia del uso de la microbiota intestinal indígena. Rev Mex Ing Quím 3:181–191

Ross R, Galvin M, McAuliffe O et al (1999) Developing applications for lactococcal bacteriocins. Anton Leeuw 76:337–346

Ross GR, Gusils C, Oliszewski R et al (2010) Effects of probiotic administration in swine. J Biosci Bioeng (6):545–549

Saarela M, Mogensen G, Fonden R et al (2000) Probiotic bacteria: safety functional and technological properties. J Biotechnol 84:197–215

Salminen S, Deighton MA, Benno Y et al (1998) Lactic acid bacteria in health and disease. In: Salminen S, Von Wright A (eds) Lactic acid bacteria: microbiology and functional aspect. Marcel Dekker, New York, pp 211–253

Santini C, Baffoni L, Gaggia F et al (2010) Characterization of probiotic strains: an application as feed additives in poultry against *Campylobacter jejuni*. Int J Food Microbiol 141:S98–S108

Savage DC (1987) Factors influencing biocontrol of bacterial pathogens in the intestine. Food Technol 41:82–87

Saxelin M, Pessi T, Salminen S (1995) Fecal recovery following oral administration of *Lactobacillus* strain GG (ATCC 53103) in gelatine capsules to healthy volunteers. Int J Food Microbiol 25:199–203

Schillinger U, Lucke FK (1989) Antimicrobial activity of *Lactobacillus sake* isolated from meat. Appl Environ Microbiol 55:1901–1906

Schneider R, Rosmini MR, Hermann M et al (2004) Identificación de bacterias lácticas componentes de la microbiota típica de los terneros criados en condiciones artificiales. Revista FAVE Ciencias Vet 3:7–15

Schwarz S, Kehrenberg C, Walsh TR (2001) Use of antimicrobial agents in veterinary medicine and food animal production. Int J Antimicrob Agents 17:431–437

Shu Q, Gill HS (2002) Immune protection mediated by the probiotic *Lactobacillus rhamnosus* HN001 (DR20™) against *Escherichia coli* O157:H7 infection in mice. FEMS Immun Med Microbiol 34:59–64

Signorini ML, Soto LP, Zbrun MV et al (2012) Impact of probiotic administration on health and faecal microbiota in young calves: a meta-analysis of randomized controlled trials of lactic acid bacteria. Res Vet Sci 93:250–258

Smiricky-Tjardes MR, Grieshop CM, Flickinger EA et al (2003) Dietary galactooligosaccharides affect ileal and total tract nutrient digestibility, ileal and fecal bacterial concentrations, and ileal fermentative characteristics of growing pigs. J Anim Sci 81:2535–2545

Soto LP, Drago S, Frizzo LS et al (2006) Efecto de hidrolizados proteicos sobre el crecimiento de *Lactobacillus casei* DSPV 318T. II Simposio Internacional de Bacterias Lácticas. Primer encuentro Red BAL Argentina, San Miguel de Tucumán, 11 al 13 de octubre. p 131

Soto LP, Frizzo LS, Bertozzi E et al (2009) Milk evaluation as growth and cold preservation medium of a probiotic inoculum for young calves. J Anim Vet Adv 8:1353–1360

Soto LP, Frizzo LS, Avataneo E et al (2011) Design of macrocapsules to improve bacterial viability and supplementation with a probiotic for young calves. Anim Feed Sci Technol 165:176–183

Stephens TP, Stanford K, Rode LM et al (2010) Effect of a direct-fed microbial on animal performance, carcass characteristics and the shedding of *Escherichia coli* O157 by feedlot cattle. Anim Feed Sci Technol 158:65–72

Terré M, Calvo MA, Adelantado C et al (2007) Effects of mannan oligosaccharides on performance and microorganism fecal counts of calves following an enhanced-growth feeding program. Anim Feed Sci Technol 137:115–125

Timmerman HM, Koning CJM, Mulder L et al (2004) Monostrain, multistrain and multispecies probiotics—a comparison of functionality and efficacy. Int J Food Microbiol 96:219–233

Timmerman HM, Mulder L, Everts H et al (2005) Health and growth of veal calves fed milk replacers with or without probiotics. J Dairy Sci 88:2154–2165

Vassalo M, Fialho ET, Oliveira AIG et al (1997) Probióticos para leitões dos 10 aos 30 kg de peso vivo. Rev Soc Bras Zoot 26:31–138

Verdonk JMAJ, Shim SB, van Leeuwen P et al (2005) Application of inulin-type fructans in animal feed and pet food. Brit J Nut 93(Suppl. 1):S125–S138

Vinderola CG, Bailo N, Reinheimer JA (2000) Survival of probiotic microflora in Argentinian yogurts during refrigerated storage. Food Res Int 33:97–102

Walker WA, Duffy LC (1998) Diet and bacterial colonization: role of probiotics and prebiotics. J Nutr Biochem 9:668–675

Weinbreck F, Bodnár I, Marco ML (2010) Can encapsulation lengthen the shelf-life of probiotic bacteria in dry products? Int J Food Microbiol 136:364–367

Zhou JS, Shu Q, Rutherfurd KJ et al (2000) Acute oral toxicity and bacterial translocation studies on potentially probiotic strains of lactic acid bacteria. Food Chem Toxicol 38:153–161

# 7 Probiotics and Prebiotics for the Health of Companion Animals

Loredana Baffoni

Companion animals, also referred as pets, are animals kept primarily as companions at home or having close daily relationship with humans. Species suitable to be companion animals include dogs; cats; horses; house rabbits; ferrets; avian pets, such as canaries, parakeets and parrots; guinea pigs; and other small mammals, reptiles, and fish. Dogs and cats are largely considered the major human companions. Both have been cohabiting with us for thousands of years becoming an important part of our life; therefore, considering this strong emotional bond, their health and wellness are essential.

Ensuring a wholesome and balanced diet is a fundamental duty of a responsible pet ownership. The primary role of diet is to provide enough nutrients to meet metabolic requirements while giving the consumer a feeling of well-being. However, beyond meeting nutritional needs, diet may modulate various functions in the body and may play detrimental or beneficial roles in some diseases (Bontempo 2005). In the last decades, the growing knowledge in canine and feline nutrition has greatly contributed to improve longevity and well-being of companion animals, and some researchers have focused their attention on the importance of dietary fibres and beneficial microorganisms in pets' diet. Fibre sources, such as beet pulp, cellulose, corn fibre, fruit fibre, rice bran and whole grains, are suitable ingredients for pet foods, and experimental evidences support their beneficial effects in improving the health status of pets (De Godoy et al. 2013).

While the knowledge on the canine and feline intestinal microbiota is still expanding, the use of probiotics and prebiotics is becoming increasingly popular for treatment and/or prevention of diseases in companion animals (Jugan et al. 2017; Di Cerbo et al. 2017; Grześkowiak et al. 2015). Nevertheless, published papers on probiotic, prebiotic or synbiotic applications are greatly limited compared to available human studies.

L. Baffoni
Department of Agricultural Sciences, University of Bologna, Bologna, Italy
e-mail: loredana.baffoni@unibo.it



D. Di Gioia, B. Biavati (eds.), *Probiotics and Prebiotics in Animal Health and Food Safety*, https://doi.org/10.1007/978-3-319-71950-4_7

A noteworthy number of probiotic products are available commercially for dogs and cats as tablets, capsules, pastes and liquid formulations. Moreover, numerous animal feeds are available on the market, claiming to contain probiotics and/or prebiotics. Incorporation of probiotics into normal feed may have the advantage of an easy, daily administration of beneficial organisms. However, more studies on specific strains for pet nutrition are necessary, and controls on product quality and misleading product labelling are required. In the past, analysis on commercial pet foods revealed a significant percentage of products not containing the organisms declared on the label or containing different species (Weese and Arroyo 2003).

Data supporting the usefulness of probiotics and prebiotics in companion animals are increasing, and it is likely that appropriate strains or formulations may exert some effects (Baillon et al. 2004; Marshall-Jones et al. 2006; O'Mahony et al. 2009; Kanakupt et al. 2011). However, trials on healthy and/or diseased animals are often questionable in terms of number of subjects, timing, dosage, and strains used, making the comparison among studies really difficult (Jugan et al. 2017).

Some investigators hypothesized that commensal organisms may exert species-specific effects, and therefore a successful canine or feline probiotic organism would ideally be derived from the gastrointestinal tract (GIT) of the animal in question (McCoy and Gilliland 2007). For this reason, different studies have focused on isolating, testing and characterizing canine- and feline-specific probiotics (McCoy and Gilliland 2007; Biagi et al. 2007; Perelmuter et al. 2008; O'Mahony et al. 2009).

The intestinal microbiota, the bacterial communities residing in the intestinal tract, consists of a balance of beneficial and potentially harmful bacteria. This microbiota is established early in life; its disruption or perturbation can result in intestinal upset and poor immune function development (Czarnecki-Maulden 2008). Therefore, it is of great importance the study of pets' microbiota composition to understand the relationship between an unbalanced microbiota and pathological symptoms and to ascertain the positive action of probiotic and prebiotic supplements. Recent advances in molecular techniques and the large use of new-generation sequencing technologies have allowed a wider knowledge on gut microbiota composition in healthy animals (Suchodolski et al. 2008; Desai et al. 2009; Ritchie et al. 2010; Handl et al. 2011; Suchodolski 2011b; Swanson et al. 2011; Garcia-Mazcorro et al. 2012; Tun et al. 2012; Hand et al. 2013) as well as the effect of diet on microbiome composition (Hang et al. 2012; Beloshapka et al. 2013; Bermingham et al. 2013; Hooda et al. 2013; Kerr et al. 2013; Deusch et al. 2014; Young et al. 2016).

## 7.1 Microbiota of Dogs and Cats

Because of anatomical and physiological differences, each intestinal compartment harbours a unique microbial ecosystem (Suchodolski et al. 2005). Microorganisms reside in specialized niches and provide specialized functions by utilizing host nutrients and, in return, providing metabolites for host uptake. Each animal harbours a unique and peculiar microbial profile (Suchodolski et al. 2004; Ritchie et al. 2010). The major differences occur at species and strain level, with typically only minor overlap of bacterial species between individual animals, while most mammals share similar bacterial phyla, order and genera. For example, a study on faecal cat microbiota showed that 84% of samples harboured *Bifidobacterium* spp.; however, each individual cat seems to have a peculiar pattern of *Bifidobacterium* species (Ritchie et al. 2010).

The stomach harbours between $10^4$ and $10^5$ cfu/g of bacteria (Kil and Swanson 2011). Bacterial counts in the duodenum and jejunum are typically low ($10^5$ cfu/g of content), but can reach up to $10^9$ cfu/mL in some dogs and cats (Johnston 1999). This concentration is considerably greater compared to that of the human duodenum, where total bacterial counts higher than $10^5$ cfu/g have been associated with the small intestinal bacterial overgrowth (SIBO) syndrome. Cats appear to have greater counts of anaerobic bacteria in their small intestine compared with dogs (Johnston et al. 1993). Ileum contains a more diverse microbiota and greater bacterial numbers ($10^7$ cfu/mL). Bacterial counts in the colon range between $10^9$ and $10^{11}$ cfu/g of content (Mentula et al. 2005; Suchodolski 2011b).

Using culture techniques, *Bacteroides*, *Clostridium*, *Lactobacillus*, *Bifidobacterium* spp. and *Enterobacteriaceae* are the predominant bacterial groups that have been identified from the canine and feline intestine. Afterwards, with the growth of molecular tools, our knowledge about the phylogenetic diversity within the canine and feline gut has been greatly expanded. Recent studies have revealed several hundred bacterial phylotypes in the canine and feline intestinal tract (Suchodolski et al. 2008; Swanson et al. 2011; Garcia-Mazcorro et al. 2012; Hand et al. 2013). The phyla *Firmicutes*, *Bacteroidetes*, *Proteobacteria*, *Fusobacteria* and *Actinobacteria* constitute more than 99% of all gut microbiota in dogs and cats. The remaining bacterial groups are represented by the phyla *Spirochaetes*, *Tenericutes*, *Verrucomicrobia*, *Cyanobacteria* and *Chloroflexi* and a few unclassified bacterial lineages. Generally, aerobic bacteria or facultative anaerobic bacteria occur in greater abundance in the small intestine, whereas anaerobes predominate in the large intestine. In the stomach, mucosa-adherent *Helicobacter* spp. predominate, followed by *Lactobacillus*, *Streptococcus* spp., and *Clostridium* spp. Ten and eleven different bacterial phyla were identified in the proximal small intestine of dogs and cats (Suchodolski 2011b), respectively. *Firmicutes* (mainly *Clostridiales* and *Lactobacillales*), *Bacteroidetes*, *Proteobacteria* and *Actinobacteria* constituted approximately 95% of sequences. *Firmicutes* (mainly *Clostridiales*), *Bacteroidetes* and *Fusobacteria* have been reported to be the predominant bacterial phyla in the colon and faeces of dogs and cats even if other phyla have been reported such as *Actinobacteria*, *Proteobacteria* and *Bacteroidetes/Chlorobi* group (Swanson et al. 2011; Deng and Swanson 2015).

*Firmicutes* is a heterogeneous bacterial phylum comprising several phylogenetically distinct *Clostridium* clusters. These clusters differ in abundance in the different intestinal tract. Clusters XIVa and IV encompass many important short-chain fatty acids (SCFA) producing bacteria (e.g. *Ruminococcus* spp., *Faecalibacterium* spp., *Dorea* spp. and *Turicibacter* spp.) and predominate in the ileum and colon. Cluster XI and Cluster I (*Clostridium perfringens* group) are the second most abundant group in the small and large intestine of dogs and cats (Ritchie et al. 2008; Suchodolski et al. 2008). Several studies have described the metabolic products generated by the canine and feline intestinal microbiota, including SCFA, lactate, ammonia, and other end products (Sunvold et al. 1995; Sparkes et al. 1998). The ability of intestinal microbes to ferment dietary products into SCFA has positive implications in GI health. Acetate, propionate and butyrate are the most abundant SCFA, constituting approximately 60, 25 and 10%, respectively, in canine and feline faeces (Sunvold et al. 1995; Barry et al. 2010). Minor components of SCFA are branched-chain fatty acids, including isobutyric acid, butyric acid and isovaleric acid, which are produced during protein degradation (Barry et al. 2010). Although marked differences have been observed in the microbiota composition among animals of the same species, the metabolic end products are quite similar. It has already been suggested for the human microbiota that a functional redundancy exists in the GI tract. Several members of the community are able to perform similar functions, and if one microbial group is displaced because of perturbations (e.g. antibiotic therapy), other members of the community are capable of maintaining a stable ecosystem functionality (Suchodolski et al. 2009).

Despite minor overlaps, observed abundance of the various bacterial groups differs between studies. For example, percentages of *Firmicutes* in faecal samples range between 25 and 95% of obtained sequences (Middelbos et al. 2010; Ritchie et al. 2010; Swanson et al. 2011; Handl et al. 2011). It is likely that these discrepancies may be due to differences in DNA extraction methods and PCR protocols in the different studies.

For example, 16S rRNA gene approaches routinely underestimate the abundance of *Actinobacteria* in intestinal samples using universal primer. The use of species-specific primers or probes for *Bifidobacterium* spp. (*Actinobacteria* phylum) usually confirms the presence of a bifidobacterial population in the intestinal tract of the majority of dogs and cats (Handl et al. 2011; Ritchie et al. 2010). Through a massive parallel 16S rRNA sequencing, a high variability could be registered in the observed *Bifidobacterium* spp. between individual dogs. Handl et al. (2011), in their pyrosequencing analysis on faecal samples, identified eight different bifidobacterial species, with *B. subtilis* and *B. bifidum* being the most prevalent. Concerning cats, Ritchie et al. (2010) used 16S rDNA libraries to study feline bifidobacteria population using group-specific primers and reported that the most prevalent phylotype shared a 98% similarity with *B. subtilis* 16S rDNA sequence and was observed in 10 cats out of 12.

Members of the order *Lactobacillales* (e.g. *Lactobacillus* spp., *Pediococcus* spp., *Enterococcus* spp., *Streptococcus* spp., *Lactococcus* spp.) seem to be highly prevalent in the duodenum, jejunum and colon of dogs (Handl et al. 2011), while they are detected in higher abundance in the proximal small intestine of cats with decreasing proportions in the colon (Ritchie et al. 2010).

## 7.2 Microbes and Gastrointestinal Disorders

The discrimination between commensal and pathogenic microbes is often difficult. Although many microbes are correlated with illness, one should determine whether a microbe is a contributor or simply one that benefits from the conditions of a diseased environment. Several potential pathogenic bacteria related to canine and

**Table 7.1** Potential pathogenic bacteria related to canine and feline gastrointestinal tract (GIT)

| | Pathogenic bacteria |
|---|---|
| GIT | *Anaerobiospirillum* spp.<br>*Bacillus cereus*<br>*Campylobacter jejuni* and *Campylobacter coli Clostridium perfringens* and *C. difficile* Enteropathogenic *Escherichia coli*<br>*Klebsiella pneumoniae*<br>*Salmonella* spp.<br>*Yersinia* spp. |

feline GIT are listed in Table 7.1. It should be noted, however, that many of these microbes are also present in healthy dogs and cats. Thus, microbial balance or activity or both appear to be more important than the mere presence of any pathogenic microbe (Kil and Swanson 2011).

Compositional changes in the small intestinal microbiota, potentially leading to changes in intestinal permeability and digestive function, have been suggested in canine small intestinal dysbiosis or antibiotic-responsive diarrhoea.

Current theories for the development of inflammatory bowel disease (IBD), both in humans and pets, propose a combination of environmental factors, the intestinal microbiota and a genetic susceptibility of the host. There is however mounting evidence that microbes play a fundamental role in the pathogenesis of IBD (Suchodolski 2011a), considering also that studies in engineered animal models, with susceptibility for inflammation, indicate that IBD develops only if bacteria are present (Packey and Sartor 2009). The cause-effect relationship between microbial alterations and inflammation is not well understood. It is suspected that intestinal inflammation causes a dysbiosis towards Gram-negative bacteria (i.e. *Proteobacteria*), and this depletion of some commensal bacterial groups may lead to a reduced capability of the intestinal microbiome to downregulate an aberrant intestinal immune response, leading to a perturbation of the intestinal balance (Sokol et al. 2008). Some pathogenic bacteria such as *Campylobacter jejuni* and *Salmonella* trigger changes in mucosal architecture and in the innate immune system, which diminish the colonization resistance of resident microbes (Stecher and Hardt 2008). In this context, researchers have hypothesized the usefulness of a probiotic and/or prebiotic therapy in order to strengthen the mucosal barrier and enhance the immune response.

Recent molecular studies, performed in dogs and cats, have underlined the differences in the intestinal microbiota between healthy animals and IBD patients. For example, dogs and cats with idiopathic small intestinal IBD were significantly enriched in *Enterobacteriaceae* compared to controls (Janeczko et al. 2008; Xenoulis et al. 2008). Another study revealed an increase in *Proteobacteria* (i.e. *Pseudomonas* spp.) in the duodenum of IBD dogs (Suchodolski et al. 2010). Similar to humans, IBD dogs showed a reduction in the proportions of *Bacteroidales* and *Clostridiales* (i.e. *Lachnospiraceae*, *Ruminococcaceae*, *Faecalibacterium spp.*); in particular, a reduced diversity within the *Clostridium* clusters XIVa and IV would indicate that these bacterial groups, important short-chain fatty acid producers, may play an important role in promoting intestinal health (Suchodolski 2011a). In general, a reduced bacterial species richness could be observed in the small intestine of IBD dogs (Xenoulis et al. 2008; Craven et al. 2009).

Compositional changes have also been registered in the large intestine of cats with chronic enteropathies. FISH analysis revealed greater microscopic counts of total bacteria, *Bifidobacterium* spp. and *Bacteroides* spp., in healthy cats, whereas cats with IBD had greater microscopic counts of *Desulfovibrio* spp., potential producers of toxic sulfides (Inness et al. 2007).

Together with a change in the microbiota composition, studies point out that feline and canine IBDs are likely associated with an immune dysregulation, as evidenced by different cytokine expression levels and Toll-like receptor (TLR) regulation in animals with chronic enteropathies (Nguyen Van et al. 2006; Janeczko et al. 2008; Luckschander et al. 2010). Toll-like receptors are crucial members of the innate immune system. They are located on cell surfaces, recognize microbe-associated molecular patterns, and activate immune responses. TLRs have been shown to be dysregulated in various dog breeds with IBD (Burgener et al. 2008; Allenspach 2011). Granulomatous colitis of Boxer dogs (i.e. a form of Crohn's disease) has recently been associated with the presence of adherent and invasive *Escherichia coli* (AIEC; Simpson et al. 2006). These AIEC isolates share similarities to AIEC isolates obtained from ileal tissues of humans with Crohn's disease. Craven et al. (2010) discovered mutations in a NADPH oxidase subunit that reduce the ability of phagocytes to eliminate intracellular pathogens, predisposing the host to chronic infections.

## 7.3 Dogs

### 7.3.1 Probiotics

Despite the large availability of probiotic, prebiotic and synbiotic formulations on the market, the existing body of literature on in vivo studies is limited, especially for synbiotic testing.

Probiotic and prebiotic administration is typically focused, in dogs, on prevention and/or treatment of enteric diseases through microbiota modulation; but in vivo studies on animals with gastrointestinal disorders are sparse and difficult to compare.

It is interesting to underline that several works focused on isolation and characterization of new putative probiotic strains from animal gastrointestinal tract prior to in vivo testing (McCoy and Gilliland 2007; O'Mahony et al. 2009; Biagi et al. 2007; Beasley et al. 2006). Other studies tested, on the contrary, commercial available probiotics obtained from different sources (e.g. *Lactobacillus rhamnosus* strain GG (LGG), *Enterococcus faecium* strain SF68®) (Weese and Anderson 2002; Benyacoub et al. 2003). Weese and Anderson (2002) in their study on adult dogs with LGG, a popular probiotic for human consumption, reported a relatively high level of the probiotic bacterium in the faeces of some dogs, while, in other animals administered with the same dose, LGG was rarely or never detected. Authors speculated that compositional differences in gastrointestinal microbiota of adult dogs could have played a role in the obtained results. Dogs with high, pre-existing population of lactic acid bacteria may be more resistant to colonization with "foreign" lactobacilli. Bacterial species may be able to limit colonization of similar organisms through stable occupation of certain environmental or nutritional niches or through the production of specific antibacterial products. The registered persistence of LGG in dogs is shorter than that reported in humans. In conclusion, LGG cannot be defined as a "canine" probiotic; however, this study demonstrated that it could be safely administered to dogs and it can survive gastrointestinal transit (Weese and Anderson 2002). Further studies regarding the efficacy of LGG in canine gastrointestinal diseases are envisaged.

*Enterococcus faecium* SF68® is a lactic acid bacterium (LAB) with inhibitory effects against important enteropathogens. Therefore, it might be useful as an antidiarrhoeal agent for pets, as already demonstrated in humans. Benyacoub et al. (2003) performed an in vivo study to assess a possible immune stimulation of SF68® in puppies. *E. faecium* supplementation succeeded in increasing faecal IgA concentrations and vaccine-specific IgG and IgA for canine distemper viral disease. Another short-term treatment with SF68® failed, on the contrary, to affect giardial cyst shedding or antigen content in dog with naturally acquired giardiasis, and it did not alter innate or adaptive immune responses (Simpson et al. 2009). Safety concerns have been raised about the use of enterococci as probiotics, for the incidence of virulence traits among enterococcal strains and their resistance to many antibiotics (Franz et al. 2011). Moreover, an in vitro study evidenced the ability of some enterococci strains to significantly enhance the adhesion of *C. jejuni* to canine mucus, making dogs in this case a potential carrier and possibly a source for human infections. However, the study also reported that LAB strains of canine origin reduced significantly the adhesion of *C. perfringens*, and this outcome stresses the importance of host-derived strains compared to foreign ones (Rinkinen et al. 2003).

Putative probiotic *Lactobacillus* and *Bifidobacterium* strains isolated and characterized from faeces of healthy dogs are listed in Box 1.

**Box 1 Putative Probiotic Species Isolated from Canine GIT**

***Lactobacillus fermentum***
***Lactobacillus salivarius***
***Lactobacillus rhamnosus***
***Lactobacillus animalis***
***Lactobacillus mucosae***
***Lactobacillus murinus/ruminis***
***Lactobacillus acidophilus***
***Bifidobacterium pseudolongum***
***Bifidobacterium animalis***

Isolated strains are usually characterized in vitro to test their resistance at low pH, bile salts and freeze-drying and to assess their antimicrobial activity against a wide range of intestinal pathogens. These are essential characteristics for a probiotic strain. The comparison among available trials is difficult because animal experiments often differ for what concerns animal number, administration methods, doses and monitored parameters. In vivo tests about persistence and/or colonization of the administered probiotic strains in dogs' gastrointestinal tract revealed that fed strains usually did not permanently colonize the intestine (Manninen et al. 2006; O'Mahony et al. 2009). In rare cases, the isolated potential probiotic strains showed a good persistence, for example, the canine *L. fermentum* AD1 that was found in canine GIT even 6 months after stopping administration in healthy dogs (Strompfová et al. 2006).

The majority of canine strains showed a good inhibitory action against intestinal pathogen. Tests in vitro gave positive outcomes for *L. animalis* LA4 and a *L. pentosus* strain against *C. perfringens* (Biagi et al. 2007; Rinkinen et al. 2003). Moreover, a *L. reuteri* strain was effective against *S. enterica* serovar Typhimurium in associative cultures (McCoy and Gilliland 2007). A canine strain, *Bifidobacterium animalis* AHC7, significantly reduced translocation of *S.* Typhimurium to both the liver and spleen in murine challenge. In the same work, *B. animalis* AHC7-fed animals showed a reduction of *C. difficile* numbers. *C. difficile* could be acquired during hospitalization of dogs and is associated with the development of diarrhoea. The elimination of *C. difficile* from the canine gut may not only improve canine gastrointestinal health but may also help reduce the risk of human infection due to owner-pet interactions (O'Mahony et al. 2009).

Concerning the animal microbiota modulation through probiotic supplementation, limited results are available. Frequently, the in vitro performances are only partially confirmed by in vivo trials on healthy animals. *L. animalis* LA4, for example, did not significantly affect faecal counts of *C. perfringens*, coliforms and enterococci in vivo. Authors speculated that the low number of animals used in this study and the high individual variability could have influenced the results (Biagi et al. 2007); however this problem has been shown in different trials. The number of total aerobes, *Bacteroides*, *E. coli*, lactobacilli or bifidobacteria were not affected by administration with *B. animalis* AHC7 (O'Mahony et al. 2009). The supplementation of *L. fermentum* AD1 in the diet, on the other end, increased significantly the number of lactic acid bacteria in canine digestive tract together with total proteins and total lipids while decreasing the concentration of glucose in the bloodstream of dogs (Strompfová et al. 2006).

Probably, the limited evidences about the modulation of the indigenous microbiota through probiotic feeding could be due to the low number of animals enrolled for the experiments but also to the different approaches used to quantify bacteria (classical microbiological methods vs molecular techniques). It could be speculated that the supplementation of a suitable prebiotic compound during probiotic treatment (synbiotic) may enhance the probiotic performances.

Two examples of clinical trials evidenced the possible use of probiotics for canine intestinal upsets. One trial evaluated the effect of a probiotic product in acute self-limiting gastroenteritis. The probiotic cocktail consisted of thermostabilized

*Lactobacillus acidophilus* and live strains of *Pediococcus acidilactici*, *Bacillus subtilis*, *Bacillus licheniformis* and *Lactobacillus farciminis*. Authors evidenced that the probiotic cocktail reduced the convalescence time of dogs with acute self-limiting diarrhoea (Herstad et al. 2010). Another trial evaluated whether a probiotic supplementation in dogs with food-responsive diarrhoea (FRD) had beneficial effects on intestinal cytokine patterns and on microbiota. Probiotic cocktail consisted of three different lyophilized *Lactobacillus* spp. strains (two *L. acidophilus* and one *L. johnsonii*). Results showed that the Canine Inflammatory Bowel Disease Activity Index, a scoring system comprising general attitude, appetite, faecal consistency, defecation frequency and vomit, decreased in all dogs after probiotic treatment. However, only mild effects were detected concerning microbiota and cytokine modulation (Sauter et al. 2006).

### 7.3.2 Prebiotics and Synbiotics

#### Prebiotics

Prebiotics may be considered as functional food ingredients. They are attracting considerable interest from pet owners and pet food manufacturers. The most common forms of prebiotics are non-digestible oligosaccharides (NDOs, polymers typically with two to ten monosaccharide units) such as mannan oligosaccharides, gluco-oligosaccharides, galactooligosaccharides, fructooligosaccharides and long-chain prebiotics such as inulin (>20 residues).

The type of supplemented prebiotic must be considered when comparing results among various experiments, and it is important to take into consideration that oligosaccharides with different average DP (degree of polymerization) can elicit different responses. Degradation of longer chains is supposed to be slower, enabling their arrival into more distal parts of the intestine (Roberfroid et al. 1998). Especially for animal feed, where different types of animals have different intestinal tract morphologies and different degrees of bacterial associations, the chain length factor could be a valuable criterion for the design of a "tailor-made" animal feed, meeting the specific requirements of different animals. An apparent effect in cats may not necessarily be repeatable in dogs or may require a different level of supplementation (Flickinger et al. 2003). During the colonic fermentation of endogenous and undigested amino acids, several putrefactive compounds are produced and are responsible for the malodour of pet faeces. Because two of the main expected effects of prebiotics are the improvement of intestinal microbial balance and the reduction of faecal odour components, these effects would be more noticeable in animals fed with a high-protein diet. The prebiotic dose to be supplemented should be carefully evaluated in order to minimize potential negative side effects. Flatulence and loose stools can occur at very high levels (i.e. >20% of dry matter) of supplementation or at moderate levels (i.e. at >10% of dry matter) in non-adapted animals. Pets that consume large, infrequent meals could potentially receive a large bolus of NDOs, eliciting adverse effects (Flickinger et al. 2003). Optimal inclusion levels in diets have yet to be established for different animal species and prebiotic compounds

(Flickinger and Fahey 2002). Most of the available in vivo trials with prebiotics include fructooligosaccharides and/or inulin (Hussein et al. 1999; Strickling et al. 2000; Flickinger et al. 2003; Propst et al. 2003; Barry et al. 2009; Beloshapka et al. 2013); however also other oligosaccharides or fibre-rich foodstuffs have been tested for their potential probiotic properties (Knapp et al. 2008; Biagi et al. 2010; Faber et al. 2011a, b; De Godoy et al. 2013).

FOS (fructooligosaccharides), scFOS (short-chain fructooligosaccharides) and inulin were used in vivo at different concentrations. These compounds have been shown to positively affect diet digestibility and intestinal characteristics of dogs, generally at a dietary inclusion of 1% or greater. Barry et al. (2009) studied, on the contrary, the effects of a low-level supplementation (0.2 and 0.4%) of inulin and scFOS on nutrient digestibility, ileal IgA concentration, stool protein catabolite concentrations and microbiota in faeces of healthy, adult dogs. The low prebiotic supplementation seemed to enhance the fermentative activity in the gut. Authors speculated that inulin and scFOS were fully fermented in the proximal colon, and the low amount of SCFA in treated animals is indicative of fructan fermentation and subsequent SCFA absorption in the ascending colon. Moreover, phenol concentration decreased linearly, while ileal IgA concentration was not affected, as expected in adult dogs with a fully developed immune system. Similar results were reported in other experiments (Grieshop et al. 2004; Verlinden et al. 2006). A decrease in stool protein catabolites, reported by Barry et al. (2009), resulted in a less offensive stool odour and was expected to be beneficial for the intestinal health while reducing the presence of potential harmful compounds. On the other end, faecal microbiota was not affected by treatment. Authors underlined that it is important to establish threshold levels at which biological responses might be expected.

Recent reviews on prebiotic efficacy did not show a strong modulation of intestinal microbiota upon prebiotic supplementation (Pinna and Biagi 2016; Kozlowska et al. 2016); however, diet formulation and way of administration are fundamental and can account for differences in experimental outcomes. Middelbos et al. (2007) reported a significantly higher bifidobacteria concentration in animals supplemented with fructooligosaccharides and cellulose compared to the cellulose treatment. Apanavicius et al. (2007) examined the effects of fructan supplementation on the immune response of weanling puppies subjected to bacterial challenge with *Salmonella* Typhimurium. Fructan supplementation increased faecal acetate, total SCFA and *Lactobacillus* spp. concentration and decreased the lack of appetite in infected animals. Moreover, supplemented puppies had also a reduced enterocyte sloughing compared to animals feeding a control diet. Still concerning microbiota modulation, some authors reported a decreased concentration of *C. perfringens* in faecal samples after administration of mannan oligosaccharides, oligofructose, lactitol and polydextrose (Strickling et al. 2000; Flickinger et al. 2003; Biagi et al. 2010; Beloshapka et al. 2012).

Interesting studies tried to investigate the prebiotic effect of nonclassical oligosaccharides and fibres. Faber et al. (2011a, b) firstly tested in vitro and then in vivo new fermentable carbohydrate sources. Their first study evaluated the hydrolytic digestibility, fermentative capability and microbiota-modulating properties of temulose molasses and four hydrolysed fractions of temulose molasses. Temulose molasses derived from the fibreboard manufacturing process. The hydrolysis of temulose molasses removes arabinose and xylose, producing a galactoglucomannan oligosaccharide (GGMO) product. This GGMO was then tested in vivo. Temulose molasses and selected fractions of temulose molasses, having different DP, were evaluated for fermentative and microbiota-modulating properties in vitro, using canine faecal inoculum. The tested substrates resulted in a significant drop in pH and produced greater concentrations of SCFA compared to the control substrates (scFOS and a yeast cell wall preparation). The temulose and its fractions also resulted in significant increase of bifidobacteria population and decrease of *E. coli* in vitro. GGMO was subsequently supplemented to adult dogs at different concentrations (0, 0.5, 1, 2, 4, and 8%). The GGMO substrate contained high concentrations of oligosaccharides, with the mannose component accounting for 35% of dry matter (DM). Faecal microbial populations were unaffected by the addiction of the GGMO substrate except for *Bifidobacterium* spp. with 8% supplementation of GGMO. The increase of nutrient digestibility and faecal SCFA concentrations, together with the decrease of crude protein (CP) digestibility, digesta pH and phenol and indole concentrations, indicates an active large bowel fermentation with GGMO supplementation. Data presented in this work provide evidence of the positive nutritional properties in vivo, but not necessarily prebiotic potential, of supplemental GGMO when incorporated in a high-quality dog food. Because of an increased concentration of mannan, continued research on its pathogen binding capability and its potential as immunomodulatory agent is needed in order to assess its efficacy as a dietary supplement for canine health and well-being (Faber et al. 2011a).

The work of Knapp et al. (2008) aimed at quantifying in vitro digestion, glycaemic and insulinaemic responses and gastrointestinal tolerance of fructose (Fruc), maltodextrin (Malt), polydextrose (Poly), pullulan (Pull), resistant starch (RS), sorbitol (Sorb) and xanthan gum (Xan) in adult dogs. Limited digestion of RS, Poly and Xan occurred with Malt having the highest area under the curve for glucose and insulin in the glycaemic tests. Gastrointestinal tolerance was examined for diets containing carbohydrates at either 100 or 200% of the adequate intake value for dietary fibre. At 100 and 200% Malt, RS and Sorb resulted in ideal faecal scores, while Pull and Xan resulted in looser stools and Poly resulted in diarrhoea. Data indicated that novel carbohydrates give different results concerning digestibility, energy content, glycaemic and insulinaemic responses and gastrointestinal tolerance. Variation in responses is due largely to the individual carbohydrate molecular structure and binding pattern. The impact on animal microbiota was not evaluated.

Biagi et al. (2010) investigated the effect, on the composition and activity of the canine intestinal microbiota, of different sources of soluble fibres, including fibres widely used by the pet food industry, and some prebiotic substances that might be considered as potential supplements for dog diets. Fourteen treatments were evaluated in vitro and then two were chosen for an in vivo trial. The treatments included different types of industrial product of FOS inulins and pectins, lactitol, glucuronic acid, chicory, beet pulp, pea hull fibre, psyllium fibre and guar gum. These compounds were added to canine faecal cultures and incubated for 24 h.

Dog faecal inoculum was able to ferment most of the substrates that were tested. Two substrates (lactitol and Pectin Classic CU201) were selected to be tested in vivo in adult dogs (10 g/kg for 30 days). These substrates were chosen considering the different beneficial properties shown during the in vitro trial: increased production of propionic (lactitol) and n-butyric acid (Pectin Classic CU201), reduced ammonia concentration (Pectin Classic CU201) and increased lactobacilli and reduced coliform viable counts (lactitol). The in vivo results confirmed the efficacy of lactitol in reducing viable count of *C. perfringens* and total coliforms, but administration failed in increasing *Lactobacillus* spp. population. Also Pectin Classic CU201 did not increase faecal lactobacilli counts, while it decreased *C. perfringens* levels (even if to a lesser extent than lactitol).

**Synbiotics**

The use of synbiotic supplements in dogs has been poorly investigated. Some in vitro trials tested the effects of the association of probiotics and prebiotics on faecal microbiota, but in vivo trials are limited. One work could be cited that was performed to determine the efficacy of fructooligosaccharides (FOS) and/or *Lactobacillus acidophilus* (LAC) in modulating the concentrations of gut microbial populations, fermentative end products and nutrients digestibility in healthy adult dogs (Swanson et al. 2002). After the trial authors concluded that the supplementation of FOS and *L. acidophilus* together as a synbiotic may prove to be beneficial because it may decrease the concentration of several faecal putrefactive compounds (biogenic amines, branched-chain fatty acids, phenols, indoles) to a greater extent compared to the single supplementation of the prebiotic or the probiotic. FOS, fed alone, appeared to enhance indices of gut health by positively altering gut microbial ecology and faecal protein catabolites, whereas the probiotic was more effective when fed in combination with FOS rather than alone; however, it is important to underline the human origin of the probiotic strain used.

An interesting approach has been used by Tzortzis et al. in their works (2003, 2004). They described the possible use of glycotechnology in order to deliberately design and synthesize synbiotics consisting of an efficient probiotic and a prebiotic tailored to enhance the growth of that specific probiotic. This approach takes into consideration the glycosidase specificity of the probiotic microorganism to design suitable oligosaccharide mixtures. A further possibility implies the use of the probiotic strain enzymes for the synthesis of an appropriate prebiotic compound, which would act as a highly selective substrate. Tzortzis et al. (2003) focused on the production of new α-galactosides using the α-galactosidase of a *Lactobacillus reuteri* strain (NCIMB 41152) isolated from canine large intestinal tissue samples. The synthesized oligosaccharide was a galactosyl melibiose mixture (GMM). The subsequent step was the study of the fermentability of these α-galactosyl oligosaccharides in a synbiotic formulation compared to the oligosaccharide alone (Tzortzis et al. 2004). The in vitro trial showed a higher increase of bifidobacteria and lactobacilli with GMM addiction compared to FOS, melibiose and raffinose. GMM, when compared to the commercial oligosaccharides, was also associated with a significant decrease in clostridia, *E. coli* and eubacteria. Furthermore, when

*L. reuteri* was added to the fermenters together with GMM, bifidobacteria and lactobacilli increase still more. This tailor-made approach could represent the successful strategy to formulate new effective synbiotic products.

A recent work of Gagné et al. (2013) on healthy sled dogs evaluated the impact of a synbiotic product on faecal microbiota composition. The synbiotic, consisting of *E. faecium* SF68®, *Bacillus coagulans*, *L. acidophilus* and several prebiotics (FOS, MOS) and vitamins (B3, B6), was administered in a placebo-controlled trial. Authors reported an increase of *Lactobacillaceae* and faecal butyrate concentration across all dogs. Faecal scores also improved compared to the control group at 5 weeks.

## 7.4 Cats

### 7.4.1 Probiotics

Minimal information exists regarding probiotic applications in cats, and few clinical studies have been performed. Because of differences in host physiology and diet, probiotic efficacy in cats cannot be extrapolated from studies in dogs. Cats are obligate carnivores and have evolved consuming a diet of prey with high protein content, low/moderate fat content and a minimal amount of carbohydrate which is usually consumed in small quantities many times a day.

The effect of a dietary supplementation of the probiotic strain *Lactobacillus acidophilus* DSM13241 was evaluated in healthy adult cats by Marshall-Jones et al. (2006). The probiotic strain was recovered from faeces, demonstrating survival through the feline gastrointestinal tract. Probiotic supplementation was associated with increased numbers of beneficial *Lactobacillus* spp. and *L acidophilus* groups in faeces and decreased numbers of *Clostridium* spp. and *Enterococcus faecalis*, indicating an altered bacterial balance in the gastrointestinal tract microbiota. A decrease of faecal pH was registered, reflecting the observed changes in the microbiota, as well as an immunomodulatory effect.

Lappin et al. (2009) evaluated the efficacy of the probiotic strain *Enterococcus faecium* SF68® in cats with chronic feline herpesvirus 1 (FHV-1) infection. This virus is frequently associated with morbidity because of recurrent ocular and respiratory clinical signs of disease, and *E. faecium* is considered an immune-enhancing probiotic. Faecal microbial diversity was maintained throughout the study in cats supplemented with SF68®, while a decrease was evidenced in control animals. Clinical results varied among individual cats, but the overall findings suggested that administration of the probiotic bacterium lessened morbidity associated with chronic FHV-1 infection, even if authors underlined the need of further studies to determine SF68® efficacy in a clinical setting.

Feline renal failure is a significant cause of morbidity and mortality in cats in the United States. For this reason, the reduction of blood urea nitrogen (BUN) and serum creatinine levels is desirable in renal failure patients; and this could be achieved through a reduced level of high biological value proteins. This dietary therapy has been shown to increase survival of feline renal failure patients. A clinician, curious about the manufacturer's claims of a multi-strain probiotic product, Kibow Biotics®, examined the efficacy of this product on azotemia in cats. Kibow Biotics® contains a mixture of bacteria consisting of *Streptococcus thermophilus*, *Lactobacillus acidophilus* and *Bifidobacterium longum* strains reported to decrease BUN and serum creatinine levels. The results showed that the probiotic mixture seemed to benefit the animals. The manufacturer's promise of decreasing azotemia appeared to be verified, and these patients did experience improved health and vitality.

The efficacy of a probiotic supplementation has also been evaluated in adult cats with a *Campylobacter*-induced diarrhoea. The study aimed at determining whether the probiotic strain *Lactobacillus acidophilus* DSM13241 was able to affect the recovery and elimination of a clinical *Campylobacter* infection. Cats were treated with antibiotics before starting the trial and then subdivided into two groups, a control group and a group supplemented with the probiotic bacterium. Results showed that probiotic supplementation significantly reduced pathogen shedding and favoured a more rapid response to antibiotic treatment (Baillon and Butterwick 2003).

Considering the scarce number of studies described in the literature, it is not simple to draw any conclusion, and new trials are envisaged to assess the possible usefulness of probiotic microorganisms in the management of feline gastrointestinal diseases.

### 7.4.2 Prebiotics and Synbiotics

#### Prebiotics

Although cats are strict carnivores and are metabolically different from dogs, the potential benefits of a prebiotic supplementation also exist for this species.

Barry et al. (2010) evaluated the impact of different fibre sources (4% cellulose, FOS or pectin) in adult cats monitoring nutrient digestibility, faecal protein catabolite concentrations and faecal microbiota concentrations. The research demonstrated that cats are able to adapt to moderate concentrations of fibre. Although protein catabolites and, thus, stool odour were increased in both FOS and pectin diets, end products of carbohydrate fermentation were also increased. The FOS treatment significantly increased the bifidobacteria population compared to pectin. Moreover, pectin diet appeared to have a harder texture than other diets; cats consuming this diet began to produce softer faeces than normal. Some of these outcomes had been previously underlined by Hesta et al. (2001). They evaluated the effect of different concentrations of oligofructose and inulin on faecal characteristics and nutrient digestibility in healthy cats. Oligofructose supplementation was added to the diet at 0.3, 6 and 9% while inulin at 0.3 and 6%. There were no significant differences regarding the macroscopical and chemical aspects of faeces between control and 0.3% supplemented groups. The higher amounts of fresh faeces were produced in the 6 and 9% FOS groups and could be due to excessive moisture but also to a reduced digestibility. Authors evidenced no significant differences between 0.3% inulin and oligofructose supplementation, although oligofructose seemed to be more easily fermentable, because of the higher concentration of SCFA in faecal samples of this group. No surveys were performed on faecal microbiota.

The objective of the study of Kanakupt et al. (2011) was to determine the effects of low-level prebiotic inclusion [0.5% scFOS, 0.5% galactooligosaccharides (GOS) and 0.5% scFOS +0.5% GOS] on nutrient digestibility, fermentative metabolite concentrations and large bowel microbial ecology of healthy adult cats. The GOS + scFOS supplementation resulted in a lower pH and higher concentrations of butyrate, valerate, acetate, total SCFA and total branched-chain fatty acids compared to single prebiotic supplementations. As expected, scFOS- and GOS-supplemented diets affected faecal microbial *Bifidobacterium* spp. concentrations, but whereas bifidobacteria populations were increased, faecal *Lactobacillus* spp., *E. coli* and *Clostridium perfringens* were not affected by dietary treatments. Authors concluded that the effects observed for the scFOS + GOS supplementation could be ascribed to the higher prebiotic concentration in the combined treatment rather than to any synergy that might exist between them.

In conclusion, data from prebiotic trials underline that oligosaccharides may be considered as nutritional interventions to improve intestinal health of cats. Concentrations higher than 0.5% should be used, but it is evident that an excessive supplementation can cause intestinal discomfort. More studies are necessary to clarify dose, length of supplementation period, type of prebiotic compounds and possible effect on intestinal microbiota.

### Synbiotic

Concerning synbiotic applications, spare data exist for cats. Biagi et al. (2013) investigated the effect of feeding a selected probiotic-prebiotic combination on the composition and metabolism of intestinal microbiota in adult cats. *B. pseudocatenulatum* strain B82 and GOS were administered to ten animals for 15 days. Results evidenced in faecal samples a reduction in ammonia concentration and an increase in bifidobacteria count, suggesting an improvement of the animal intestinal health.

Rishniw and Wynn (2011) investigated the effect of a commercial synbiotic product on diseased cats. Chronic kidney disease (CKD) is a common geriatric feline disorder with high morbidity and mortality. Specific bacteria, capable of metabolizing urea, creatinine, indoles, phenol and nitrosamine into nontoxic metabolites, have been selected to treat CKD. Azodyl is a commercial product, an enteric-coated capsule containing three different microorganisms (*Streptococcus thermophilus*, *Lactobacillus acidophilus* and *Bifidobacterium longum* strains) and a prebiotic source (psyllium husk). The capsule releases its content within the ileocolic region. This particular trial aimed to demonstrate the inefficiency of this product to reduce azotemia when sprinkled onto food instead of orally administered. This was probably due to a reduced survival of the bacterial strains in feed, underlining the importance of microencapsulation technology to ensure an easier and effective probiotic and prebiotic supplementation in feed.

## Conclusions

Although not so many data exist on the effects of probiotic and prebiotic supplementation in companion animals, studies performed mainly on cats and dogs underline that this approach is promising on healthy animals. Studies concerning the effects on pathogen loads are also sparse. The problems related to companions animals are basically three: (1) commercial probiotic formulations often are not rigorous in the declaration of the microorganism contained in the products; therefore it is not clear which strain(s) are being administered; (2) the strains used in probiotic formulation are often poorly characterized; and (3) the knowledge about gut microbiota is not as detailed as it is for other animals, including humans. The three points are related, and when the gut microbial composition will be elucidated, it will become clearer which strains are more suitable in formulations. This will help companies in the design and production of new targeted formulations that, hopefully, will be based on the synbiotic approach, considering the good results obtained up to now.

## References


Allenspach K (2011) Clinical immunology and immunopathology of the canine and feline intestine. Vet Clin North Am Small Anim Pract 41(2):345–360

Apanavicius CJ, Powell KL, Vester BM, Karr-Lilienthal LK, Pope LL, Fastinger ND, Wallig MA, Tappenden KA, Swanson KS (2007) Fructan supplementation and infection affect food intake, fever, and epithelial sloughing from Salmonella challenge in weanling puppies. J Nutr 137(8):1923–1930

Baillon ML, Butterwick RF (2003) The efficacy of a probiotic strain, *Lactobacillus acidophilus* DSM13241, in the recovery of cats from clinical *Campylobacter* infection. J Vet Int Med 17:416–419

Baillon ML, Marshall-Jones ZV, Butterwick RF (2004) Effects of probiotic *Lactobacillus acidophilus* strain DSM13241 in healthy adult dogs. Am J Vet Res 65(3):338–343

Barry KA, Hernot DC, Middelbos IS, Francis C, Dunsford B, Swanson KS, Fahey GC Jr (2009) Low-level fructan supplementation of dogs enhances nutrient digestion and modifies stool metabolite concentrations, but does not alter fecal microbiota populations. J Anim Sci 87(10):3244–3252

Barry KA, Wojcicki BJ, Middelbos IS, Vester BM, Swanson KS, Fahey GC Jr (2010) Dietary cellulose, fructooligosaccharides, and pectin modify fecal protein catabolites and microbial populations in adult cats. J Anim Sci 88(9):2978–2987

Beasley SS, Manninen TJ, Saris PE (2006) Lactic acid bacteria isolated from canine faeces. J Appl Microbiol 101(1):131–138

Beloshapka AN, Dowd SE, Suchodolski JS, Steiner JM, Duclos L, Swanson KS (2013) Fecal microbial communities of healthy adult dogs fed raw meat-based diets with or without inulin or yeast cell wall extracts as assessed by 454 pyrosequencing. FEMS Microbiol Ecol 84(3):532–541

Beloshapka N, Wolff AK, Swanson KS (2012) Effects of feeding polydextrose on faecal characteristics, microbiota and fermentative end products in healthy adult dogs. Br J Nutr 108(4):638–644

Benyacoub J, Czarnecki-Maulden GL, Cavadini C, Sauthier T, Anderson RE, Schiffrin EJ, von der Weid T (2003) Supplementation of food with *Enterococcus faecium* (SF68) stimulates immune functions in young dogs. J Nutr 133(4):1158–1162

Bermingham EN, Young W, Kittelmann S, Kerr KR, Swanson KS, Roy NC, Thomas DG (2013) Dietary format alters fecal bacterial populations in the domestic cat (Felis catus). Microbiology 2(1):173–181

Biagi G, Cipollini I, Pompei A, Zaghini G, Matteuzzi D (2007) Effect of a Lactobacillus animalis strain on composition and metabolism of the intestinal microflora in adult dogs. Vet Microbiol 124:160–165

Biagi G, Cipollini I, Grandi M, Zaghini G (2010) Influence of some potential prebiotics and fibre-rich foodstuffs on composition and activity of canine intestinal microbiota. Anim Feed Sci Technol 159:50–58

Biagi G, Cipollini I, Bonaldo A, Grandi M, Pompei A, Stefanelli C, Zaghini G (2013) Effect of feeding a selected combination of galacto-oligosaccharides and a strain of *Bifidobacterium pseudocatenulatum* on the intestinal microbiota of cats. Am J Vet Res 74(1):90–95

Bontempo V (2005) Nutrition and health of dogs and cats: evolution of petfood. Vet Res Commun 29(Suppl 2):45–50

Burgener IA, König A, Allenspach K, Sauter SN, Boisclair J, Doherr MG, Jungi TW (2008) Upregulation of toll-like receptors in chronic enteropathies in dogs. J Vet Intern Med 22(3):553–560

Craven M, Acland GM, Mezey JG, Boyko AR, Wang W, Meurs K, McDonough SP, Simpson KW (2010) W1250 Genome-Wide Association Scan reveals polymorphisms in the P67phox Subunit (Ncf2) of the NADPH Oxidase Complex in boxer dogs with adherent and invasive *E. coli*-associated granulomatous colitis: a potential model of chronic granulomatous disease. Gastroenterology 138(5):S683. (Abstract)

Craven M, Dowd SE, McDonough SP, Simpson KW (2009) High throughput pyrosequencing reveals reduced bacterial diversity in the duodenal mucosa of dogs with IBD. J Vet Intern Med 23:731. (Abstr)

Czarnecki-Maulden GL (2008) Effect of dietary modulation of intestinal microbiota on reproduction and early growth. Theriogenology 70(3):286–290

De Godoy MR, Kerr KR, Fahey GC Jr (2013) Alternative dietary fiber sources in companion animal nutrition. Forum Nutr 5(8):3099–3117

Deng P, Swanson KS, (2015) Gut microbiota of humans, dogs and cats: current knowledge and future opportunities and challenges. Br J Nutr 113 (S1):S6–S17

Deusch O, O'Flynn C, Colyer A, Morris P, Allaway D, Jones PG, Swanson KS (2014) Deep Illumina-based shotgun sequencing reveals dietary effects on the structure and function of the fecal microbiome of growing kittens. PLoS One 9(7):e101021

Desai AR, Musil KM, Carr AP, Hill JE (2009) Characterization and quantification of feline fecal microbiota using cpn60 sequence-based methods and investigation of animal-to-animal variation in microbial population structure. Vet Microbiol 137(1–2):120–128

Di Cerbo A, Morales-Medina JC, Palmieri B, Pezzuto F, Cocco R, Flores G, Iannitti T (2017) Functional foods in pet nutrition: focus on dogs and cats. Res Vet Sci 112:161–166

Faber TA, Hopkins AC, Middelbos IS, Price NP, Fahey GC Jr (2011a) Galactoglucomannan oligosaccharide supplementation affects nutrient digestibility, fermentation end-product production, and large bowel microbiota of the dog. J Anim Sci 89(1):103–112

Faber TA, Bauer LL, Price NP, Hopkins AC, Fahey GC (2011b) In vitro digestion and fermentation characteristics of temulose molasses, a coproduct of fibreboard production, and select temulose fractions using canine fecal inoculum. J Agric Food Chem 59(5):1847–1853

Flickinger EA, Fahey GC Jr (2002) Pet food and feed applications of inulin, oligofructose and other oligosaccharides. Br J Nutr 87(Suppl 2):S297–S300

Flickinger EA, Van Loo J, Fahey GC Jr (2003) Nutritional responses to the presence of inulin and oligofructose in the diets of domesticated animals: a review. Crit Rev Food Sci Nutr 43(1):19–60

Franz CM, Huch M, Abriouel H, Holzapfel W, Gálvez A (2011) Enterococci as probiotics and their implications in food safety. Int J Food Microbiol 151(2):125–140

Gagné JW, Wakshlag JJ, Simpson KW, Dowd SE, Latchman S, Brown DA, Brown K, Swanson KS, Fahey GC (2013) Effects of a synbiotic on fecal quality, short-chain fatty acid concentrations, and the microbiome of healthy sled dogs. BMC Vet Res 9(1):246

Garcia-Mazcorro JF, Dowd SE, Poulsen J, Steiner JM, Suchodolski JS (2012) Abundance and short-term temporal variability of fecal microbiota in healthy dogs. Microbiology 1(3):340–347

Grieshop CM, Flickinger EA, Bruce KJ, Patil AR, Czarnecki-Maulden GL, Fahey GC Jr (2004) Gastrointestinal and immunological responses of senior dogs to chicory and mannan-oligosaccharides. Arch Anim Nutr 58:483–493

Grześkowiak Ł, Endo A, Beasley S, Salminen S (2015) Microbiota and probiotics in canine and feline welfare. Anaerobe 34:14–23

Hand D, Wallis C, Colyer A, Penn CW (2013) Pyrosequencing the canine faecal microbiota: breadth and depth of biodiversity. PLoS One 8(1):e53115

Handl S, Dowd SE, Garcia-Mazcorro JF, Steiner JM, Suchodolski JS (2011) Massive parallel 16S rRNA gene pyrosequencing reveals highly diverse fecal bacterial and fungal communities in healthy dogs and cats. FEMS Microbiol Ecol 76(2):301–310

Hang I, Rinttila T, Zentek J, Kettunen A, Alaja S, Apajalahti J, Harmoinen J, de Vos WM, Spillmann T (2012) Effect of high contents of dietary animal-derived protein or carbohydrates on canine faecal microbiota. BMC Vet Res 8(1):90

Herstad HK, Nesheim BB, L'Abée-Lund T, Larsen S, Skancke E (2010) Effects of a probiotic intervention in acute canine gastroenteritis—a controlled clinical trial. J Small Anim Pract 51(1):34–38

Hesta M, Janssens GP, Debraekeleer J, De Wilde R (2001) The effect of oligofructose and inulin on faecal characteristics and nutrient digestibility in healthy cats. J Anim Physiol Anim Nutr (Berl) 85(5–6):135–141

Hooda S, Boler BMV, Kerr KR, Dowd SE, Swanson KS (2013) The gut microbiome of kittens is affected by dietary protein: carbohydrate ratio and associated with blood metabolite and hormone concentrations. Br J Nutr 109(9):1637–1646

Hussein HS, Flickinger EA, Fahey GC Jr (1999) Petfood applications of inulin and oligofructose. J Nutr 129(S7):1454S–1456S

Inness VL, McCartney AL, Khoo C, Gross KL, Gibson GR (2007) Molecular characterisation of the gut microflora of healthy and inflammatory bowel disease cats using fluorescence in situ hybridisation with special reference to Desulfovibrio spp. J Anim Physiol Anim Nutr (Berl) 91(1–2):48–53

Janeczko S, Atwater D, Bogel E, Greiter-Wilke A, Gerold A, Baumgart M, Bender H, McDonough PL, McDonough SP, Goldstein RE, Simpson KW (2008) The relationship of mucosal bacteria to duodenal histopathology, cytokine mRNA, and clinical disease activity in cats with inflammatory bowel disease. Vet Microbiol 128(1–2):178–193

Johnston KL (1999) Small intestinal bacterial overgrowth. Vet Clin North Am Small Anim Pract 29:523–550

Johnston KL, Lamport A, Batt RM (1993) An unexpected bacterial flora in the proximal small intestine of normal cats. Vet Rec 132:362–363

Jugan MC, Rudinsky AJ, Parker VJ, Gilor C (2017) Use of probiotics in small animal veterinary medicine. J Am Vet Med Assoc 250(5):519–528

Kanakupt K, Vester Boler BM, Dunsford BR, Fahey GC Jr (2011) Effects of short-chain fructooligosaccharides and galactooligosaccharides, individually and in combination, on nutrient digestibility, fecal fermentative metabolite concentrations, and large bowel microbial ecology of healthy adults cats. J Anim Sci 89(5):1376–1384

Kerr KR, Forster G, Dowd SE, Ryan EP, Swanson KS (2013) Effects of dietary cooked navy bean on the fecal microbiome of healthy companion dogs. PLoS One 8(9):e74998

Kil DY, Swanson KS (2011) Companion animals symposium: role of microbes in canine and feline health. J Anim Sci 89(5):1498–1505

Knapp BK, Parsons CM, Swanson KS, Fahey GC Jr (2008) Physiological responses to novel carbohydrates as assessed using canine and avian models. J Agric Food Chem 56(17):7999–8006

Kozłowska I, Marć-Pieńkowska J, Bednarczyk M (2016) Beneficial Aspects of Inulin Supplementation as a Fructooligosaccharide Prebiotic in Monogastric Animal Nutrition–A Review. Annals Anim Sci, 16(2)315–331

Lappin MR, Veir JK, Satyaraj E, Czarnecki-Maulden G (2009) Pilot study to evaluate the effect of oral supplementation of Enterococcus faecium SF68 on cats with latent feline herpesvirus 1. J Feline Med Surg 11(8):650–654

Luckschander N, Hall JA, Gaschen F, Forster U, Wenzlow N, Hermann P, Allenspach K, Dobbelaere D, Burgener IA, Welle M (2010) Activation of nuclear factor-kappaB in dogs with chronic enteropathies. Vet Immunol Immunopathol 133(2–4):228–236

Manninen TJ, Rinkinen ML, Beasley SS, Saris PE (2006) Alteration of the canine small-intestinal lactic acid bacterium microbiota by feeding of potential probiotics. Appl Environ Microbiol 72(10):6539–6543

Marshall-Jones ZV, Baillon ML, Croft JM, Butterwick RF (2006) Effects of *Lactobacillus acidophilus* DSM13241 as a probiotic in healthy adult cats. Am J Vet Res 67(6):1005–1012

McCoy S, Gilliland SE (2007) Isolation and characterization of *Lactobacillus* species having potential for use as probiotic cultures for dogs. J Food Sci 72(3):M94–M97

Mentula S, Harmoinen J, Heikkila M, Westermarck E, Rautio M, Huovinen P, Kononen E (2005) Comparison between cultured small-intestinal and fecal microbiotas in Beagle dogs. Appl Environ Microbiol 71:4169–4175

Middelbos IS, Vester Boler BM, Qu A, White BA, Swanson KS, Fahey GC Jr (2010) Phylogenetic characterization of fecal microbial communities of dogs fed diets with or without supplemental dietary fiber using 454 pyrosequencing. PLoS One 5(3):e9768

Middelbos IS, Fastinger ND, Fahey GC Jr (2007) Evaluation of fermentable oligosaccharides in diets fed to dogs in comparison to fiber standards. J Anim Sci 85(11):3033–3044

Nguyen Van N, Taglinger K, Helps CR, Tasker S, Gruffydd-Jones TJ, Day MJ (2006) Measurement of cytokine mRNA expression in intestinal biopsies of cats with inflammatory enteropathy using quantitative real-time RT-PCR. Vet Immunol Immunopathol 113(3–4):404–414

O'Mahony D, Murphy KB, MacSharry J, Boileau T, Sunvold G, Reinhart G, Kiely B, Shanahan F, O'Mahony L (2009) Portrait of a canine probiotic *Bifidobacterium*—from gut to gut. Vet Microbiol 139:106–112

Packey CD, Sartor RB (2009) Commensal bacteria, traditional and opportunistic pathogens, dysbiosis and bacterial killing in inflammatory bowel diseases. Curr Opin Infect Dis 22:292–301

Perelmuter K, Fraga M, Zunino P (2008) In vitro activity of potential probiotic *Lactobacillus murinus* isolated from the dog. J Appl Microbiol 104(6):1718–1725

Pinna C, Biagi G (2016) The Utilisation of Prebiotics and Synbiotics in Dogs. Ital J Anim Sci 13(1):3107

Propst EL, Flickinger EA, Bauer LL, Merchen NR, Fahey GC Jr (2003) A dose-response experiment evaluating the effects of oligofructose and inulin on nutrient digestibility, stool quality, and fecal protein catabolites in healthy adult dogs. J Anim Sci 81(12):3057–3066

Rinkinen M, Jalava K, Westermarck E, Salminen S, Ouwehand AC (2003) Interaction between probiotic lactic acid bacteria and canine enteric pathogens: a risk factor for intestinal *Enterococcus faecium* colonization? Vet Microbiol 92(1–2):111–119

Rishniw M, Wynn SG (2011) Azodyl, a synbiotic, fails to alter azotemia in cats with chronic kidney disease when sprinkled onto food. J Feline Med Surg 13(6):405–409

Ritchie LE, Burke KF, Garcia-Mazcorro JF, Steiner JM, Suchodolski JS (2010) Characterization of fecal microbiota in cats using universal 16S rRNA gene and group-specific primers for *Lactobacillus* and *Bifidobacterium* spp. Vet Microbiol 144(1–2):140–146

Ritchie LE, Steiner JM, Suchodolski JS (2008) Assessment of microbial diversity along the feline intestinal tract using 16S rRNA gene analysis. FEMS Microbiol Ecol 66:590–598

Roberfroid MB, Van Loo JA, Gibson ER (1998) The bifidogenic nature of chicory inulin and its hydrolysis products. J Nutr 128:11–19

Sauter SN, Benyacoub J, Allenspach K, Gaschen F, Ontsouka E, Reuteler G, Cavadini C, Knorr R, Blum JW (2006) Effects of probiotic bacteria in dogs with food responsive diarrhoea treated with an elimination diet. J Anim Physiol Anim Nutr (Berl) 90(7–8):269–277

Simpson KW, Rishniw M, Bellosa M, Liotta J, Lucio A, Baumgart M, Czarnecki-Maulden G, Benyacoub J, Bowman D (2009) Influence of *Enterococcus faecium* SF68 probiotic on giardiasis in dogs. J Vet Intern Med 23(3):476–481

Simpson KW, Dogan B, Rishniw M, Goldstein RE, Klaessig S, McDonough PL, German AJ, Yates RM, Russell DG, Johnson SE, Berg DE, Harel J, Bruant G, McDonough SP, Schukken YH (2006) Adherent and invasive *Escherichia coli* is associated with granulomatous colitis in boxer dogs. Infect Immun 74(8):4778–4792

Sokol H, Pigneur B, Watterlot L, Lakhdari O, Bermudez Humaran LG, Gratadoux JJ, Blugeon S, Bridonneau C, Furet JP, Corthier G, Grangette C, Vasquez N, Pochart P, Trugnan G, Thomas G, Blottiere HM, Dore J, Marteau P, Seksik P, Langella P (2008) *Faecalibacterium prausnitzii* is an anti-inflammatory commensal bacterium identified by gut microbiota analysis of Crohn disease patients. Proc Natl Acad Sci U S A 105:16731–16736

Sparkes AH, Papasouliotis K, Sunvold G, Werrett G, Clarke C, Jones M, Gruffydd-Jones TJ, Reinhart G (1998) Bacterial flora in the duodenum of healthy cats, and effect of dietary supplementation with fructo-oligosaccharides. Am J Vet Res 59:431–435

Stecher B, Hardt WD (2008) The role of microbiota in infectious disease. Trends Microbiol 16:107–114

Strickling JA, Harmon DL, Dawson KA, Gross KL (2000) Evaluation of oligosaccharide addition to dog diets: influences on nutrient digestion and microbial populations. Anim Feed Sci Technol 86:205–219

Strompfová V, Marcináková M, Simonová M, Bogovic-Matijasić B, Lauková A (2006) Application of potential probiotic *Lactobacillus fermentum* AD1 strain in healthy dogs. Anaerobe 12(2):75–79

Suchodolski JS (2011a) Companion animals symposium: microbes and gastrointestinal health of dogs and cats. J Anim Sci 89(5):1520–1530

Suchodolski JS (2011b) Intestinal microbiota of dogs and cats: a bigger world than we thought. Vet Clin North Am Small Anim Pract 41(2):261–272

Suchodolski JS, Xenoulis PG, Paddock CG, Steiner JM, Jergens AE (2010) Molecular analysis of the bacterial microbiota in duodenal biopsies from dogs with idiopathic inflammatory bowel disease. Vet Microbiol 142(3–4):394–400

Suchodolski JS, Dowd SE, Westermarck E, Steiner JM, Wolcott RD, Spillmann T, Harmoinen JA (2009) The effect of the macrolide antibiotic tylosin on microbial diversity in the canine small intestine as demonstrated by massive parallel 16S rRNA gene sequencing. BMC Microbiol 9:210

Suchodolski JS, Camacho J, Steiner JM (2008) Analysis of bacterial diversity in the canine duodenum, jejunum, ileum, and colon by comparative 16S rRNA gene analysis. FEMS Microbiol Ecol 66:567–578

Suchodolski JS, Ruaux CG, Steiner JM, Fetz K, Williams DA (2004) Application of molecular fingerprinting for qualitative assessment of small-intestinal bacterial diversity in dogs. J Clin Microbiol 42:4702–4708

Suchodolski JS, Ruaux CG, Steiner JM, Fetz K, Williams DA (2005) Assessment of the qualitative variation in bacterial microflora among compartments of the intestinal tract of dogs by use of a molecular fingerprinting technique. Am J Vet Res 66:1556–1562

Sunvold GD, Hussein HS, Fahey GC Jr, Merchen NR, Reinhart GA (1995) In vitro fermentation of cellulose, beet pulp, citrus pulp, and citrus pectin using fecal inoculum from cats, dogs, horses, humans, and pigs and ruminal fluid from cattle. J Anim Sci 73:3639–3648

Swanson KS, Dowd SE, Suchodolski JS, Middelbos IS, Vester BM, Barry KA, Nelson KE, Torralba M, Henrissat B, Coutinho PM, Cann IK, White BA, Fahey GC Jr (2011) Phylogenetic and gene-centric metagenomics of the canine intestinal microbiome reveals similarities with humans and mice. ISME J 5(4):639–649

Swanson KS, Grieshop CM, Flickinger EA, Bauer LL, Chow J, Wolf BW, Garleb KA, Fahey GC Jr (2002) Fructooligosaccharides and *Lactobacillus acidophilus* modify gut microbial populations, total tract nutrient digestibilities and fecal protein catabolite concentrations in healthy adult dogs. J Nutr 132(12):3721–3731

Tun HM, Brar MS, Khin N, Jun L, Hui RKH, Dowd SE, Leung FCC (2012) Gene-centric metagenomics analysis of feline intestinal microbiome using 454 junior pyrosequencing. J Microbiol Methods 88(3):369–376

Tzortzis G, Jay AJ, Baillon ML, Gibson GR, Rastall RA (2003) Synthesis of α-galactooligosaccharides with α-galactosidase from *Lactobacillus reuteri* of canine origin. Appl Microbiol Biotechnol 63(3):286–292

Tzortzis G, Goulas AK, Baillon ML, Gibson GR, Rastall RA (2004) *In vitro* evaluation of the fermentation properties of galactooligosaccharides synthesised by alpha-galactosidase from *Lactobacillus reuteri*. Appl Microbiol Biotechnol 64(1):106–111

Verlinden A, Hesta M, Hermans JM, Janssens GPJ (2006) The effects of inulin supplementation of diets with or without hydrolysed protein sources on digestibility, faecal characteristics, haematology and immunoglobulins in dogs. Br J Nutr 96:936–944

Weese JS, Anderson ME (2002) Preliminary evaluation of *Lactobacillus rhamnosus* strain GG, a potential probiotic in dogs. Can Vet J 43(10):771–774

Weese JS, Arroyo L (2003) Bacteriological evaluation of dog and cat diets that claim to contain probiotics. Can Vet J 44(3):212–216

Xenoulis PG, Palculict B, Allenspach K, Steiner JM, Van House AM, Suchodolski JS (2008) Molecular-phylogenetic characterization of microbial communities imbalances in the small intestine of dogs with inflammatory bowel disease. FEMS Microbiol Ecol 66(3):579–589

Young W, Moon CD, Thomas DG, Cave NJ, Bermingham EN (2016) Pre-and post-weaning diet alters the faecal metagenome in the cat with differences vitamin and carbohydrate metabolism gene abundances. Sci Rep 6:34668

# 8 Probiotic Applications for Finfish Aquaculture

Ana Rodiles, Mark D. Rawling, David L. Peggs, Gabriella do Vale Pereira, Sam Voller, Rungtawan Yomla, Benedict T. Standen, Peter Bowyer, and Daniel L. Merrifield

## 8.1 Introduction

Aquaculture is the farming of aquatic organisms including finfish, crustaceans, molluscs, aquatic plants, algae, amphibians, some reptiles and other organisms (such as echinoderms and tunicates). The production of these organisms is practised in fresh, brackish and marine water environments of all climates across the globe, from tropical equatorial regions to within the Arctic Circle.

Aquaculture is the fastest-growing sector of the agribusiness industry, and, although growth has slowed over the past two decades, aquaculture production (excluding aquatic plants and algae) has more than doubled from 32.4 million tonnes in 2000 to 73.8 million tonnes in 2014 (FAO 2016). Furthermore, expansion has consistently exceeded population growth rate in recent years and is therefore seen as a solution to meet an ever-increasing global demand for seafood. In contrast, global capture fisheries have plateaued, and many wild fish stocks have collapsed (FAO 2016). The latest data show that in 2014, aquaculture contributed 44% of total global fishery production. It is predicted that aquaculture production will surpass capture fisheries in 2021, and its input to global food fish supply is expected

A. Rodiles
Trouw Nutrition, Amersfoort, The Netherlands

M.D. Rawling • G. do Vale Pereira • S. Voller • R. Yomla • D.L. Merrifield (✉)
School Biological and Marine Sciences, Plymouth University, Plymouth, UK
e-mail: Daniel.Merrifield@plymouth.ac.uk

D.L. Peggs
Skretting ARC, Stavanger, Norway

B.T. Standen
Biomin Holding GmbH, Getzersdorf, Austria

P. Bowyer
Bell Farms, Schell City, MO, USA



D. Di Gioia, B. Biavati (eds.), *Probiotics and Prebiotics in Animal Health and Food Safety*, https://doi.org/10.1007/978-3-319-71950-4_8

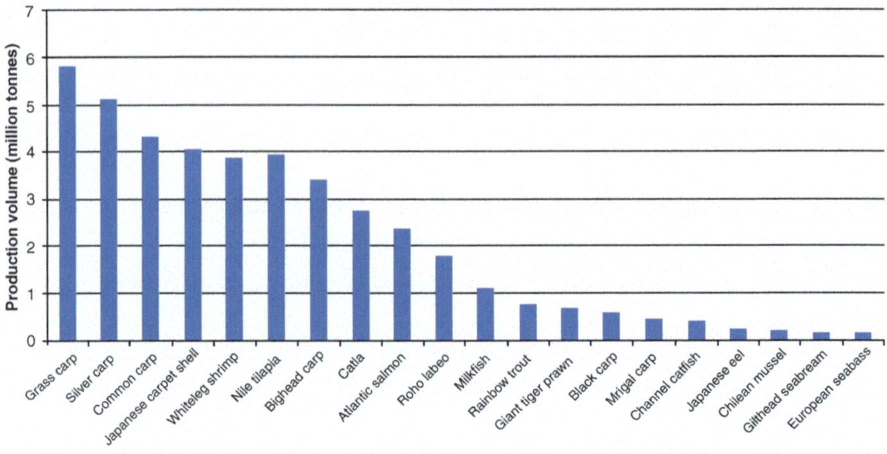


**Fig. 8.1** Production volumes of some of the major aquaculture species (excluding aquatic plants and algae) in 2015 (FAO 2016)

to reach 52% in 2025 (FAO 2016). However, in 2014 the aquaculture sector overtook wild-caught fish contributions in the supply of fish for human consumption (FAO 2016). In comparison to terrestrial meat, farmed fish production has long exceeded sheep and goat meat production volumes, and in 2011 a significant milestone was reached when farmed fish production surpassed beef production (Larsen and Roney 2013).

A total of 543 farmed species (including 362 finfishes, 104 molluscs, 62 crustaceans, 9 aquatic invertebrates and 6 amphibians and reptiles) were registered with production data by FAO in 2014. Figure 8.1 shows the production volumes some of the main farmed species. The majority of farmed fish are derived from freshwater systems (57.7%) (FAO 2016). This includes carps, barbels and other cyprinids (28 million tonnes), tilapias (5.3 million tonnes) and other miscellaneous species (9.1 million tonnes) such as catfishes (FAO 2016) which are farmed across most continents (Fig. 8.1). Mariculture (inclusive of diadromous and marine fish) currently represents only 9.8% of total finfish production; however, its corresponding financial value is estimated at 21% of total farmed fish value due to a number of species which are highly valued by consumers (FAO 2014). These include salmonid species (3.4 million tonnes) of which Atlantic salmon (*Salmo salar*) represents approximately two thirds of salmonid production (FAO 2015). The farming of this species is now a large, well-established industry in the North Atlantic as well as the South-Eastern Pacific. Localised mariculture industries have also thrived, as is the case in the Mediterranean with species such as the European seabass (*Dicentrarchus labrax*) and gilt-head seabream (*Sparus aurata*), which have grown side by side from approximately 5k tonnes in 1990 to nearly 160k tonnes in 2014 (FAO 2016) (Fig. 8.1). Rapid biological and technological advancements have also led to the emergence of newly cultured species over the past decade. Such examples are red drum (*Sciaenops ocellatus*) and cobia (*Rachycentron canadum*) which have increased in production from less than 3k tonnes in the early 2000s to 72k and 40k tonnes, respectively, in 2014 (FAO 2016). The recent closure of the bluefin tuna (*Thunnus orientalis*) life cycle, and subsequent prompt efforts to industrialise the farming of this species, further epitomises the global effort to expand the culture of prized marine food fish.

However, with a rapid increase in production has come a steep learning curve in culture practices and pathology. Aquatic environments are naturally rich in nutrient and pathogen loads, which can fluctuate heavily under seasonal, meteorological and anthropological influence. This can often be exacerbated by intensive production methods, aggravating the potential for disease outbreak amongst stock. Limited historical and scientific knowledge has led this young industry to seek reliable methods of mitigating and treating pathological threats.

Furthermore, an ever-increasing pressure is being placed upon the industry to implement sustainable practices on a socioeconomic and environmental level so as to secure responsible growth and maintenance of the industry. This includes a move away from using marine ingredients in aquafeeds and a reduction in the use of traditional pharmaceuticals in disease prevention, particularly the use of antibiotic growth promoters. However, the range of environments and climates, alongside the vast array of species and their respective immunological and physiological characteristics, means that solutions are complex in this diverse category of livestock farming.

## 8.2 Comparative Physiology and Immunology of Finfish

From an evolutionary perspective, fish (*Teleostei*) are considered as the earliest class of vertebrates having both innate and adaptive immunity. The immune system operates at the crossroads between the innate and adaptive responses and is habituated to the environment and the poikilothermic nature of the fish (Tort et al. 2003). The aquatic environment is highly antigenic, and thus the external barriers of the fish such as the skin, gills and digestive tract play an important role in controlling potential infectious routes. Such protective barriers are reinforced by the production of mucus. Mucus contains a number of humoral soluble compounds, such as lectins, pentraxins, lysozymes, complement proteins, antibacterial peptides and immunoglobulin (IgM, IgT/IgZ), which have an important role in inhibiting the entry of pathogens (for reviews see Foey and Picchietti 2014; Esteban and Cerezuela 2015; Koppang et al. 2015; Salinas and Parra 2015).

An increasing body of evidence, both from mammalian and fish studies indicates that the innate (non-specific) and the acquired (adaptive) immune systems operate synergistically to combat disease. Innate responses in vertebrates and invertebrates are thought to precede the adaptive responses in so much that the innate responses activate and determine the nature of the adaptive response. Thus, they cooperate to maintain homeostasis during development, growth and following tissue damage (see Fig. 8.2; Fearon and Locksley 1996; Fearon 1997).

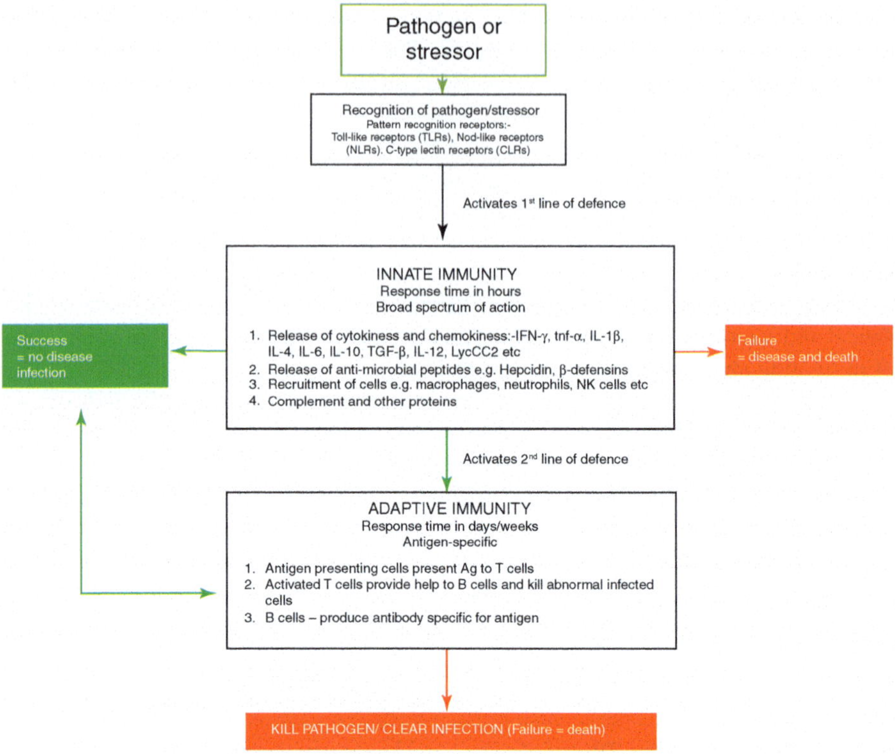


**Fig. 8.2** Summary of the immune responses of finfish

The innate immune system is an evolutionary ancient system characterised by its non-specific nature. It is mediated by germ-line encoded parameters, namely, pattern recognition proteins or receptors (PRP/R). These parameters identify conserved molecular patterns called pathogen-, microbe- and damage-associated molecular patterns (PAMPs, MAMPs, DAMPs, respectively) which are associated with microbes and inherent danger signals from malignant tissues or apoptotic cells (Medzhitov and Janeway 2002). Typical PAMPs include polysaccharides, glycoproteins such as bacterial lipopolysaccharides (LPS), peptidoglycans, DNA CpG motifs and virus-associated double-stranded RNA (dsRNA) (Janeway 1989; Medzhitov and Janeway 2002). The advantage of the innate system, through the process of being inducible by external molecules, allows for a rapid response, which has been tailored by environmental factors and pathogenic associations. As a result, the specificity of the innate defence is an inheritable trait that provides a preliminary line of defence (Medzhitov and Janeway 1998; Carroll and Janeway 1999; Du Pasquier 2001, 2004; Tort et al. 2003; Alvarez-Pellitero 2008). In fish, the innate immune system is commonly divided into three compartments: epithelial/mucosal barrier, the cellular components and humoral components. Figure 8.3 provides a schematic overview of the gut-associated lymphoid tissues (GALT) of a typical teleost. Several

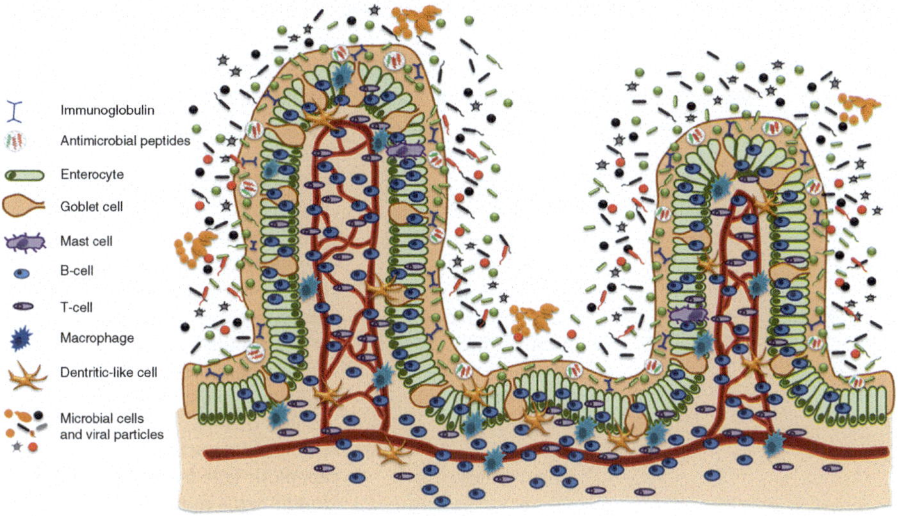


**Fig. 8.3** Schematic representation of the teleost intestine, demonstrating the main cells which comprise the GALT. Note: not to scale

important differences are apparent when compared to the GALT of *Aves* and *Mammalia*; these include a lack of lacteal vessels, an absence of Peyer's patches and associated lymphoid follicles, the possible absence of M cells and an absence of mesenteric lymph nodes. As such, teleost GALT is more of a diffuse collection of cells, rather than the structured tissues found in terrestrial vertebrates.

In the wild, fish have a well-developed and complex innate system that may be constitutive or responsive (Ellis 2001; Magnadottir 2010). In contrast, in a fish farm or a fish tank, the infection pressure is much greater due to the physical constraints. Upon infection, systemic innate immune responses can provide an early defence against the pathogen; however, in most cases pathogens are adept at evading these responses and infecting weak or immuno-compromised fish (Magnadottir 2010). Consequently the immune response of the fish operates at two distinct levels: local and systemic (Table 8.1).

The adaptive immune system is a relatively recent evolutionary development, first appearing in jawed vertebrates about 400–500 million years ago (Tort et al. 2003). The key components in the evolution of the adaptive system are the appearance of the thymus, the B and T lymphocytes and the RAG (recombination activation gene) enzymes, which through the process of gene rearrangement can generate the observed diversity of the immunoglobulin superfamily (B- and T-cell receptors and the major histocompatibility complex). Unlike the innate system, the components of the adaptive system are not germ-line encoded; however, it has an impressive capacity to recognise and respond to very specific structures presented by pathogens (Agrawal et al. 1998). This results in an unlimited diversity of pathogen recognition, and so the specific activity reflects the disease history of the individual.

**Table 8.1** The components and functions of the immune system of fish. Adapted from Schley and Field (2002)

| Immune system | Type of defence | Physical components | Modes of action |
|---|---|---|---|
| Innate immune system | Physical barriers | Skin, gills, scales, mucus membranes | Prevent the entry of antigens from entering systemic circulation, e.g. pathogenic bacteria, parasites |
| | Cell-mediated barriers | Phagocytic cells, e.g. neutrophils, macrophages<br>Inflammatory cells, e.g. mast cells, basophils<br>Natural killer cells<br>Complement system<br>Interferons/ Mx- proteins<br>Transferrin<br>Chemokines | Phagocytosis, secretion and activation, cytokine production, T-lymphocyte stimulation<br>Release of inflammatory mediators, e.g. histamine, prostaglandins<br>Induce apoptosis of infected or malignant cells. Synthesise and secrete IFN-γ<br>Complement activation. Cause apoptotic cell death<br>Inhibit virus replication<br>Chelates iron inhibits growth of bacteria. Activates macrophages<br>Activate/recruit other cells to site of infection |
| | Humoral-mediated barriers (soluble factors) | Acute phase protein<br>Lytic enzymes<br>Antiproteases<br>Antimicrobial peptides | Promote the repair of damaged tissues<br>Modulation of surface charge of bacteria to facilitate phagocytosis<br>Restrict bacteria to growth *in vivo*<br>Induce precipitation and agglutination reactions. Activate complement. Induce cytokine release |
| Adaptive immune response | B lymphocytes<br>T lymphocytes | Plasma cells<br>CD4+ T cells<br>Th1 cells<br>Th2 cells<br>Th17<br>Th22<br>CD8+ T cells | Secrete antibodies<br>Induce activation of lymphocytes<br>Production of IFN-γ, promote intracellular cell-mediated responses<br>Promote humoral (antibody) responses and clear parasitic infections<br>Production of IL-17 and antimicrobial peptides, control extracellular bacterial infections<br>Production of IL-22 and antimicrobial peptides, control extracellular bacterial infections<br>Cytotoxic action—Destroy infected and malignant cells. Suppress activity of lymphocytes |

Note that not all cell types and soluble factors are present in all teleosts

The adaptive responses of fish are predominated by humoral IgM antibody responses which are recognised, typically, to be slower to develop when compared to the mammalian counterparts (Ellis 2001; Magnadottir 2010; Trichet 2010). So, when confronted by a highly variable and antigenic environment, the fish immune response is predominated by a broader range of innate responses characterised by a lack of antigen specificity and memory compensating for a relatively slow reacting and adaptive immune response. Although IgM is present in the mucus secretions which coat the epithelial surfaces, IgT is the specialised mucosal isotype in teleosts, which is analogous to the IgA of terrestrial vertebrates. IgT is transcytosed across the epithelium and released into the mucus with a secretory protein, which affords some protection against the harsh conditions present in, for example, the alimentary tract (Magadan et al. 2015).

## 8.3 The Gastrointestinal Tract of Fishes

The gastrointestinal (GI) tracts of fish have evolved to varying degrees of specialisation to suit a number of different niches. The anatomy and physiology of the fish GI tract depends to a great extent on their diet; carnivores, herbivores, detritivores, algivores and omnivores (De Silva and Anderson 1995) display considerable variations in alimentary tract morphology and function. These physiological differences include the presence or absence of pharyngeal teeth or gizzards, the presence or absence of a stomach, the stomach shape and size, the presence or absence of pyloric caeca, the number of pyloric caeca, the intestinal length and its degree of looping and motility (Kapoor et al. 1975). These adaptations reflect the fact that fish are a very diverse group of vertebrates able to process a multitude of foodstuffs. The pancreas and epithelial cells secrete endogenous enzymes into the lumen; however, fermentation processes may be involved in the degradation of specific nutrients from plants and algae in many teleosts (Ray and Ringø 2014). Many fish species do not fit neatly or completely into distinct dietary classifications, however, and depending on feed availability and life cycle, fish may display different feeding strategies during their life span (Olsen and Ringø 1997).

Generally, the GI tract can be divided into distinct regions: foregut (mouth, gill arch, oesophagus, stomach and pyloric caeca), anterior intestine, mid-intestine and posterior intestine; however, there are some deviations on this classification which depend largely on the dietary habits of the species (Harder 1975; Løkka et al. 2013; Ray and Ringø 2014). Some species have evolved pharyngeal teeth or gizzards for the grinding of ingested food (e.g. common carp (*Cyprinus carpio*) and milkfish (*Chanos chanos*)), respectively. The stomach temporarily stores ingested food and releases hydrochloric acid and trypsinogen to initiate the digestive process. The pH may be as low as 2 or 3 in many species and is effective at reducing the viable microbial load in the chyme. In the absence of the stomach, many species have developed a saclike structure called the intestinal bulb, or *pseudogaster*, which performs this process (Fänge and Grove 1979; Olsen and Ringø 1997). Other species lack such structures, and it is thought that an acid phase in the digestion process is absent in these species. The pyloric caeca are fingerlike projections located in the anterior part of the intestine acting as extensions of this organ. They are not present in all fish but in some species can account for as much as 70% of the gut (Wulff et al. 2012). There is also a large variation in the size and number of the pyloric caeca between the species which possess these structures. It is thought that the functions of the pyloric caeca are to increase absorptive surface area of the intestine and thus aid in the digestive process (Ray and Ringø 2014). Beyond the stomach/intestinal bulb and pyloric regions, the intestine is a simple cylindrical structure which continues to the anus. This organ is the primary site of digestion of feed and absorption of nutrients; it also plays a crucial role in the water-electrolyte balance and endocrine regulation, as well as supporting metabolism and immunity (Ringø et al. 2003).

Within the fish intestine, where the pH is generally between 7 and 9, alkaline enzymes, bile salts, bicarbonate, antimicrobial substances and mucus are secreted into the lumen (Ray and Ringø 2014). The mucus, as well as the dietary components, in the chyme serve as substrates which support microbial life (Ray and Ringø 2014). In general, the relative intestinal length correlates to the feeding habit of the fish. In carnivorous fish, the length of the gut is approximately equal to or slightly less than the total body length, whilst in herbivorous and detritivorous species, the gut can exceed 20 times that of the total body length (Parameswaran et al. 1974; Olsen and Ringø 1997). This is, however, a general rule with many exceptions. Indeed, the relative gut length may change as a consequence of transferring from a carnivorous diet to a herbivorous diet as is the case with *Labeo gonius* and *Labeo calbasu* (Parameswaran et al. 1974; Sinha 1976). As in mammals, differences in the relative intestinal length reflect the nature and nutritional value of the food being processed (Clements and Raubenheimer 2005). The generally longer intestine of herbivorous fish enlarges the absorptive surface area, increasing the retention time in order to enhance the utilisation of foods with relatively poor nutritional value (Olsen and Ringø 1997). The passage rate and residence time from the stomach to the anus depend on several variables including temperature, stress, meal, pellet and fish size (Smith 1980). It is accepted that diet plays a crucial role in determining the intestinal microbial community composition and activity; however, the effect of the digestion speed, the gut microbiota and the influence on the interactions between the host and microbes are still largely unknown (Ray and Ringø 2014).

## 8.4 The Gut Microbiota of Fishes

The microbiomes of fish are dominated by complex and diverse communities of *Bacteria* and to a lesser extent yeasts, *Archaea*, viruses and protists which inhabit the skin, gills and GI tract. In larval fish, the microbiota becomes established following first feeding, initially comprised by microbes from the egg surfaces (the epibiota), rearing water and first feed. Ontologically, this precedes full activation of the adaptive immune response and immunological memory. Once fully established, the abundance of the gut microbiota is typically several orders of magnitude greater than the microbial communities inhabiting the skin, gills, rearing water or aquafeeds (for review see: Romero et al. 2014). Despite a growing body of knowledge derived from hundreds of research studies published in scientific journals, our knowledge of the composition, activities and functions of the gut microbiota of fish is somewhat primitive compared to that of the gut microbiota of terrestrial animals. Worse still, our understanding of the microbial communities inhabiting the skin and gills lags far behind what we know of the gut microbiota (Merrifield and Rodiles 2015).

Much of our knowledge of the gut microbiota has been derived from studies using culture-based or DNA barcoding methods (e.g. using DGGE). Such studies have yielded important findings, including evidence of the sensitivity of the microbiota to a number of biotic and abiotic factors. However, considering that the gut microbiota of finfish are mostly uncultivable under routine culture conditions (Romero et al. 2014) and that DNA barcoding methods only detect the dominant operational taxonomic units (OTUs) and provides only a semi-quantitative analysis, our understanding of the true diversity and abundance of the microbiota, and the extent that biotic and abiotic factors impact them, has been somewhat limited. The continual decreasing cost of sequencing and improvements of bioinformatics tools has led to a new wave of research on the gut microbiota of fish, which, using high-throughput sequencing approaches to generate 16S rRNA libraries, has extended our knowledge of the "rare biosphere" of the fish gut and provided fascinating insights into the effects of diet on these communities (Apper et al. 2016; Falcinelli et al. 2015, 2016; Standen et al. 2015). Such studies have also demonstrated variability of bacterial communities, as well as core communities, across different GI sites within a species (Gajardo et al. 2016), as well as across different life histories and biogeography (Llewellyn et al. 2015). The rapid proliferation of studies in the last 5 years which have used high-throughput sequencing analysis is most welcomed, but the information available from such studies is highly fragmented with inconsistencies in the methods used (e.g. sequencing platforms, bioinformatics pipelines, 16S V regions) and the often ambiguous description of the sample type. This piecemeal approach prevents an overarching understanding of the microbiomes of fish which can only be rectified by coordinated and concerted research efforts using standardised and consistent analytical approaches.

It is increasingly clear that the microbiomes of fish are intimately involved in multiple aspects of nutrition and disease, as well as host development at the larval stages. The microbiome actively contributes to the digestive process, and a wide range of microbes capable of producing extracellular digestive enzymes have been isolated from the gut of fish. Many of these enzymes are often enzymes that the host is unable to produce (or may only produce in low concentrations), such as cellulase, chitinase and phytase (Ray et al. 2012). Microbial fermentation processes could also aid host digestive function, especially in herbivorous, omnivorous and detritivorous fish species. Indeed, the presence of short-chain fatty acids (SCFAs), including acetate and to a lesser extent propionate and butyrate, have been described in the intestine of fish (Clements et al. 1994; Clements and Choat 1995; Mountfort et al. 2002). These SCFAs are rapidly absorbed from the gut lumen and supply energy either directly to the enterocytes or to other organs via the vascular system. These fermentation products may also increase the solubility of the minerals by decreasing the pH in the gut lumen and can make the intestinal tract an unfavourable environment for opportunistic pathogens (Merrifield and Rodiles 2015). The bacterial flora has also been reported to produce vitamins. For example, *Cetobacterium somerae* produces copious amounts of vitamin B12 (cobalamin) and has been identified in the gut of various fish (Sugita et al. 1991; Tsuchiya et al. 2008; Larsen et al. 2014). Vitamin B12 is involved in erythrocyte development and fatty acid metabolism (Lin et al. 2010), amongst other things, and the contribution of *C. somerae* to the vitamin B12 requirements of fish is inferred by the fact that fish such as Nile tilapia *Oreochromis niloticus* and rainbow trout *Oncorhynchus mykiss* where this bacterium is commonly present in the GI tract typically have no dietary requirement for vitamin B12. In contrast, species such as the channel catfish *Ictalurus punctatus* and Japanese eel *Anguilla japonica*, where the bacterium is not commonly found in the GI tract, require the dietary provision of vitamin B12 (Tsuchiya et al. 2008).

The importance of the microbiota in terms of host development and immune status has been presented in several gnotobiotic and germ-free studies. The seminal work of Rawls et al. (2004) revealed that 212 genes in the GI tract were regulated by the microbial communities; these genes were involved in numerous processes including immunity (e.g. *Saa1*, *Crp*, *C3* and *Socs3*), cell division and DNA replication (e.g. minichromosome maintenance genes and *Pcna*) and nutrition (e.g. genes involved in lipid metabolism, *Cpt1a*, *Ctp1b* and *Fbp1*). Similar studies reveal that the absence of the microbiota retards gut development, function and immune status, as revealed by reduced levels of enteroendocrine cells and goblet cells, a lack of brush border intestinal alkaline phosphatase activity (an enzyme which detoxifies bacterial endotoxin, amongst other things), reduced epithelial cell turnover, immature enterocyte glycan patterns and a loss of epidermal integrity (Bates et al. 2006). Other studies demonstrate that the microbiome is also important in mediating barrier function (i.e. excluding foreign pathogens), through competition for adhesion sites and nutrients and via the production of various antimicrobial compounds (for reviews see Merrifield et al. 2014; Romero et al. 2014; Merrifield and Rodiles 2015).

## 8.5 Probiotics Used in Aquaculture

The aqueous environment that surrounds aquatic animals can support a rich community of microbes. This affords an alternative delivery mechanism since the rearing water could be used to supply beneficial microbes to the host. The possibility of using beneficial microbes to improve the microbial population, or chemical quality of the rearing water, is therefore a unique opportunity for exploitation. Such approaches have led to debate around the definition of the term probiotic when used for aquatic organisms (Merrifield et al. 2010a, b). Applications of microbes to improve the chemical (i.e. the breakdown of toxins) or microbial (i.e. the reduction of known pathogens) quality of the rearing water are described as bioremediation or biocontrol applications, respectively. Though some scientists refer to these applications as probiotics, others restrict the term to incidences whereby water-based provision (or dietary provision) of beneficial microbes leads to colonisation of the gut of the target organism.

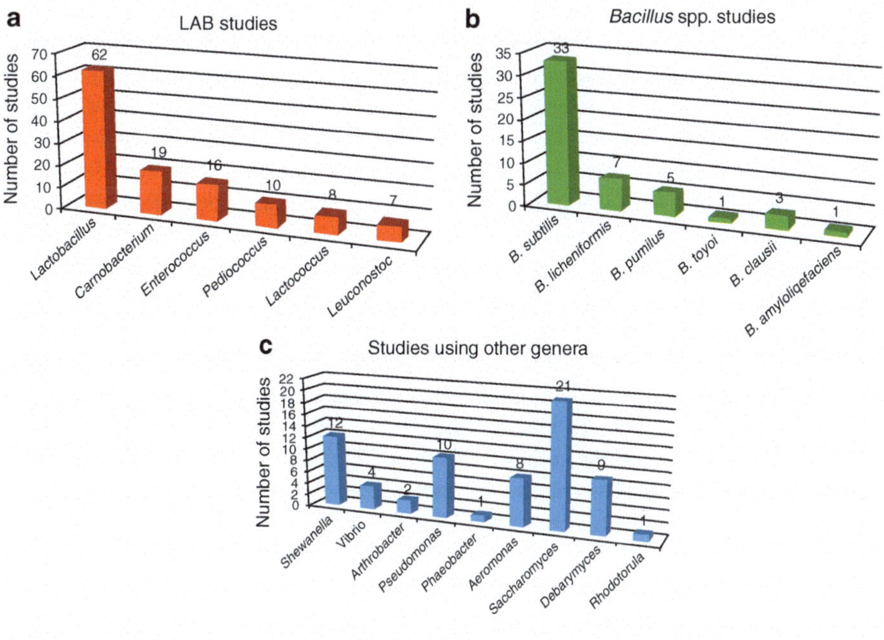


**Fig. 8.4** Abundance of scientific studies (from the reviews of Merrifield et al. 2010a, Dimitroglou et al. 2011; Carnevali et al. 2014; Lauzon et al. 2014) in fish distributed by probiotic species investigated. *A* = LAB, *B* = *Bacillus* spp. and *C* = other genera

Research on the application of probiotic microorganisms for aquatic species began in earnest more than two decades ago. Since this time, the body of research has grown substantially with hundreds of papers now available, covering the effects of probiotics on fish growth performance, feed conversion, gut morphology, immune status, disease resistance, stress and fecundity (for reviews see: Merrifield et al. 2010a; Dimitroglou et al. 2011; Carnevali et al. 2014; Lauzon et al. 2014). These review articles cover around 200 *in vivo* fish probiotic studies (refer to Fig. 8.4) and reveal that, as of 2014, studies have been carried out in over 20 fish species, with the most well-researched probiotic genera being *Lactobacillus* (>60 studies), *Bacillus* (>40 studies) and *Saccharomyces* (>20 studies). With a rapidly growing body of knowledge which reveals the potential benefits of probiotics for important aquaculture species, it has become increasingly common to use probiotics, or other microbial modifiers, in aquaculture practices.

### 8.5.1 Probiotic Colonisation and Modulation of the Intestinal Microbiota

Of all the functions and modes of actions of probiotics that have been reported in aquaculture, the capacity of a probiont to colonise the intestinal tract and positively modulate the fish gut microbiome is considered the most important. Probiotics use a variety of mechanisms to compete with endogenous microbes in order to establish populations in the intestine; these include production of inhibitory compounds, competition for chemicals or available energy, competition for adhesion sites, inhibition of virulence gene expression or disruption of quorum sensing (Merrifield et al. 2010a, b; Merrifield and Carnevali 2014). Numerous studies have reported that probiotics can survive the upper GI tract of fish; thereafter, they are able to populate the lumen (as components of the allochthonous microbiota) or the mucus or epithelial surfaces (as components of the autochthonous microbiota). Readers with a specific interest are referred to the review of Merrifield and Carnevali (2014). In brief, the impact of the probiotic on the gut microbiota can lead to a multitude of possible outcomes, including elevated LAB levels (Ferguson et al. 2010; Jatoba et al. 2011; Standen et al. 2013, 2016), elevated total viable counts (Ridha and Azad 2012), decreased total viable counts (Jatoba et al. 2011), decreased presumptive pathogen levels (Jatoba et al. 2011; Del'Duca et al. 2013) and altered microbial diversity (Ramos et al. 2013). High-throughput sequencing studies have revealed that the relative abundance of different taxa or OTUs is affected by probiotic feeding (Falcinelli et al. 2015, 2016). Such inconsistent, and sometimes contradictory, effects on the gut microbiota are a result of the different resident microbiota present in different fish species, different probiotic feeding regimes and different fish rearing conditions. Irrespective of these factors, it is clear that continued provision of probiotic feeding is required to maintain the implanted probiotic population, with several studies revealing that the probiotic abundance within the intestine decreases to non-detectable levels within 2–3 weeks after the cessation of feeding (Merrifield and Carnevali 2014).

### 8.5.2 Probiotic Benefits Reported in Finfish

The mucus layer in the GI tract provides a physical, mechanical and chemical barrier against pathogenic insults. It contains mucins, of which *Muc2*, *Muc2-like*, *Muc13* and *I-Muc* appear to be the most important in the alimentary mucus of teleosts (although not all are present in all teleosts), which help to bind and trap pathogens. It also contains various antimicrobial peptides and antibodies. Combined with a continuous turnover and sloughing of the surface layer and replenishment from goblet cells, it provides an effective first barrier that potential pathogens have to negotiate. Probiotic provision has been observed to modulate intestinal mucus characteristics and the attachment success of pathogens to the intestinal mucus of fish. For example, several studies have revealed, *in vitro*, that probiotics may retard pathogen adhesion to, or growth within, fish intestinal mucus (Chabrillón et al. 2005; Balcázar et al. 2008). Through histological analysis of the intestine, multiple studies have demonstrated elevated goblet cells in the intestine of probiotic-fed fish, which has been interpreted as being indicative of elevated mucus production (Standen et al. 2013, 2016; Reda and Selim 2015). Further, there is evidence that probiotics may also be able to increase the lysozyme activity of the intestinal mucus of rainbow trout (Newaj-Fyzul et al. 2007). Taken together, these findings provide clear evidence that probiotics have the potential to enhance the protective role of fish intestinal mucus.

The complex host-microbe interactions, which occur at the intestinal barrier, are only partly described in fish, and the mechanisms involved therein are poorly understood. However, fish are known to share certain molecules and immune processes with mammals where the depth of knowledge of this topic is far greater. The expression of PRRs allows for the detection of microbes at the mucosal interface. Perhaps the best characterised receptors in fish are those belonging to the toll-like receptor (TLR) and the intracytoplasmic Nod-like receptor families. These receptors are involved in the recognition of PAMPs, MAMPs or commensal-associated molecular patterns (CAMPs). TLR recognition triggers a series of molecular pathways which include adaptor molecules, such as Myd88, and the subsequent production of the transcription factor NFκB which leads to the production of cytokines including those involved in the inflammatory responses, for example, tumour necrosis factor-α (TNF-α), interleukin-1β (IL-1β), interleukin-8 (IL-8) and interleukin-10 (IL-10) (Tlaskalova-Hogenova et al. 2005).

A number of studies assessing the effects of probiotics on fish immunity have reported microbial-associated immune responses in the GI tract. These include elevated intestinal TLR (Standen et al. 2016), TNF-α (Liu et al. 2013; Standen et al. 2013, 2016), IL-1β (Liu et al. 2013; Standen et al. 2016) and IL-8 (Pérez-Sánchez et al. 2011) mRNA levels in probiotic-fed fish. On the contrary, Picchietti et al. (2009) demonstrated probiotic (*Lactobacillus delbrueckii*)-induced lower intestinal transcript levels of IL-β as well as trends towards lower IL-10, Cox-2 and transforming growth factor-β (TGF-β). Standen et al. (2016) observed concomitant increased intestinal transcripts of both pro-inflammatory (TNF-α and IL-1β) and anti-inflammatory (TGF-β and IL-10) cytokines when feeding a multispecies probiotic product to tilapia. Similar pro-inflammatory and anti-inflammatory signals were observed in the spleen and kidney of probiotic-fed rainbow trout by Panigrahi et al. (2007), and Liu et al. (2013) revealed a bacterial species-dependent, and time-dependent, effect of probiotics on the intestinal expression of TNF-α, IL-1β and TGF-β genes. In addition, and indeed likely in response to and subsequently also contributing to, such changes in immune regulatory gene expression in the intestine, probiotics can stimulate an increase in the number of intestinal Ig + cells, acidophilic granulocytes, T cells (Picchietti et al. 2007, 2009) and total intraepithelial leucocytes (Standen et al. 2013, 2015, 2016).

Beyond the localised intestinal responses, many studies have reported increased systemic or peripheral immune responses including elevated serum lysozyme activity (Ferguson et al. 2010; Wang et al. 2008; Telli et al. 2014), serum alternative complement activity (Wang et al. 2008; Pirarat et al. 2006, 2011), serum myeloperoxidase content (Wang et al. 2008; Zhou et al. 2010), serum bactericidal activity (Pirarat et al. 2011; Abdel-Tawwab 2012), peripheral leucocyte levels (Ferguson et al. 2010; Eissa and Abou-ElGheit 2014), peripheral Ig levels (Ridha and Azad 2012), respiratory burst activity (Aly et al. 2008; Wang et al. 2008; Zhou et al. 2010; Iwashita et al. 2015), phagocytic activity (Ridha and Azad 2015) and modulated expression of cytokine genes in the lymphoid organs (Pérez-Sánchez et al. 2011; Pirarat et al. 2011; Liu et al. 2013) of fish fed probiotic-supplemented diets. With clear potential to improve both mucosal and systemic immune responses, it is therefore not surprising that there are a large number of studies which have reported improved disease resistance of probiotic-fed fish. Such benefits have been observed with a wide variety of probiotics, in numerous fish species, and against a range of bacterial pathogens, viruses and ectoparasites. Since disease resistance studies are considered by many as the ultimate validation of probiotic efficacy, there are a large number of fish studies on this topic. The review of Lauzon et al. (2014) summarised all of the available peer-reviewed literature on the effects of probiotics on the disease resistance of cold water fish, and of the 43 *in vivo* salmonid challenge studies available at the time, at least one of the probiotic treatment regimens was able to reduce mortalities in 40 of the studies (93%). This demonstrates a clear potential for probiotics to improve fish disease resistance, but this statistic should be viewed with caution because studies which fail to induce improved disease resistance are less likely to be published. Likewise, there are a large number of studies which did not observe the aforementioned localised or systemic immunological benefits. Readers with a specific interest in this topic are referred to Merrifield et al. (2010a), Dimitroglou et al. (2011), Carnevali et al. (2014) and Lauzon et al. (2014).

A number of studies have investigated the impact of probiotics on the ultrastructure of the intestine. Several of these studies have reported that dietary probiotics can improve the uniformity, density and/or length of the microvilli comprising the apical brush border in the intestine of a number of fish species (Sáenz de Rodrigáñez et al. 2009; Merrifield et al. 2010b; Standen et al. 2015; Falcinelli et al. 2016). In addition, a plethora of studies have revealed nutritional benefits as a consequence of dietary probiotic provisions. Such benefits include elevated intestinal enzyme activities. Examples include elevated intestinal protease, amylase and cellulase activities in grass carp fed *B. coagulans* (Wang 2011), amylase activities in grass carp fed *Rhodopseudomonas palustris* and *Lb. acidophilus* (Wang 2011) and lipase, protease and amylase activities in the intestine of common carp fed *Bacillus* sp. and photosynthetic bacteria (Yanbo and Zirong 2006). In turbot, increased protein degradation in the distal intestine was observed when fed *Vibrio proteolyticus* supplemented diets, resulting in higher nitrogen digestibility and higher ammonia contents and an elevated fraction of smaller soluble proteins in the intestine (De Schrijver and Ollevier 2000).

Several recent studies using zebrafish larvae provide novel insight into other mechanisms that are involved in modulating growth, nutrient utilisation and metabolism of probiotic-fed fish (Falcinelli et al. 2015, 2016). Feeding of *L. rhamnosus* to larval zebrafish modulated host lipid processing by the downregulation of genes involved in cholesterol and triglyceride metabolism (*fit2*, *agpat4*, *dgat2*, *mgll*, *hnf4a*, *scap* and *cck*) which resulted in decreasing larval total body cholesterol and triglyceride content and elevated fatty acid levels (Falcinelli et al. 2015). These changes resulted in elevated zebrafish larval growth performance. The application of *L. rhamnosus* has also been reported to upregulate the expression of genes involved in elevating blood glucose levels (*nucb2a*, *Glp-1* and *insulin*) and genes involved in suppressing appetite (*leptin* and *mc4r*) (Falcinelli et al. 2016). Concomitantly, genes involved in enhancing appetite (*cb1* and *npy*) were downregulated, larval whole-body glucose levels were decreased and larval appetite was reduced, as evidenced by lower feed intake. It is not yet clear how suppression of appetite correlates with improved growth performance and digestive function, neither is it clear if such an effect can be extended to other probiotic species or fish species. Nonetheless, from such studies it is becoming increasingly clear that the gut microbiota, and probiotics, play key roles in the digestive processes of fish, which have only been partly described to date. These include modulation of gene networks involved in glucose metabolism, lipid metabolism and appetite, as well as improvements in the brush border morphology, yielding higher potential absorptive surface area and elevated intestinal digestive enzyme activities. Probiotic feeding may also enhance skeletal development of fish larvae, with studies reporting that probiotics participate in the regulation of genes involved in osteocyte formation, such as *Mapk1/3* in zebrafish (Maradonna et al. 2013) and *bglap* in European seabass (Lamari et al. 2013). These are likely mechanisms that underpin improved host growth performance, which has frequently been reported in probiotic-fed fish (Yan et al. 2016; Munir et al. 2016; Standen et al. 2016; Hamdan et al. 2016).

Results from a growing number of studies have also demonstrated the potential positive effects of probiotics on improving reproductive performance. For example, studies have revealed enhanced fertilisation rates, viable egg abundance and egg maturation rates, of probiotic-fed fish (Ghosh et al. 2007; Gioacchini et al. 2010a, b, c, 2011, 2012, 2013; Giorgini et al. 2010; Lombardo et al. 2011). Several of these studies have begun to reveal the mechanisms which underpin these effects, which include higher gonadal somatic indices, enhanced oocyte germinal vesicle breakdown, elevated responsiveness of oocytes to maturation inducing hormone and modulations in the expression of genes involved in reproduction. These studies are reviewed by Gioacchini et al. (2014).

**Conclusions**

Despite the deep body of research which demonstrates the possible benefits of utilising probiotics in fish rearing (as summarised in Sect. 8.5.2), there are an equal, or greater, number of studies which reveal a lack of effect, either positive or negative, when applying probiotics to fish. Further, there are often difficulties in obtaining reproducible outcomes, with some studies using the same probiotics and the same target fish species but obtaining differing results. This can partially be explained by the differences in probiotic feeding regime, basal diets used, fish life stage and culture conditions. It is becoming clear that biogeography, life history, seasonality and diet are factors that influence the composition or activity of the microbiomes of fishes. Since the probiotic concept is based on improving the gut microbiome by transplanting a population of beneficial microbes, the inter- and intraspecies resident host gut microbiota variations under different trial conditions are likely to influence the efficacy of probiotics and thus hamper reproducibility. It is difficult therefore, although not impossible, to find probiotic strains that have the versatility to work across multiple fish species, rearing conditions and life stages. One strain which has a well-documented level of success and reproducibility is *Pediococcus acidilactici* CNCM 18/5MA, which is sold under the brand name Bactocell® (Lallemand SAS, France). At present this is the only strain authorised for use in aquaculture as a probiotic in the EU. An alternative strategy is to use a combination of probiotic strains that have complimentary, or synergistic, modes of action and benefits, in order to maximise the efficacy for different fish species and for use under different conditions and life histories. A good example is the Aquastar® product range produced by Biomin GmbH (Austria) which contains strains of *B. subtilis*, *L. reuteri*, *P. acidilactici* and *Enterococcus faecium*. Both of these commercial products have a rich and diverse range of well-documented benefits across many fish and crustacean species. In contrast, there are a large number of products available on the market which lack credibility due to a lack of scientific data to support claimed benefits and some spurious products which do not contain the species or concentrations claimed on the product labels and marketing materials (Nimrat and Vuthiphandchai 2011).

Moving forward, spurious products must be removed from the market by regulatory authorities and market forces, and sustained research efforts must be made to increase the reproducibility of the benefits of the efficacious products on the market. Since the probiotic must populate the intestinal tract of the host species, and through competition and antagonism with endogenous microbial communities, favourably modulate the host microbiota, we must increase our understanding of the composition and functionality of the gut microbiota of fishes. Gaining a better understanding of the microbiomes of fishes at the larval stages will help to improve the efficacy of probiotic intervention with live feeds and starter feeds. Further understanding of the *normal* microbiomes of fish with emphasis on biogeography, life history, host genotypes, seasonality and other factors, will help to ascertain which probiotic regimes are appropriate for a given species, at a given life stage or stage in the production cycle. Armed with such information, it should be easier to design appropriate probiotic strategies and may lead to better tailored application solutions. The use of probiotics (including biocontrol and bioremediation applications) has a bright future in the farming of aquatic animals, and there are many further opportunities to be exploited. These include expanding probiotic applications to new and emerging fish species, optimising probiotic regimes to the ever-evolving dietary formulations which contain lower levels of marine ingredients and higher levels of nontraditional ingredients and improving technologies for easier inclusion of viable probiotics into aquafeeds.

## References

Abdel-Tawwab M (2012) Interactive effects of dietary protein and live bakery yeast, *Saccharomyces cerevisiae* on growth performance of Nile tilapia, *Oreochromis niloticus* (L.) fry and their challenge against *Aeromonas hydrophila* infection. Aquac Int 20:317–331

Agrawal A, Eastman QM, Schatz DG (1998) Transposition mediated by RAG1 and RAG2 and its implications for the evolution of the immune system. Nature 394:744–751

Alvarez-Pellitero P (2008) Fish immunity and parasite infections: from innate immunity to immunoprophylactic prospects. Vet Immunol Immunopathol 126(3–4):171–198

Aly SM, Mohamed MF, John G (2008) Effect of probiotics on the survival, growth and challenge infection in *Tilapia nilotica* (*Oreochromis niloticus*). Aquac Res 39:647–656

Apper E, Weissman D, Respondek F, Guyonvarch A, Baron F, Boisot P, Rodiles A, Merrifield DL (2016) Hydrolysed wheat gluten as part of a diet based on animal and plant proteins supports good growth performance of Asian seabass (*Lates calcarifer*), without impairing intestinal morphology or microbiota. Aquaculture 453:40–48
Balcázar JL, Vendrell D, de Blas I, Ruiz-Zarzuela I, Múzquiz JL, Gironés O (2008) Characterization of probiotic properties of lactic acid bacteria isolated from intestinal microbiota of fish. Aquaculture 278:188–191
Bates JM, Mittge E, Kuhlman J, Baden KN, Cheesman SE, Guillemin K (2006) Distinct signals from the microbiota promote different aspects of zebrafish gut differentiation. Dev Biol 297(2):374–386
Carnevali O, Sun Y-Z, Merrifield DL, Zhou Z, Picchietti S (2014) Probiotic applications in temperate and warm water fish species. In: Merrifield DL, Ringø E (eds) Aquaculture nutrition: gut health, probiotics and prebiotics. Wiley, London, pp 185–222
Carroll MC, Janeway CA (1999) Innate immunity. Curr Opin Immunol 11(1):11–12
Chabrillón M, Rico RM, Balebona MC, Moriñigo MÁ (2005) Adhesion to sole, *Solea senegalensis* Kaup, mucus of microorganisms isolated from farmed fish, and their interaction with Photobacterium damselae subsp. piscicida. J Fish Dis 28:229–237
Clements K, Choat H (1995) Fermentation in tropical marine herbivorous fishes. Physiol Zool 68:355–378
Clements KD, Raubenheimer D (2005) Feeding and nutrition. In: Evans DH, Claiborne JB (eds) The physiology of fishes, 3rd edn. CRC Press, Gainesville, pp 47–82
Clements KD, Gleeson VP, Slaytor M (1994) Short-chain fatty acid metabolism in temperate marine herbivorous fish. J Comp Physiol B 164(5):372–377
De Schrijver R, Ollevier F (2000) Protein digestion in juvenile turbot (*Scophthalmus maximus*) and effects of dietary administration of *Vibrio proteolyticus*. Aquaculture 186:107–116
De Silva SS, Anderson TA (1995) Fish nutrition in aquaculture. Chapman & Hall, London
Del'Duca A, Evangelista Cesar D, Galuppo Diniz C, Abreu PC (2013) Evaluation of the presence and efficiency of potential probiotic bacteria in the gut of tilapia (*Oreochromis niloticus*) using the fluorescent in situ hybridization technique. Aquaculture 388–391:115–121
Dimitroglou A, Merrifield DL, Carnevali O, Picchietti S, Avella M, Daniels C, Güroy D, Davies SJ (2011) Microbial manipulations to improve fish health and production-a Mediterranean perspective. Fish Shellfish Immunol 30(1):1–16
Du Pasquier L (2001) The immune system of invertebrates and vertebrates. Comp Biochem Physiol B: Biochem Mol Biol 129(1):1–15
Du Pasquier L (2004) Innate immunity in early chordates and the appearance of adaptive immunity. C R Biol 327(6):591–601
Eissa N, Abou-ElGheit E (2014) Dietary supplementation impacts of potential non-pathogenic isolates on growth performance, hematological parameters and disease resistance in Nile tilapia (*Oreochromis niloticus*). J Vet Adv 4:712–719
Ellis AE (2001) Innate host defense mechanisms of fish against viruses and bacteria. Dev Comp Immunol 25(8–9):827–839
Esteban MA, Cerezuela R (2015) Fish mucosal immunity: skin. In: Beck BH, Peatman E (eds) Mucosal health in aquaculture. Wiley, London, pp 67–92
Falcinelli S, Picchietti S, Rodiles A, Cossignani L, Merrifield DL, Taddei AR, Maradonna F, Olivotto I, Gioacchini G, Carnevali O (2015) *Lactobacillus rhamnosus* lowers zebrafish lipid content by changing gut microbiota and host transcription of genes involved in lipid metabolism. Sci Rep 5
Falcinelli S, Rodiles A, Unniappan S, Picchietti S, Gioacchini G, Merrifield DL, Carnevali O (2016) Probiotic treatment reduces appetite and glucose level in the zebrafish model. Sci Rep 6
Fänge R, Grove D (1979) Digestion. In: Hoar WS, Randall DJ, Brett JR (eds) Fish physiology, vol VIII. Academic Press, New York, pp 161–260
FAO (2014) The state of world fisheries and aquaculture 2014. Rome. 223 pp
FAO (2015) Coping with climate change—the roles of genetic resources for food and agriculture. Rome. http://www.fao.org/3/a-i3866e.pdf

FAO (2016) The state of world fisheries and aquaculture 2016. Contributing to food security and nutrition for all. Rome. 200 pp
Fearon DT (1997) Seeking wisdom in innate immunity. Nature 388(6640):323–324
Fearon DT, Locksley RM (1996) The instructive role of innate immunity in the acquired immune response. Science 272(5258):50–53
Ferguson RM, Merrifield DL, Harper GM, Rawling MD, Mustafa S, Picchietti S, Balcazar JL, Davies SJ (2010) The effect of *Pediococcus acidilactici* on the gut microbiota and immune status of on-growing red tilapia (*Oreochromis niloticus*). J Appl Microbiol 109(3):851–862
Foey A, Picchietti S (2014) Immune defences of teleost fish. In: Merrifield DL, Ringø E (eds) Aquaculture nutrition: gut health, probiotics and prebiotics. Wiley, London, pp 14–52
Gajardo K, Rodiles A, Kortner TM, Krogdahl Å, Bakke AM, Merrifield DL, Sørum H (2016) A high-resolution map of the gut microbiota in Atlantic salmon (*Salmo salar*): a basis for comparative gut microbial research. Sci Rep 6
Ghosh S, Sinha A, Sahu C (2007) Effect of probiotic on reproductive performance in female livebearing ornamental fish. Aquac Res 38:518–526
Gioacchini G, Maradonna F, Lombardo F, Bizzaro D, Olivotto I, Carnevali O (2010a) Increase of fecundity by probiotic administration in zebrafish (*Danio rerio*). Reproduction 140:953–959
Gioacchini G, Bizzaro D, Giorgini E, Ferraris P, Sabbatini S, Carnevali O (2010b) Oocytes maturation induction by *Lactobacillus rhamnosus* in *Danio rerio*: *in vivo* and *in vitro* studies. Hum Reprod 25:205–206
Gioacchini G, Giorgini E, Ferraris P, Tosi G, Bizzaro D, Silvi S, Carnevali O (2010c) Could probiotics improve fecundity? *Danio rerio* as case of study. J Biotechnol 150:59–60
Gioacchini G, Lombardo F, Merrifield DL, Silvi S, Cresci A, Avella MA, Carnevali O (2011) Effects of probiotic on zebrafish reproduction. J Aquac Res Dev S1. https://doi.org/10.4172/2155-9546.S1-002
Gioacchini G, Giorgini E, Merrifield LD, Hardiman G, Borini A, Vaccari L, Carnevali O (2012) Probiotics can induce follicle maturational competence: the *Danio rerio* case. Biol Reprod 86:65
Gioacchini G, Dalla Valle L, Benato F, Fimia GM, Nardacci R, Ciccosanti F, Piacentini M, Borini A, Carnevali O (2013) Interplay between autophagy and apoptosis in the development of Danio rerio follicles and the effects of a probiotic. Reprod Fertil Dev 25:1115–1125. https://doi.org/10.1071/RD12187
Gioacchini G, Giorgini E, Vaccari L, Carnevali O (2014) Can probiotics affect reproductive processes of aquatic animals? In: Merrifield DL, Ringø E (eds) Aquaculture nutrition: gut health, probiotics and prebiotics. Wiley, London, pp 328–346
Giorgini E, Conti C, Ferraris P, Sabbatini S, Tosi G, Rubini C, Vaccari L, Gioacchini G, Carnevali O (2010) Effects of *Lactobacillus rhamnosus* on zebrafish oocytes maturation: an FT-IR imaging and biochemical analysis. Anal Bioanal Chem 398:3063–3072
Hamdan AM, El-Sayed AFM, Mahmoud MM (2016) Effects of a novel marine probiotic, Lactobacillus plantarum AH 78, on growth performance and immune response of Nile tilapia (*Oreochromis niloticus*). J Appl Microbiol 120:1061–1073
Harder W (1975) Anatomy of fishes. E. Schweizerbart'sche, Stuttgart, Germany
Iwashita MK, Nakandakare IB, Terhune JS, Wood T, Ranzani-Paiva MJ (2015) Dietary supplementation with *Bacillus subtilis*, *Saccharomyces cerevisiae* and *Aspergillus oryzae* enhance immunity and disease resistance against *Aeromonas hydrophila* and *Streptococcus iniae* infection in juvenile tilapia Oreochromis niloticus. Fish Shellfish Immunol 43:60–66
Janeway CA Jr (1989) Approaching the asymptote? Evolution and revolution in immunology. Cold Spring Harb Symp Quant Biol 54 Pt 1:1–13
Jatoba A, Vieira FD, Buglione-Neto CC, Mourino JLP, Silva BC, Seiftter WQ, Andreatta ER (2011) Diet supplemented with probiotic for Nile tilapia in polyculture system with marine shrimp. Fish Physiol Biochem 37(4):725–732
Kapoor BG, Smit H, Verighina IA (1975) The alimentary canal and digestion in fish. In: Young CM, Russell FS (eds) Advances in marine biology, vol 13. Academic Press, New York, pp 109–213

Koppang EO, Kvellestad A, Fischer U (2015) Fish mucosal immunity: gill. In: Beck BH, Peatman E (eds) Mucosal health in aquaculture. Wiley, London, pp 93–134

Lamari F, Castex M, Larcher T, Mazurais D, Castex M, Bakhrouf A, Gatesoupe F-J (2013) Comparison of the effects of the dietary addition of two lactic acid bacteria on the development and conformation of sea bass larvae, *Dicentrarchus labrax*, and the influence on associated microbiota. Aquaculture 376–379:137–145

Larsen J, Roney JM (2013) Farmed fish production overtakes beef. 2013. Plan B updates. Earth Policy Institute, Washington, DC

Larsen AM, Mohammed HH, Arias CR (2014) Characterization of the gut microbiota of three commercially valuable warmwater fish species. J Appl Microbiol 116(6):1396–1404

Lauzon HL, Pérez-Sánchez T, Merrifield DL, Ringø E, Balcázar JL (2014) Probiotic applications in cold water fish species. In: Merrifield DL, Ringø E (eds) Aquaculture nutrition: gut health, probiotics and prebiotics. Wiley, London, pp 253–289

Lin Y-H, Wua J-Y, Shiaua S-Y (2010) Dietary cobalt can promote gastrointestinal bacterial production of vitamin B12 in sufficient amounts to supply growth requirements of grouper, *Epinephelus malabaricus*. Aquaculture 302(1–2):89–93

Liu W, Ren P, He S, Xu L, Yang Y, Gu Z, Zhou Z (2013) Comparison of adhesive gut bacteria composition, immunity, and disease resistance in juvenile hybrid tilapia fed two different *Lactobacillus* strains. Fish Shellfish Immunol 35:54–62

Llewellyn MS, McGinnity P, Dionne M, Letourneau J, Thonier F, Carvalho GR, Creer S, Derome N (2015) The biogeography of the Atlantic salmon (*Salmo salar*) gut microbiome. Short communication. ISME J 10:1280–1284

Løkka G, Austbø L, Falk K, Bjerkås I, Koppang EO (2013) Intestinal morphology of the wild Atlantic salmon (*Salmo salar*). J Morphol 274(8):859–876

Lombardo F, Gioacchini G, Carnevali O (2011) Probiotic-based nutritional effects on killifish reproduction. Fish Aquac J FAJ-33

Magadan S, Sunyer OJ, Boudinot P (2015) Unique features of fish immune repertoires: particularities of adaptive immunity within the largest group of vertebrates. Pathogen-host interactions: antigenic variation vs somatic adaptations. Results Probl Cell Differ 57:235–264

Magnadottir B (2010) Immunological control of fish diseases. Mar Biotechnol (NY) 12(4):361–379

Maradonna F, Gioacchini G, Falcinelli S, Bertotto D, Radaelli G, Olivotto I, Carnevali O (2013) Probiotic supplementation promotes calcification in *Danio rerio* larvae: a molecular study. PLoS One 8(12):e83155

Medzhitov R, Janeway CA Jr (1998) Innate immune recognition and control of adaptive immune responses. Semin Immunol 10:351–353

Medzhitov R, Janeway CA Jr (2002) Decoding the patterns of self and non-self by the innate immune system. Sci Signal 296:298

Merrifield DL, Carnevali O (2014) Probiotic modulation of the Gut microbiota of fish. In: Merrifield DL, Ringø E (eds) Aquaculture nutrition: gut health, probiotics and prebiotics. Wiley, London, pp 169–222

Merrifield DL, Dimitroglou A, Foey A, Davies SJ, Baker RTM, Bogwald J, Castex M, Ringø E (2010a) The current status and future focus of probiotic and prebiotic applications for salmonids. Aquaculture 302:1–18

Merrifield DL, Harper G, Baker RTM, Ringø E, Davies SJ (2010b) Possible influence of probiotic adhesion to intestinal mucosa on the activity and morphology of rainbow trout (*Oncorhynchus mykiss*) enterocytes. Aquac Res 41:1268–1272

Merrifield DL, Rodiles A (2015) The fish microbiome and its interactions with mucosal tissues. In: Beck BH, Peatman E (eds) Mucosal health in aquaculture. Wiley, London, pp 273–295

Mountfort DO, Campbell J, Clements KD (2002) Hindgut fermentation in three species of marine herbivorous fish. Society 68(3):1374–1380

Munir MB, Hashim R, Chai YH, Marsh TL, Nor SAM (2016) Dietary prebiotics and probiotics influence growth performance, nutrient digestibility and expression of immune regulatory gene in snakehead (*Channa striata*) fingerlings. Aquaculture 460:59–68

Newaj-Fyzul A, Adesiyunz AA, Mutani A, Ramsudhag A, Brunt J, Austin B (2007) *Bacillus subtilis* AB1 controls *Aeromonas* infection in rainbow trout (*Oncorhynchus mykiss*, Walbaum). J Appl Microbiol 103:1699–1706

Nimrat S, Vuthiphandchai V (2011) *In vitro* evaluation of commercial probiotic products used for marine shrimp cultivation in Thailand. Afr J Biotechnol 10:4643–4650

Olsen RE, Ringø E (1997) Lipid digestibility in fish: a review. Recent Res Dev Lipid Res 1:199–264

Panigrahi A, Kiron V, Satoh S, Hirono I, Kobayashi T, Sugita H, Puanghaew J, Aoki T (2007) Immune modulation and expression of cytokines genes in rainbow trout *Oncorhynchus mykiss* upon probiotic feeding. Dev Comp Immunol 31:372–382

Parameswaran S, Selvaraj C, Radhakrishnan S (1974) Observations on the biology of *Labeo gonius* (Hamilton). Indian J Fish 21:54–75

Pérez-Sánchez T, Balcázar JL, Merrifield DL, Carnevali O, Gioacchini G, de Blas I, Ruiz-Zarzuela I (2011) Expression of immune-related genes in rainbow trout (*Oncorhynchus mykiss*) induced by probiotic bacteria during *Lactococcus garvieae* infection. Fish Shellfish Immunol 31(2):196–201

Picchietti S, Mazzini M, Taddei AR, Renna R, Fausto AM, Mulero V, Carnevali O, Cresci A, Abelli L (2007) Effects of administration of probiotic strains on GALT of larval gilthead seabream: immunohistochemical and ultrastructural studies. Fish Shell Immunol 22:57–67

Picchietti S, Fausto AM, Randelli E, Carnevali O, Taddei AR, Buonocore F, Scapigliati G, Abelli L (2009) Early treatment with *Lactobacillus delbrueckii* strain induces an increase in intestinal T-cells and granulocytes and modulates immune-related genes of larval *Dicentrarchus labrax* (L.) Fish Shellfish Immunol 26:368–376

Pirarat N, Kobayashi T, Katagiri T, Maita M, Endo M (2006) Protective effects and mechanisms of a probiotic bacterium *Lactobacillus rhamnosus* against experimental *Edwardsiella tarda* infection in tilapia (*Oreochromis niloticus*). Vet Immunol Immunopathol 113:339–347

Pirarat N, Pinpimai K, Endo M, Katagiri T, Ponpornpisit A, Chansue N (2011) Modulation of intestinal morphology and immunity in Nile tilapia (*Oreochromis niloticus*) by *Lactobacillus rhamnosus* GG. Res Vet Sci 91:92–97

Ramos MA, Weber B, Gonçalves JF, Santos GA, Rema P, Ozório ROA (2013) Dietary probiotic supplementation modulated gut microbiota and improved growth of juvenile rainbow trout (*Oncorhynchus mykiss*). Comp Biochem Physiol A Mol Integr Physiol 166(2):302–307

Rawls JF, Samuel BS, Gordon JI (2004) Gnotobiotic zebrafish reveal evolutionarily conserved responses to the gut microbiota. Proc Natl Acad Sci U S A 101:4596–4601

Ray AK, Ringø E (2014) The gastrointestinal tract of fish. In: Merrifield DL, Ringø E (eds) Aquaculture nutrition: gut health, probiotics and prebiotics. Wiley, London, pp 1–13

Ray AK, Ghosh K, Ringø E (2012) Enzyme-producing bacteria isolated from fish gut: a review. Aquac Nutr 18:465–492

Reda RM, Selim KM (2015) Evaluation of *Bacillus amyloliquefaciens* on the growth performance, intestinal morphology, hematology and body composition of Nile tilapia, Oreochromis niloticus. Aquac Int 23(1):203–217

Ridha MT, Azad IS (2012) Preliminary evaluation of growth performance and immune response of Nile tilapia *Oreochromis niloticus* supplemented with two putative probiotic bacteria. Aquac Res 43(6):843–852

Ridha MT, Azad IS (2015) Effect of autochthonous and commercial probiotic bacteria on growth, persistence, immunity and disease resistance in juvenile and adult Nile tilapia *Oreochromis niloticus*. Aquac Res 47:2757–2767

Ringø E, Olsen RE, Mayhew TM, Myklebust R (2003) Electron microscopy of the intestinal microflora of fish. Aquaculture 227:395–415

Romero J, Ringø E, Merrifield DL (2014) The gut microbiota of fish. In: Merrifield DL, Ringø E (eds) Aquaculture nutrition: gut health, probiotics and prebiotics. Wiley, London, pp 75–100

Sáenz de Rodrigáñez MA, Diaz-Rosales P, Chabrillon M, Smidt H, Arijo S, Leon-Rubio JM, Alarcon FJ, Balebona MC, Morinigo MA, Cara JB, Moyano FJ (2009) Effect of dietary administration of probiotics on growth and intestine functionality of juvenile Senegalese sole (*Solea senegalensis*, Kaup 1858). Aquac Nutr 15:177–185

Salinas I, Parra D (2015) Fish mucosal immunity: intestine. In: Beck BH, Peatman E (eds) Mucosal health in aquaculture. Wiley, London, pp 135–170

Schley PD, Field CJ (2002) The immune-enhancing effects of dietary fibres and prebiotics. Br J Nutr 87(2):221–230

Sinha GM (1976) Comparative morphology, anatomy and histology of the alimentary canal of an Indian freshwater major carp, *Labeo calbasu* (Hamilton) during the different life history stages in relation to food and feeding habits. Anat Anz 139:348–362

Smith LS (1980) Digestion in teleost fishes. Lectures presented at the FAO/UNDP Training Course in fish feeding technology, ACDP/REP/80/11:3–17

Standen BT, Rawling MD, Davies SJ, Castex M, Foey A, Gioacchini G, Carnevali O, Merrifield DL (2013) Probiotic *Pediococcus acidilactici* modulates both localised intestinal- and peripheral immunity in tilapia (*Oreochromis niloticus*). Fish Shellfish Immunol 35:1097–1104

Standen BT, Rodiles A, Peggs DL, Davies SJ, Santos GA, Merrifield DL (2015) Modulation of the intestinal microbiota and morphology of tilapia, *Oreochromis niloticus*, following the application of a multi-species probiotic. Appl Microbiol Biotechnol 99(20):8403–8417

Standen BT, Peggs DL, Rawling MD, Foey A, Davies SJ, Santos GA, Merrifield DL (2016) Dietary administration of a commercial mixed-species probiotic improves growth performance and modulates the intestinal immunity of tilapia, *Oreochromis niloticus*. Fish Shellfish Immunol 49:427–435

Sugita H, Miyajima C, Deguchi Y (1991) The vitamin B12-producing ability of the intestinal microflora of freshwater fish. Aquaculture 92:267–276

Telli GS, Ranzani-Paiva MJ, Dias Dde C, Sussel FR, Ishikawa CM, Tachibana L (2014) Dietary administration of Bacillus subtilis on hematology and non-specific immunity of Nile tilapia *Oreochromis niloticus* raised at different stocking densities. Fish Shellfish Immunol 39:305–311

Tlaskalova-Hogenova H, Tuckova L, Mestecky J, Kolinska J, Rossmann P, Stepankova R, Kozakova H, Hudcovic T, Hrncir T, Frolova L, Kverka M (2005) Interaction of mucosal microbiota with the innate immune system. Scand J Immunol 62(1):106–113

Tort L, Balasch JC, Mackenzie S (2003) Fish immune system. A crossroads between innate and adaptive responses. Inmunología 22(3):277–286

Trichet VV (2010) Nutrition and immunity: an update. Aquac Res 41:356–372

Tsuchiya C, Sakata T, Sugita H (2008) Novel ecological niche of *Cetobacterium somerae*, an anaerobic bacterium in the intestinal tracts of freshwater fish. Lett Appl Microbiol 46(1):43–48

Wang Y (2011) Use of probiotic *Bacillus coagulans, Rhodopseudomonas palust*ris and *Lactobacillus acidophilus* as growth promoters in grass carp (*Ctenopharyngodon idella*) fingerlings. Aquac Nutr 17:e372–e378

Wang YB, Tian ZQ, Yao JT, Li WF (2008) Effect of probiotics, *Enterococcus faecium*, on tilapia (*Oreochromis niloticus*) growth performance and immune response. Aquaculture 277:203–207

Wulff T, Petersen J, Nørrelykke MR, Jessen F, Nielsen HH (2012) Proteome analysis of pyloric ceca: a methodology for fish feed development? J Agric Food Chem 60:8457–8464

Yan YY, Xia HQ, Yang HL, Hoseinifar SH, Sun YZ (2016) Effects of dietary live or heat-inactivated autochthonous *Bacillus pumilus* SE5 on growth performance, immune responses and immune gene expression in grouper *Epinephelus coioides*. Aquac Nutr 22:698–707

Yanbo W, Zirong X (2006) Effect of probiotics for common carp (*Cyprinus carpio*) based on growth performance and digestive enzyme activities. Anim Feed Sci Technol 127:283–292

Zhou X, Tian Z, Wang Y, Li W (2010) Effect of treatment with probiotics as water additives on tilapia (*Oreochromis niloticus*) growth performance and immune response. Fish Physiol Biochem 36:501–509

# Probiotics for Honeybees' Health

# 9

Francesca Gaggìa, Loredana Baffoni, and Daniele Alberoni

> *Honey bees are social insects and their activities within and outside the hive have been described over the centuries since they are a combination of organization, intelligence and sensitivity, starting from the ritualized body movements to the their capacity to "sampling" the environment and smell the odour of the food source.*
>
> Menzel (1993)

## 9.1 The Importance of Honeybee

Honeybee is certainly one of the most familiar flying insect of terrestrial habitats. Honeybee belongs to the order Hymenoptera, family Apidae, and is a member of the genus *Apis*. The center of origin is presumably Southeast Asia where most of the species are found. Mainly, they are limited in range to tropical and montane zones in Southeast and South Asia, but two species have far broader ranges, e.g., *A. mellifera* and *A. ceranae*. Ten species are generally recognized within the genus *Apis* (Engel 1999; Arias and Sheppard 2005). Phylogenetic analyses based on nuclear and mitochondrial DNA markers strongly support a cluster into three distinct groups: cavity-nesting bees (*A. mellifera*, *A. cerana*, *A. koschevnikovi*, *A. nulensis*), giant bees (*A. dorsata*, *A. laboriosa*, *A. binghami*, *A. nigrocincta*), and dwarf bees (*A. florea*, *A. andreniformis*) (Arias and Sheppard 2005; Raffiudin and Crozier 2007). In this chapter, we will focus on the western honeybee *A. mellifera*, which is

F. Gaggìa (✉) • L. Baffoni • D. Alberoni
Department of Agricultural Sciences, University of Bologna, Bologna, Italy
e-mail: francesca.gaggia@unibo.it





D. Di Gioia, B. Biavati (eds.), *Probiotics and Prebiotics in Animal Health and Food Safety*, https://doi.org/10.1007/978-3-319-71950-4_9

the most widely distributed honeybee in the world because of its great honey-harvesting potential. The native distribution of *A. mellifera* includes Africa, Europe, and Western Asia, and molecular dating suggests that the population expanded into this range around one million years ago. Conflicting hypotheses have been proposed for the origin of this expansion (Middle East and Africa), although a recent work put *A. mellifera* closer to the only other *Apis* species, which are all restricted to Asia (Wallberg et al. 2014). The species includes 25 subspecies or geographic races described by morphometric and molecular analysis and grouped into evolutionary branches based on their morphological similarities.

It expanded its range into Europe and Asia as the Ice Age glaciers retreated, and it has been spread by humans to the Americas, Australia, and Hawaii. *A. mellifera* has also been introduced through much of the range occupied by *A. cerana*, including Japan and China.

Honeybees have an extremely elaborate social life, fulfilling the requirement of the "superorganism"; the honeybee colony "superorganism" consists of individual, groups, and hive components, complete with a large repertoire of socially interactive and homeostatic behaviors (Hölldobler et al. 2009). They typically live in colonies with intra-colonial homeostasis, consisting of a single queen, approximately 10–30 thousand "sterile" female workers, and from zero to a few thousand males, depending on the time of year (Page and Peng 2001). Food is stored in designated areas of the nest, and the workers use glandular secretions to feed the brood. Division of labor is well developed and pheromone regulated (Moritz and Southwick 1992).

Honeybees are critically important in the environment, sustaining biodiversity and providing essential pollination for a wide range of crops and wild plants (EFSA 2017). The Food and Agriculture Organization of the United Nations (FAO) estimates that of the 100 crop species that provide 90% of food worldwide, 71 are pollinated by bees (Copping 2013). The majority of crops grown in the European Union depend on insect pollination. The annual monetary value of pollination has been estimated to be billions of dollars (Hedtke et al. 2015). They contribute to human health and well-being directly through the production of honey, which is produced by honeybees from the nectar they gather, and other food and feed supplies such as pollen, wax, propolis, and royal jelly, as dietary supplements and ingredients in food (Ajibola et al. 2012). They can also be considered important bioindicators of environmental pollution (Celli and Maccagnani 2003).

Beekeeping is the art and science of rearing, breeding, and managing honeybee colonies in artificial hives for economic benefits (Ikediobi et al. 1985; Morse 1989). The most common species utilized for this purpose is *Apis mellifera* of which about 25 subspecies of economic importance occur in Europe, Middle East of Asia, and Africa (Leven et al. 1997).

Beekeeping is an ancient tradition, and honeybees have been kept in Europe for several millennia.

In recent years, a growing interest has been reported for the urban beekeeping practice as a fascinating rewarding pastime, which allows people to increase biodiversity, produce local foods, and reconnect with nature (Moore and Kosut 2013).

**Fig. 9.1** Hives in organic management

Given the importance of honeybees in the ecosystem and the food chain, and given the multiple services they provide to humans, their protection is pivotal. Beyond the essentiality of honeybee for a balanced vitamin and antioxidant-rich diet, honeybee is vital for the mankind for their contribution to biodiversity and to some extent to human survival.

Extensive losses of honeybee colonies in recent years are becoming a major cause of concern. Unfortunately, they continuously face threats (diseases, climate change, and management practices); managed honeybees are highly social, frequent a multitude of environmental niches, and continually share food, conditions that promote the transmission of parasites and pathogens (Fig. 9.1).

## 9.2 Biotic and Abiotic Stresses

Although managed honeybee colonies are continuously increasing over the last 55 years, colony populations have significantly decreased in many European and North American countries (Aizen and Harder 2009), as a result of several incoming stressors (agrochemicals, pathogens, climate change) and socioeconomic reasons (Potts et al. 2010; VanEngelsdorp and Meixner 2010).

There is still a huge gap between pollinator demand and honeybee colony supply because the area planted with insect-pollinated crops increased more than the number of honeybee colonies (Aizen and Harder 2009; Breeze et al. 2014). At the same time, cultivation of crops, relying on insects for pollination, has increased (Aizen and Harder 2009).

In the last decade, a special attention has arisen toward "colony collapse disorder" (CCD) in the USA with the alarming claims of media, describing the dramatic demise of honeybee colonies, a world pollinator crisis, and the spectra of massive human starvation. Colony losses have exceeded 90% in some locations, and loss of pollination services has had major impacts on some fruit and vegetable production. Nevertheless, in the twentieth century, several honeybee losses were already registered (Oldroyd 2007). Symptoms were very close to those observed in the USA, and consequent losses of colonies were also traced throughout the world, but a clear explanation of the main causes was never found. Surely, viruses (i.e., Israeli acute paralysis virus) and the mite *Varroa destructor* were involved since the broad patterns of CCD coincide with continents with different pressures from *V. destructor*. In addition, during the same period of the CCD, a new parasite was moving all over the world, *Nosema ceranae*, jumping from its host, the Asian honeybee *Apis cerana*, to the western *A. mellifera*, causing gradual depopulation and copious colony death (Higes et al. 2008). Moreover, in such a dramatic moment, the attention was also addressed to the agrochemicals, above all the neonicotinoid insecticide imidacloprid, which are over employed in the American agriculture system. Overall, researchers concluded that no single driver could emerge as the definitive cause of the phenomenon and that CCD was a multifactorial syndrome. Bees were all sick, but each colony seemed to suffer from a different combination of such diseases.

As reported by Vanengelsdorp and Meixner (2010), honeybee can die in many ways, and CCD is just one of them. Finally, since a great genetic variability exists both in honeybee host and pathogens, the symptoms and causes of colony losses may well be different in different regions (Neumann and Carreck 2010).

Concerning the abovementioned pesticides, in 2013, the EU imposed a temporary ban on the use of the three key neonicotinoids on some crops. However, the new proposals are for a complete ban on their use in fields, with the only exception being for plants entirely grown in greenhouses (EFSA 2013). Monitoring schemes on pesticide effects are currently ongoing in some member states to provide more insights into the acute effects of pesticides on honeybees. The effects of pesticide drifting during treatment were addressed in the "APEnet" project (Apenet, 2011), which mentioned the case of the fatal powdering of bees in flight with particulates of neonicotinoid seed coating, the implications of humidity (Girolami et al. 2012), and the lethal aerial powdering of honeybees with neonicotinoids from fragments of maize seed coat (Marzaro et al. 2011). Moreover, some reports of experimental studies describe an interaction between *N. ceranae* and other stressors (e.g., chronic bee paralysis virus (CBPV), black queen cell virus (BQCV), or imidacloprid) that can lead to elevating honeybee mortality (Alaux et al. 2010; Doublet et al. 2015).

The situation is different with honeybee colony losses (i.e., the death of colonies), which mainly occur during the winter season (winter losses of honeybee colonies). These winter losses do not follow a general pattern. In some countries and some winters, losses are high (above 15%), sometimes even catastrophic (above 30%), but they are not always and everywhere high and unusual and catastrophic. The emerging picture is that the losses reported by beekeepers to the media are always much higher than the losses counted by official inspectors in the course of nationwide monitoring programs or surveys (see the official reports available under http://ec.europa.eu/food/animals/live_animals/bees/study_on_mortality/index_en.htm). While in the winter 2012–2013, at least the Northern part of Europe experienced high winter losses, in the winter 2013–2014, the losses were below 15% in all participating member states except for Sweden (15.4%) and in some member states even below 10% or 5%. This is far from being an alarming situation. In addition, such losses are not a problem for a normal beekeeper who replaces lost colonies easily by nucs made during the bee season.

While it is impossible to identify a single factor, which can account for all colony losses in all regions of the world over a given time period, it is clear that several biological and environmental factors acting alone or in combination have the potential to cause premature colony mortality by adversely affecting colony health and life span. Among these factors, certain honeybee diseases and parasites have been shown to play a significant role in increased honeybee colony mortality and in the described colony losses.

In the following paragraph, a list and a brief description of the main pathogens, affecting honeybee health, will be listed.

## 9.3 Pathogens Affecting Honey Bee

### 9.3.1 Brood Pathogens

*Melissococcus plutonius* is the causative agent of the European foulbrood (EFB) affecting honeybee larvae in the western Apis mellifera. However, the bacterium can also infect and kill the brood of the Eastern honeybee (*Apis ceranae*) and the Himalayan honeybee (*Apis laboriosa*) (Bailey 1974; Allen et al. 1990). *M. plutonius* is a lanceolate non-spore-forming coccus with a close phylogenetic relationship to the genus *Enterococcus* (Cai and Collins 1994). Bacterial cells are ingested with contaminated food and invade the midgut where they reproduce, assimilating the larval food. Infected larvae can die before or after capping from starvation (Bailey 1983), or they may successfully pupate and form normal or undersized adults. Following infection, secondary invaders, like *Paenibacillus alvei* and *Enterococcus faecalis*, are involved in the decomposition of the larval remains. Dead larvae are found twisted around the walls of the cell or stretched out lengthways. These larvae turn yellow and then brown and finally decompose, adopting a grayish black color (Forsgren 2010). Although symptomatology is rather well described, many aspects of the pathogenesis, transmission and control of *M. plutonius* are poorly understood and remain elusive (Genersch 2010). In a recent work performed in our laboratory, we evidenced that honeybee larvae were affected by EFB, with the presence of an atypical *Paenibacillus* strain (*P. dendritiformis*) as a new putative second invader, which presumably conferred a different symptomatology to the diseased brood (Gaggia et al. 2015).

EFB did not create serious problems in many European countries since many infected and diseased colonies spontaneously recovered from the disease (Bailey 1968). Nevertheless, a dramatic increase in the incidence of EFB has been recently observed, in particular in the United Kingdom, Switzerland (Wilkins et al. 2007; Roetschi et al. 2008), and Norway (Dahle et al. 2011).

*Paenibacillus larvae* is a Gram-positive, spore-forming bacillus that causes the American foulbrood (AFB) (Genersch et al. 2006), which contaminate the first instar larvae leading to its death after the cell capping. AFB is not only fatal to single honeybee larvae, but leads to the collapse of the entire colonies. In addition, AFB is highly contagious, and the spores are extremely tenacious.

As for the EFB, the infection originates from the ingestion of food contaminated with spores; once in the midgut, spores germinate, and the vegetative cells reproduce and invade the hemocoel (Davidson 1973; Bailey and Ball 1991), by synthesizing highly active extracellular proteases (Hrabák and Martínek 2007). In the second stage, the larvae become a brownish, semifluid, glue-like colloid (ropy stage) releasing a putrid smell. The ropy aspect (dead larvae adhere and form a thread span when touched with a wooden stick) confirmed the presence of AFB. Finally, the larva remains dry down to a hard scale (foulbrood scale), which tightly adheres to the lower cell wall. The scales contain millions of spores, which could distribute the infection for many years within and between colonies (Bailey and Ball 1991).

For both foulbroods, antibiotics are used by some beekeepers (especially in the USA and other non-European countries), leading to concerns over antibiotic resistance, collateral losses of beneficial microbes, and the risks of antibiotic residues in honey and pollen destined for human consumption.

The fungus *Ascosphaera apis* is responsible for the chalkbrood disease; larvae are infected by ingesting fungal spores that germinate in the digestive tract. The subsequent mycelial growth is lethal to the larvae. Dead larvae and pupae desiccated, forming mummies that contain millions of spores and that are highly infectious (Aronstein and Murray 2010). *A. apis* is responsible for large economic losses, particularly in combination with other pathogens such as *Nosema apis* (Aydin et al. 2006), *N. ceranae*, and *V. destructor* (Hedtke et al. 2008).

### 9.3.2 *Nosema apis* and *Nosema ceranae*

Adult honeybees host two parasites belonging to the fungal phylum *Microsporidia*—*Nosema apis* and *Nosema ceranae*—both of which have received extensive attention, in particular *N. ceranae*, which moved, in the last decades, from their natural Asiatic host (*Apis cerana*) to the European one, finding fertile ground for its development (Higes et al. 2008; Rosenkranz et al. 2010). Recently, it became evident that *N. ceranae* is also widespread in the *A. mellifera* population throughout the world, particularly in countries with temperate climate (Paxton et al. 2007; Giersch et al. 2009; Higes et al. 2007). Due to its distribution, and severity, it is now considered one of the major health problems both in individual honeybees (Paxton et al. 2007; Antúnez et al. 2009) and in whole colonies (Higes et al. 2008).

As obligate intracellular parasites, the *Microsporidia* invade epithelial cells of the adult midgut and undergo repeated cell divisions to produce new infectious spores. These infections often result in heavy parasite loads, tens of millions of spores per bee (Forsgren and Fries 2010), which lead to an increase of the nutritional requirement, morbidity, and mortality of the bee host (Martín-Hernández et al. 2011).

*N. apis* is mainly characterized by dysentery, dilated abdomens, brown fecal marks on combs and the front of the hives, sick or dead bees in the vicinity of the hives, and a decrease in brood production and in the size of bee colony, particularly in spring. *N. ceranae* caused death of individuals and colonies not preceded by any visible symptoms. The microsporidium develops exploiting the host cell mitochondria (Chen et al. 2009; Higes et al. 2007), inducing a severe energetic stress and competing directly for key nutrients and energy resources. The infection firstly causes increased food consumption (Martín-Hernández et al. 2011), immune suppression (Antúnez et al. 2009), degeneration of gut epithelial cells, shortened life spans (Higes et al. 2007) and a decrease on population size and loss of adult bees. It has also been suggested that *N. ceranae* induces significantly higher mortality than *N. apis* (Paxton et al. 2007; Martín-Hernández et al. 2009; Higes et al. 2010). Considering the different symptomatology, the members of a recent international meeting assigned two different clinical patterns: nosemosis type A caused by *N. apis* and nosemosis type C caused by *N. ceranae* (COLOSS Workshop 2009).

Evidences show that *N. ceranae*, due to epithelial lesions, increases the susceptibility to other pathogens, in particular viruses (Higes et al. 2008). In addition, the exposure to sublethal concentration of neonicotinoids in immature bees significantly enhanced the number of spore production per bee (Vidau et al. 2011). Nowadays, the antibiotic Fumagilin-B (dicyclohexylammonium salt) is the only available compound to treat *N. ceranae* infection; however, it is no longer licensed in the EU states, and recent reports provide controversial results about its efficacy and its effects related to residues in honey (Lopez et al. 2008; Williams et al. 2008).

### 9.3.3 Spiroplasmosis

Spiroplasmas are small, helical, and motile eubacteria and are descendants of Gram-positive bacteria that lack a cell wall (Regassa and Gasparich 2006). *Spiroplasma melliferum* and *Spiroplasma apis* are two pathogens of adult honeybee that have been identified in Western honeybees (Clark 1977; Mouches et al. 1982), but infection has been also reported in Asia and the USA. Pathogenesis occurs when the organisms breach the gut barrier and invade the hemolymph, causing a systemic infection that can ultimately lead to fatal disease in the bee. Spiroplasma infections are much more difficult to recognize and diagnose than the foulbrood diseases, hindering the ability to monitor bacterial abundance and impact on the beekeeping industry. They remain interesting targets for study, owing to their seasonal abundance in honeybee colonies, which is presumably tied to flowering cycles of specific plants that act as transmission sites (Clark 1982).

The main groups of protists infecting honeybee have been neglected for many years due to different reasons, e.g., obscure pathology, low detectability, difficulty in culturing, and absence of genetic markers. Nowadays, the research community is focusing its attention on trypanosomes (*Crithidia mellificae* and the recent strain San Francisco), gregarines (*Apycystis bombi*), and amoeba (*Malpighamoeba mellificae*).

*C. mellificae* and gregarines colonize the hindgut and midgut, respectively. *C. mellificae* produces encrustations on the gut epithelia surface, and gregarines attach to the epithelia and absorb nutrients, creating tissue damage and reducing nutrient absorption by the bee. However, their role in honeybee health and distribution in the world is not well understood; colonies seem more susceptible in tropical climates. Trypanosomes have probably a cosmopolitan distribution since *C. mellificae* has been reported in Australia, China, France, Japan, Switzerland, and the USA (Ravoet et al. 2013). The related species *C. bombi*, also reported from Asian honeybees, has seriously affected the survival of bumble bees under stress conditions (Brown et al. 2000; Li et al. 2012). Recently, complex dynamic immune responses to *C. mellificae* infection were reported, with a distinct response when individuals were infected with *C. mellificae* and *N. ceranae* simultaneously (Schwarz and Evans 2013). In addition, an association between both pathogens was reported in the USA (Runckel et al. 2011). Gregarines infecting other bees and social wasps inhibit foraging, reduce fecundity, and increase queen mortality. After its detection in honeybees in Finland (Lipa and Triggiani 1996), *A. bombi* was also reported in honeybees in Japan (Morimoto et al. 2013) and Argentina (Plischuk et al. 2011).

The amoeba *Malpighamoeba mellificae* infects adult bees in temperate to tropical regions. The ingested cysts develop into trophozoites and invade the Malpighian tubules, degrading their tissues. As the amoebae replicate, they pack the lumen of the tubules, forming up to 500,000 cysts per bee that are shed through the feces. The damaged tubules are unable to carry out their physiological function bringing bees to death (Lipa and Triggiani 1996). Associated with spring dwindling of bee colonies, *M. mellificae* is also linked with dysentery symptoms in adult bees and the tendency of infected bees to "disappear inexplicably" from the hive (Prell 1926).

### 9.3.4 *Varroa Destructor*

*Varroa destructor* is a mite parasite of honeybees. Originally, a parasite of the Asiatic honeybee *Apis cerana* performed a host shift in the early 1970s to the European honeybee *Apis mellifera*. Where and how this switch occurred is unclear (Rosenkranz et al. 2010), anyhow since then the parasite has crossed the globe, and it is considered endemic in all the beehives of the globe. To date only Australia and few north European territories (Åland Islands and Isle of Man) result as *V. destructor*-free areas.

Varroa is feeding on the hemolymph of larvae and adult bees, thus weakening the insect. But this doesn't seem to be the determinant factor leading to the colony collapse. Indeed, varroa infect bees with a relevant number of viruses like deformed wing virus (DWV), chronic bee paralysis virus (CBPV), black queen cell virus (BQCV), and sacbrood virus (SBV). To date 16–18 truly unique viruses (24 if considering the variants) have been identified as pathogenic for bees (De Miranda et al. 2013).

Different approaches have been used to eliminate the varroa parasite from the hives. Upon its arrival in Europe, several acaricides were used to control its proliferation, but an inevitable development of multiple resistances led to commercial withdrawal of the majority of them. Nowadays, only few active ingredients result active like amitraz, coumaphos, and fluvalinate. More recently beekeepers focused their efforts on organic approaches, using organic acids like oxalic, formic, and lactic acids together with comb trapping methods. Also, essential oils and physical approaches like drone brood excision or brood heating are playing a relevant role. As the last approach in the parasite control, a number of research centers and beekeepers tried to develop varroa-resistant bees, with different approaches. Worthy to mention here is the development of the varroa-sensitive hygiene (VSH) behavior.

Nevertheless, even if eradication of the parasites from a beehive is possible, a free colony status does not last long. Indeed, varroa reinfestation occurs due to a permanent exchange of mites between foragers, or drones enter foreign colonies, voluntarily, by drifting or by robbing (Goodwin et al. 2006). Still nowadays, varroosis can be classified in the top list of destabilizing biotic factors for honeybees.

## 9.4 Digestive and Excretory Systems in *Apis mellifera*

The alimentary canal (Fig. 9.2) of honeybee extends from the mouth to the anus where the waste material is excreted. The esophagus is the connection between the mouth and the rest of the digestive system in the abdomen (through the thorax).

The posterior end opens into the crop or honey stomach an expandable bag holding (a) honey ingested in the hive and used for energy during the flight and (b) nectar and/or water collected in the field for transport back to the nest. More generally, the crop represents the microbial intersection of food sharing, food storage, and the pollination environment. The pH of the crop is highly acidic but also varies in accordance with the pH of ingested food products. There is a special structure called proventriculus near the end of the crop, which has sceleritized toothlike structures, and also muscles and valves. These structures prevent most of the liquid crop contents from passing through the ventriculus or midgut and allow the removal of pollen grains in the nectar. The proventriculus also allows filtering out particles from 0.5 to 100 μm in diameter, resulting in the partial stop of spores of *Nosema* sp. and *Paenibacillus larvae* (Peng and Marston 1986). Moreover, it prevents the contamination of the crop with enzymes and microbes from the more posterior midgut. The valve is open during the feeding, thus allowing the honey to go from the honey stomach to the ventriculus.

The contents of the crop can be spit back into cells, or fed to other workers (trophallaxis), as the case of nectar collected by foragers. Most of the nutrients from

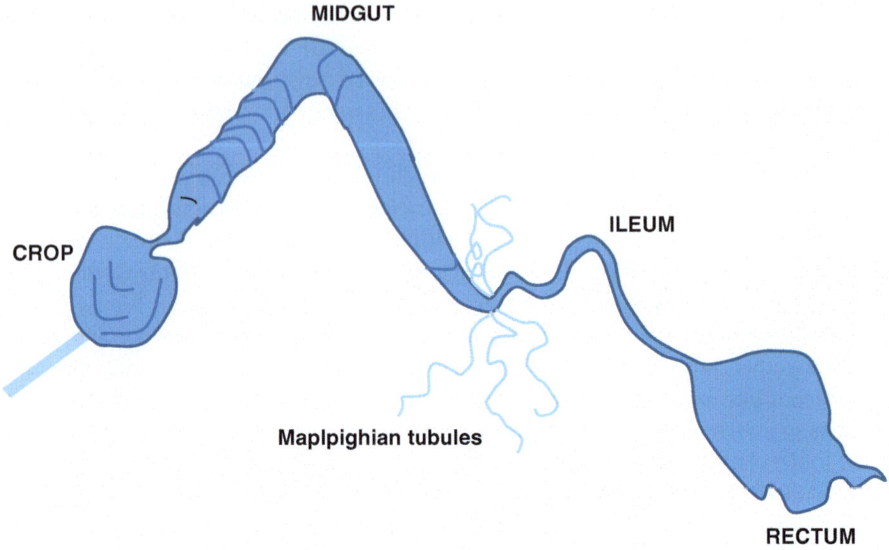


**Fig. 9.2** Representation of the honeybees' digestive system

digested feed are absorbed through the walls of the ventriculus (midgut), which is the functional stomach of bees, where most of the digestion and adsorption take place. Digestive enzymes work across a range of pH, but the optimum is pH 8. Thus, the proventriculus and the drastic change in pH between the crop and the midgut define two major microbial niches, one coevolved with liquid transfer and food storage and the other coevolved to reside in the enzymatically active and relatively nutrient-rich midgut. Malpighian tubules are small strands of tubes attached near the end of ventriculus and function as the kidney, by removing the liquid nitrogenous waste (in the form of uric acid, not as urea as in humans) from the hemolymph, and the uric acid forms crystals. The undigested material (pollen husks, dead cells, and fat globules) moves through the pyloric valve into the hindgut for excretion; the hindgut is divided into two compartments: the anterior ileum, a narrow tube with six longitudinal invaginations, and the rectum, a larger saclike compartment. During winter, the rectum expands considerably to hold waste material since bees do not defecate in the hive and wait for warm flying weather in the spring.

## 9.5 Composition of the Honeybee Gut Microbiota

The molecular tools and the new methods of DNA sequencing allowed researchers to investigate the gut microbiota of *A. mellifera*, giving a more consistent picture of its composition and role in insect health compared to culture-dependent methodologies.

In the past several microorganisms (*Bacillus* spp., Enterobacteriaceae, *Bifidobacterium* spp.), together with molds and yeasts, were identified from honeybee guts by using culture-based techniques (Gilliam and Valentine 1976; Gilliam

1997; Scardovi and Trovatelli 1969). Molds, particularly the genera *Penicillium* and *Aspergillus*, were commonly found in the alimentary canal of worker honeybees (Gilliam et al. 1974, 1977), and yeast presence appeared to be an indicator of stress conditions in honeybees (Gilliam 1997). Still today, plate count isolation and further identification allow the recovery of new species which could only be detected by traditional microbiology. This is the case of recently characterized gut bacterial species such as *Gilliamella apicola, Snodgrassella alvi, Frischella perrara, Lactobacillus kullabergensis, L. kimbladii, L. helsingborgensis, L. mellis, L. mellifer*, and *L. melliventris* (Kwong and Moran 2013; Olofsson et al. 2014).

Globally, the composition and function of the microbial community inhabiting the alimentary tract are closely related to the physiological changes and nutritional regimes associated with honeybee age and tasks. Foragers consume almost exclusively nectar and honey to meet the metabolic demands of flying (Winston 1987), while nurse bees eat large quantities of stored pollen to meet the nutritional demands for synthesizing and secreting royal jelly for larvae and other adults (Anderson et al. 2011). Investigations allowed establishing that a "core" bacterial community has coevolved with the honeybee over millions of years and now represents a relatively stable and constitutive component of healthy bees independent of geography.

A recent study showed that honey crop (honey stomach) of foragers was dominated by *Lactobacillus* with the dominance of *L. kunkeei* and Alpha 2.2 (*Acetobacteraceae*) but also contained a small number of less abundant Enterobacteriaceae that likely have their origins in the pollination environment (Corby-Harris et al. 2014). Other studies based on culture-dependent methods evidenced a crop microbiota composed of several bacterial species within the genera *Lactobacillus* and *Bifidobacterium* (Vásquez et al. 2012) with new identified lactobacilli species (Olofsson et al. 2014). The probiotic properties of these bacteria are notably recognized in vertebrates where *Lactobacillus* and *Bifidobacterium* strains exert beneficial activities within the gut microbiota (Gaggìa et al. 2010).

The estimates counts ranks from $10^2$ to $10^5$; however, the number varies numerically across seasons with the flowers visited and with the health status of bees. *Acetobacteraceae* and *L. kunkeei* thrive in sugar-rich, acidic environments such as the crop, beebread, and honey and are considered core hive bacteria, as they are associated with nurse workers and developing larvae (Anderson et al. 2013). The crop is a central organ in the honeybee's food production (beebread and honey) and food storage, and all the isolated bacteria exert important function, e.g., exopolysaccharide and antimicrobial compound production, biofilm formation, fermentation activities, and inhibition of spoilage microorganisms (Olofsson and Vásquez 2008; Forsgren 2010).

Finally, crop samples were also found to contain the core gut microbiota to some degree; on average, the gut-specific taxa in the greatest abundance in the forager crop corresponded to *Lactobacillus* (Firm 5), *Gilliamella apicola* (Gamma1), and *Snodgrassella alvi* (Beta).

The midgut contains relatively few bacteria, which are most concentrated at the distal region, adjoining the hindgut. Unlikely, the hindgut houses a large bacterial community dominated by eight major bacterial groups (Moran 2015): two *Alphaproteobacteria*

(*Bartonellaceae* and *Acetobacteraceae*), two *Gammaproteobacteria* (*Gilliamella apicola* and *Frischella perrara*), two members of the phylum *Firmicutes* with different species of lactobacilli (Firm 4, Firm 5), one *Betaproteobacteria* (*Snodgrassella alvi*), and one species of the genus *Bifidobacterium* (*B. asteroides*).

## 9.6 The Importance of the Gut Microbiota on Bee Health

The importance of gut-dwelling microbial communities in bees has become appreciated only recently, following the repetitive colony losses registered worldwide due to abiotic and biotic stressors, which led researchers to better understand the role of both gut symbiotic and pathogenic microbial interactions, since they are strictly related to food storage and the pollination environment. As for humans and animals, the understanding of the beneficial nature of insect-microbial systems is fundamental to investigate the effect of the microbial communities on host nutrition and pathogen defense. Thanks to the advanced molecular techniques and metagenomics, the human gut microbiota has revealed a huge number of bacterial genes (100 times more the number of genes found in the host), which strongly influence the physiological and biochemical activities of the host. The works on *Drosophila melanogaster* have given a picture of the molecular dialog between the microbiota and the insect gut. There is evidence of the role of gut microorganisms in supporting the immune system, influencing the epithelial homeostasis, promoting life span and larval growth upon food scarcity, and driving the host mating preference (Brummel et al. 2004; Ryu et al. 2008; Buchon et al. 2009; Sharon et al. 2010; Storelli et al. 2011). Shin et al. (2011) showed that acetic acid from the gut commensal bacterium, *Acetobacter pomorum*, modulates insulin/insulin-like growth factor signaling (IIS) in *Drosophila* to regulate host homeostatic programs controlling developmental rate, body size, energy metabolism, and intestinal stem cell activity. Among studies performed on insects, Dillon and Charnley (2002) reported in the desert locust *Schistocerca gregaria* the contribution of gut microbiota to host defense against pathogens by producing antimicrobial phenolic compounds and synthesizing key components of the locust cohesion pheromone. In healthy individuals of *D. melanogaster*, the immune system allows the dominance of two acetic acid bacteria (AAB) strains (*Acetobacter pomorum* and *Commensalibacter intestini*), which suppress the proliferation of the gut pathogen *Gluconobacter morbifer* by competition, which is a gut apoptosis inducer (Ryu et al. 2008; Crotti et al. 2010). A decreased presence of potentially pathogenic *Pseudomonads* spp. and a higher mating fitness were observed in the Mediterranean fruit fly males *Ceratitis capitata* fed with a diet enriched of *Klebsiella oxytoca* live cells, following irradiation (Ben-Ami et al. 2010).

A balanced gut microbiota constitutes an efficient barrier against pathogen colonization, produces metabolic substrates (e.g., vitamins and short-chain fatty acids), and actively exchanges regulatory signals with the host that primes and instructs mucosal immunity (Gaggìa et al. 2010). Although insects harbor a smaller number of symbionts, the honeybee gut microbiota displays high affinity with that of mammals (Kwong and Moran 2013). The huge number of bacterial symbionts, inhabiting selected niches along the whole tract of the gut (from honey crop to the rectum) are host-adapted species, which contribute to host defense, nutrition, and physiology (Hamdi et al. 2011).

Additionally, the honeybee gut microbiota exists at two major levels: within the relatively digestive tract and throughout the hive that houses the developing young and food stores. The majority of commensal gut bacteria are vital for the maintenance of homeostasis and health both in the single insect and into the hive, considering that activities such as trophallaxis and cleaning behavior led bees to partially share their microbial consortium.

The concept of symbiosis, in which both microbial and host elements work synergistically to maintain proper nutrition, health, and immunity, may be more important in social insects where both elements, compared to solitary insects, are often highly coevolved (Vásquez et al. 2012). The most explicative example of coevolution derives from the significant contribution to host protection provided by the interaction of the gut microbiota with the humoral and systemic immunity that is associated to the defense strategies in eusocial insects, whose genome has significantly fewer immune genes than expected (Evans et al. 2006). A balanced gut microbiota is necessarily associated with bee health since it provides countless enzymatic activities to break down the complex sugars of the honeybees' diet. Some studies evidenced that the lactobacilli and bifidobacteria community (LAB) in the crop vary numerically across seasons with the flowers visited by bees and with the health status of bees (Olofsson and Vásquez 2008). Cox-Foster et al. (2007) demonstrated a high relative abundance of the $\gamma$-proteobacterial taxa in the bees from CCD-affected hives than in the healthy ones, while the presence of *Firmicutes* and *Alphaproteobacteria*, mainly represented by taxa related to the genus *Lactobacillus* and AAB, respectively, was dramatically reduced in diseased bees. In three species of wild bumble bees, a low presence of *S. alvi* and *G. apicola* strains was associated with a higher incidence of the pathogen *Crithidia* spp. (Cariveau et al. 2014). *Snodgrassella* and *Gilliamella* form biofilm-like layers on the epithelium of the longitudinal invaginations of the ileum; *Snodgrassella* is in direct association with the host tissue followed by a thick layer of *Gilliamella*. Studies on gene functions showed significant enrichment in the categories of several activities associated with the formation of the biofilm on the gut epithelial surface and with the host interaction (Engel et al. 2012). The microbial community of *Bombus*, which is dominated by *Gilliamella* and *Snodgrassella*, seems to protect the insect against a trypanosome (Koch and Schmid-Hempel 2011), suggesting a possible role of the biofilm as a protective layer against parasite invasion. Gut symbionts are continuously involved in the bioconversion and preservation of pollen material, nectar, honey, and beebread. Vásquez and Olofsson (2009) suggested that LAB from the honeybee stomach belonging to the genera *Lactobacillus* and *Bifidobacterium* are involved in the fermentation process of beebread and may be responsible for improving the nutritive value by vitamin production. As reported by Engel et al. (2012), a wide genetic variation can be observed within different bacterial species involved in food processing, e.g., carbohydrate metabolism and pollen wall demolition, thus reflecting divergent niche adaptation within the gut of honeybees.

In conclusion, the above mentioned findings showed that the interaction host-symbionts goes beyond a mere nutritional complementation of the host diet. Honeybee gut symbionts counteract bee pathogens and parasites, enhance bee immunity, and improve aspects related to host physiology, behavior, reproduction, and evolution. Consequently, microorganisms could be a key element in managing and preserving honeybee health status toward biotic and abiotic stressors.

## 9.7 Beneficial Bacteria or Probiotic Bacteria?

LAB has been widely studied in animals and humans because of their probiotic properties, which have led to their well-built commercial exploitation in food, feed, and pharmaceutical market (Gaggìa et al. 2010, 2011; Tontou et al. 2015). The findings that a component of the honeybee gut microbiota was represented by lactobacilli and bifidobacteria have increased the interest of scientists in looking for similarity and analogy with the probiotic bacteria widely investigated in humans and animals. They are Gram-positive, acid-tolerant, facultative, and/or strictly anaerobic bacteria and produce lactic and acetic acid as the major metabolic end product of carbohydrate fermentation. LAB are well known for the production of antimicrobial peptide. They are normal inhabitants of the gastrointestinal tract of many insects, and their presence in the honeybee digestive system has been consistently reported in the literature (Olofsson and Vásquez 2008; Baffoni et al. 2016; Gaggia et al. 2015; Moran 2015). The bee's digestive system represents an optimal niche for LAB, which obtained from the bee's diet suitable substrates for their growth. The in vitro antagonistic activity toward bee pathogens due to organic acids and antimicrobial peptides (*M. plutonius*, *P. larvae*, *N. ceranae*) is well documented (Audisio et al. 2011b; Wu et al. 2013; Yoshiyama and Kimura 2009; Maggi et al. 2013; Baffoni et al. 2016). Saraiva et al. (2015) found in the gut microbiota of honeybee a relative high presence of genes involved in the biosynthesis of streptomycin and secondary metabolites, which could be associated with protection functions against exogenous microorganisms. Moreover, among *Lactobacillus*, novel species has been recently identified (Olofsson et al. 2014), thus extending the beneficial potentiality of these bacteria.

Another group of interesting bacterial species is represented by the acetic acid bacteria (AAB) that are a large group of obligate aerobic Gram-negative bacteria within the *Alphaproteobacteria* clade, commonly found in association with various kinds of sugar matrices. AAB of the genera *Gluconobacter*, *Acetobacter*, *Gluconacetobacter*, and *Saccharibacter* have been reported as symbionts of bees (Crotti et al. 2010). Among these, the sugar-loving and flower-associated *Gluconobacter* spp. are among the predominant bacterial groups in bees. Mohr and Tebbe (2006) isolated from the honeybee's gut about 100 bacterial strains belonging to different bacterial divisions. All isolates of the *Alphaproteobacteria* were AAB, closely related to *Gluconobacter oxydans* or *Saccharibacter floricola*, an osmophilic bacterium previously isolated from pollen.

Lactic acid bacteria and AAB show interesting properties like the capability to grow and tolerate acidic pH, to produce organic acids, and to metabolize different sugars. These features explain the effectiveness of LAB and AAB in colonizing the sugar-rich digestive system of bees and suggest a potential for inhibiting the growth of acid-sensitive pathogenic bacteria. Taking into account that treatments with formic, lactic, and acetic acids are widely employed by beekeepers to prevent pathogen infections, and, in the light of the final products of their metabolism, LAB and AAB may represent natural protecting bee symbionts of considerable importance (Olofsson and Vásquez 2008). It is of common use to describe these microorganisms as "probiotics"; however, the scientific community and authorities of the field have still not yet drawn up a list of properties/characteristics that identified the probiotic concept as in humans and animals. The transfer of the probiotic concept from vertebrates to invertebrates still requires further considerations, and several questions still need to be investigated and debated; in particular, we referred to the origin of the strains and the knowledge of their genome since their diffusion in the environment could be considered a risk.

Modulation of the honeybee gut microbiota by supplementation of selected bacterial strains has risen a special attention since it represents strategies to improve the health status of colonies, in terms of productivity and boosting the presence of beneficial microorganisms within the bee gut of new-generation bees. In the next section, an overview of the main published applications will be reported (Table 9.1).

**Table 9.1** Overview of beneficial microorganism applications for the treatment of the main honeybee pathogens

| Honeybee disease and infection dose | Microorganisms/metabolites | Source | Reported effect(s) | References |
|---|---|---|---|---|
| *P. larvae* (AFB) $10^3$ and $10^4$ spores/mL | *L. kunkeei, L. mellis, L. kimbladii, L. kullabergensis, L. helsingborgensis, L. melliventris, L. apis, L. mellifer, B. asteroides* and *B. coryneforme* ($10^7$ bacteria/mL) | Honey crop | Reduced larvae mortality | Forsgren (2010) |
| Natural diseased larvae | Iturin-like peptides from *B. amyloliquefaciens* LBM 5006 (800 AU/mL) | Native soil of Brazilian Atlantic Forest | In vitro: Bactericidal effect and cell lysis<br>In vivo: No effect | Benitez et al. (2012) |
| Not described | *B. thuringiensis* HD110, *B. laterosporus* BMG65. | Honeybee gut | Reduced larvae mortality | Hamdi and Daffonchio (2011) |
| *M. plutonius* (EFB) $10^7$–$10^6$–$10^5$ bacteria/mL | *L. kunkeei, L. mellis, L. kimbladii, L. kullabergensis, L. helsingborgensis, L. melliventris, L. apis, L. mellifer, B. asteroides,* and *B. coryneforme* ($10^7$ bacteria/mL) | Honey crop | Reduced larvae mortality | Vásquez et al. (2012) |

(continued)

**Table 9.1** (continued)

| Honeybee disease and infection dose | Microorganisms/metabolites | Source | Reported effect(s) | References |
|---|---|---|---|---|
| *N. Ceranae*<br>First trial: $10^5$ spores/10 μL<br>Second trial: $10^3$ spores/10 μL | – Surfactins S1–S2 (2000 and 5000 AU/mL from *B. subtilis* Mori2 and C4<br>– Bacteriocins B1–B2 (102,400 and 25,600 AU/mL) from *E. avium* DSMZ17511 and *E. faecium* CRL1385 | Honey<br>Beebread<br>Chicken crop | Decreased pathogen intensity (S2) | Porrini et al. (2010) |
| Diseased bees | Organic acids from *L. johnsonii* CRL1647:<br>Lactic acid (138 nM)<br>Phenyl-lactic acid (0.3 nM)<br>Acetic acid (38 nM) | Honeybee gut | Decreased spore counts | Maggi et al. (2013) |
| First trial: $10^4$ spores/μL<br>Second trial: Natural infection | *L. kunkeei* Dan39,<br>*L. plantarum* Dan91 and<br>*L. johnsonii* Dan92,<br>*B. asteroides* DSM 20431,<br>*B. coryneforme* C155, *B. indicum* C449 ($10^6$–$10^7$ cfu/mL of sugar syrup) | Honeybee gut | Reduced spore load | Baffoni et al. (2016) |
| *Nosema* spp.<br>$10^3$ spores/μL | *P. apium* C6 ($10^6$ cfu/500 μL) | Second instar larvae | Reduced spore detection | Corby-Harris et al. (2014) |
| Diseased bees | *L. johnsonii* CRL1647 ($10^5$ cfu/mL) | Honeybee gut | Reduced spore detection | Audisio et al. (2015) |
| Diseased bees | $10^5$ spores/mL of *Bacillus subtilis* Mori2 spores | Honey | Reduced spore detection | Sabaté et al. (2012) |

## 9.8 Application of Beneficial Microorganisms

Experiments envisaging the administration of beneficial bacteria to honeybees are diverse and sometime confusing. The main target is often to counteract the most widespread pathogens affecting both larvae and adults since in vitro tests evidenced interesting host protection properties by directly stimulating the bee's immune system and inhibiting pathogens through competitive exclusion and antimicrobial compound production (organic acids and secondary metabolites, e.g., bacteriocins and lipopeptides). Strains are usually isolated from honeybee crop/gut or from the environment; the use of formulations for animal and human consumption is also considered but disputable.

Applications addressed to infected larvae showed a significant reduction of larvae mortality after supplementation of different beneficial bacteria. However, data could result in misleading conclusions, since a reduction of larvae mortality, although statistically significant, is of little biological relevance because the colony will probably succumb to the disease, although it might take 1 or 2 weeks longer. Artificial infections with pathogens at high concentrations have a strong impact on the colony, and it would be preferable to observe lower doses, which simulate a natural infection process.

Forsgren 2010 applied a mixture of beneficial bacteria isolated from honey crop—*L. kunkeei*, *L. mellis*, *L. kimbladii*, *L. kullabergensis*, *L. helsingborgensis*, *L. melliventris*, *L. apis*, *L. mellifer*, *B. asteroides*, and *B. coryneforme*—with a final concentration of $10^7$ bacteria/mL. Infection in honeybee larvae was performed with two different spore concentrations of *P. larvae*.

In the detail, the LAB mixture was supplemented with sugar syrup, both in combination with *P. larvae* at the time of spore inoculum and 48 h postinfection. Results showed the positive effect of LAB supplementation only in the group challenged with the highest dose of *P. larvae* with a significant reduction of larvae mortality, from 70 to 55%. Hamdi and Daffonchio (2011) used a probiotic mixture composed by *Bacillus thuringiensis* HD110, *Brevibacillus laterosporus* BMG65, and *Saccharibacter* spp. The efficacy was proved on *P. larvae*-infected larvae, and the experiments showed that the addition of the bacterial mix to the diet decreased the mortality level from 70% in the control to 22% in larvae fed with the microorganism mix.

A single laboratory assay was performed in *Apis mellifera* (Vásquez et al. 2012) to evaluate the impact of beneficial bacteria against EFB. The LAB strains tested by Forsgren 2010 were orally administered to honeybee larvae challenged with *M. plutonius* at three concentrations ($10^7$, $10^6$, and $10^5$ bacteria/mL). Likewise, the obtained results do not prove the efficacy of the strategy since the reduced mortality between 10 and 20%, although significant, does not resolve the disease. Moreover, if to some extent an efficacy could be demonstrated in laboratory conditions, the "natural open field" situation could display different results, since multiple variables influence the life within the hive. Therefore, it could be interesting to investigate the efficacy of the LAB mixture in infected larvae with a lower dose of the pathogen and perform the treatments as preventive measure before the infection step to better simulate a natural infection process.

In adult honeybees an emergent pathogen affecting bee health is *Nosema ceranae*, identified as a microsporidium multiplying within gut cells without relevant symptoms during infection (see details in Higes et al. 2010). It has been associated with reduced honeybee life span and colony weakening (Goblirsch et al. 2013). Application of beneficial bacteria is mainly performed in plastic cages under laboratory conditions with newly emerging honeybees infected with the pathogen. Many issues can be argued about the use of cage experiments. Although the laboratory assessment allows the standardization of the variables and the direct observation of the introduced perturbations (e.g., diet change, pathogen inoculation, beneficial microorganisms, pesticides), most of the behavioral and social interactions both inside and outside the hive are lacking. Moreover, this confinement can also introduce stress factors and influence the experiment itself.

In all reported experiments, the biological relevance of spore reduction (less than 1 log) is questionable since the spore numbers remain high. Sabaté et al. (2012) and Audisio et al. (2015) observed a decrease in the amount of spores in field conditions in honeybees orally fed for several months with strains isolated from the gut of healthy insects, namely, *B. subtilis* Mori2 and *L. johnsonii* CRL1647. The decrease in *Nosema* incidence observed by Sabaté et al. (2012) was only evident in September and October when a slight spore increase was observed in the control group. When the control group showed a physiological decrease in the spore number, no relevant reduction was observed in the treated groups. Corby-Harris et al. (2014) observed a reduction of the spore load in honeybee adults originating from larvae fed with pollen patty mixed with an inoculum of *Parasaccharibacter apium* C6, but it can be argued if this observed reduction could be effective. Moreover, the authors did not specify the species used for the infection step, which is of pivotal importance since the infection process and symptomatology are different. Similarly, Baffoni et al. (2016) observed a significant decrease of *N. ceranae* in infected honeybees orally fed with a mixture of *Lactobacillus* and *Bifidobacterium* strains. However, the ~1 log reduction observed in challenged and treated insects is irrelevant since the spore number remained high and honeybees would surely die. However, in the same experiments, the authors also evidenced a significant reduction in spore load in honeybees exposed to a low natural infection. In this particular case, a hypothetical protective effect, contrasting the low infection rate, might be considered of biological relevance since it could be useful to contain the advance of the infection. Unfortunately, the experiment was performed in cages, and it should be envisaged to confirm the hypothetical effect of the beneficial bacteria also in open field.

An interesting approach to study *N. ceranae*-host interactions comes from Gisder and Genersch (2015). The authors developed a cell culture model by using the lepidopteran cell line IPL-LD 65Y, from *Lymantria dispar*, which was susceptible to *N. ceranae* infection and could support the entire microsporidium life cycle. By this approach, the authors tested several molecules for cytotoxicity and inhibition of *N. ceranae* intracellular development and demonstrated the efficacy of some of them.

Beneficial microorganisms are also applied to positively influence the hive productivity, and some results showed the significant increase of the brood area, honeybee numbers, and honey production (Audisio and Benítez-Ahrendts 2011a; Sabaté et al. 2012; Alberoni et al. 2015). In particular, Alberoni et al. (2015) also analyzed by NGS the change of the gut microbiota, and a clear increase of bifidobacteria and *Acetobacteraceae* was evidenced in treated honeybees after supplementation of lactobacilli and bifidobacteria. Both bacterial groups are important endosymbionts of the bee gut and have significant implications related to host nutrition physiology and protection. However, further investigations are necessary to better focus at gut level how this modulation would affect the host-gut microbe interaction.

As already mentioned, the use of beneficial bacteria commercially exploited in humans and animals has also been tested. An improved wax gland cell development was observed by Pătruică et al. (2012), following the supplementation of organic acids and a probiotic product containing *Lactobacillus* and *Bifidobacterium* spp. Both individually and in combination, they positively influenced the number, the morphology, and the diameter of the wax cells. Surprisingly, Andrearczyk et al. (2014) found an increase of *Nosema* spp. infection, following administration in both winter and summer bees of a probiotic product recommended for animals. Ptaszyńska et al. (2016) observed an increased mortality rate in *Nosema*-infected honeybees fed with the human probiotic *Lactobacillus rhamnosus*, both as preventive measure and along the infection. The authors argued that the increased infection was associated with a pH reduction of the honeybee midgut, because of the metabolic activity of the supplemented microorganism. However, this consideration relies on previous data (Ptaszyńska et al. 2013), where this association is not clearly and statistically demonstrated and further investigations are necessary to better understand such interactions. However, the use of these strains is controversial since it is preferable to select and use microorganisms from the honeybee gut, possessing the immense pool of genes for host interaction.

The production of antimicrobial compounds by gut symbionts for host protection is another interesting topic. A recent genomic analysis of 13 LAB strains, isolated from the honey crop, put in evidence that most of them produced extracellular proteins of known/unknown function related with antimicrobial action, host interaction, or biofilm formation. In particular, a putative novel bacteriocin with 51% homology with helveticin J was detected in *L. helsingborgensis* Bma5N (Butler et al. 2013). At the same time, it has to be said that some strains did not evidence any "antimicrobial function," thus confirming the high variability among the gut microorganisms inhabiting the same niches. Vásquez et al. (2012) analyzed the interaction of some LAB symbionts with the honey crop by SEM and fluorescence microscopy. The resulting images evidenced biofilm formation and structures resembling extracellular polymeric substances (EPS), which are known to be involved in host protection/colonization and cellular recognition (Flemming and Wingender 2010). A further support comes from the work of Ellegaard et al. (2015), which evidences at genome level the presence of gene clusters associated with the biosynthesis of cell wall polysaccharides in both "Firm 4" and "Bifido" groups (Ellegaard et al. 2015). Martinson et al. (2012) reported, in honeybee workers, the presence of genes in *G. apicola* and *S. alvi* encoding a relevant number of functions related to biofilm formation and host interaction (type IV pili, outer membrane proteins, and secretion), whose expression could be relevant for the establishment of a micro-niche harsh to pathogen colonization. Finally, the *Bacillaceae* family includes several spore-forming bacteria, isolated from the bee gut and from the hive environment, showing a strong antibacterial activity against bee pathogens. In this case inhibition activity was mainly due to the production of different classes of lipopeptides (Alippi and Reynaldi 2006; Lee et al. 2009; Sabaté et al. 2009; Yoshiyama and Kimura 2009).

Applications of antimicrobials that could be active against different pathogens are emerging, since it has the advantage of being less invasive. One of the first attempts, performed by Porrini et al. (2010), assessed the effect of four different antimicrobial metabolites: two surfactins (S1 and S2) from *B. subtilis* Mori4 and *B. subtilis* C4 and two bacteriocins from *Enterococcus avium* DSMZ17511 and *Enterococcus faecium* CRL1385 (B1 and B2). The performed trials – divergent for

*N. ceranae* spore inoculum, metabolites concentration, and administration period—revealed a significant reduction of spore concentration only for surfactin S2. Likely, Maggi et al. (2013) successfully tested, in hives naturally infected with *N. ceranae*, a pure metabolite from *L. johnsonii* CRL1647, mainly composed by lactic acid (five times at intervals of 5 days) and coupled in the last treatment with fumagillin. The analysis per individual bee showed a significant decrease of spore counts in treated hives compared to a control, where a regular increase along the experiment was observed. The decrease was observed both before and after the fumagillin application, thus showing a synergistic effect with the antibiotic treatment. Irrespective of the results, the partial standardization of the experiment by choosing sister queens has to be positively pointed out.

Research in this topic is still far to conclude that beneficial microorganisms could actually limit pathogen widespread and support honeybee health and the hive productivity, even if the preliminary results are promising. Nowadays, beekeepers too often rely on subspecies hybrids, with the false hope to increase disease resistance, but the resistance mechanisms against bee pathogens/parasites are usually a result of a coevolution in local ecosystems (Ruottinen et al. 2014). The available applications offer to some extent a picture of the positive influence of these microorganisms on bee health. However, the main issue is how the modulation of the honeybee gut microbiota could influence the composition of the gut microbiota itself and also host immunity and physiology. The widespread of microorganisms into the environment is undoubtedly a dangerous terrain that needs to be deeply investigated to minimize risk associated with bio-treatments.

## References


Aizen MA, Harder LD (2009) The global stock of domesticated honey bees is growing slower than agricultural demand for pollination. Curr Biol 19(11):915–918. https://doi.org/10.1016/j.cub.2009.03.071

Ajibola A, Chamunorwa JP, Erlwanger KH (2012) Nutraceutical values of natural honey and its contribution to human health and wealth. Nutr Metab (Lond) 20:9–61. https://doi.org/10.1186/1743-7075-9-61

Alaux C, Brunet JL, Dussaubat C, Mondet F, Tchamitchan S, Cousin M, Brillard J, Baldy A, Belzunces LP, Le Conte Y (2010) Interactions between Nosema microspores and a neonicotinoid weaken honeybees (Apis mellifera). Environ Microbiol 12(3):774–782. https://doi.org/10.1111/j.1462-2920.2009.02123.x

Alberoni D, Baffoni L, Gaggia F, Ryan P, Murphy K, Ross RP, Biavati B, Stanton C, Di Gioia D (2015) Administration of lactobacilli and bifidobacteria on Apis mellifera L. beehives to increase health of the bee super-organism. In: Microbial diversity 2015, the challenge of complexity, Perugia, pp 107–108

Alippi AM, Reynaldi FJ (2006) Inhibition of the growth of Paenibacillus larvae, the causal agent of American foulbrood of honeybees, by selected strains of aerobic spore-forming bacteria isolated from apiarian sources. J Invertebr Pathol 91(3):141–146. https://doi.org/10.1016/j.jip.2005.12.002

Allen MF, Ball BV, Underwood BA (1990) An isolate of Melissococcus pluton from Apis laboriosa. J Invertebr Pathol 55:439–440. https://doi.org/10.1016/0022-2011(90)90090-S

Anderson KE, Sheehan TH, Eckholm BJ, Mott BM, DeGrandi-Hoffman G (2011) An emerging paradigm of colony health: microbial balance of the honey bee and hive (Apis mellifera). Insectes Sociaux 58(4):431–444

Anderson KE, Sheehan TH, Mott BM, Maes P, Snyder L, Schwan MR, Walton A, Jones BM, Corby-Harris V (2013) Microbial ecology of the hive and pollination landscape: bacterial associates from floral nectar, the alimentary tract and stored food of honey bees (Apis mellifera). PLoS One 8(12):e83125. https://doi.org/10.1371/journal.pone.0083125

Andrearczyk S, Kadhim MJ, Knaga S (2014) Influence of a probiotic on the mortality, sugar syrup ingestion and infection of honeybees with Nosema spp. under laboratory assessment. Med Weter 70:762–765

Antúnez K, Martín-Hernández R, Prieto L, Meana A, Zunino P, Higes M (2009) Immune suppression in the honey bee (*Apis mellifera*) following infection by *Nosema ceranae* (Microsporidia). Environ Microbiol 11:2284–2290. https://doi.org/10.1111/j.1462-2920.2009.01953.x

APENET (2011) Effects of coated maize seed on honey bees. Report based on results obtained from the second year (2011) activity of the APENET project. http://www.reterurale.it/apenet

Arias MC, Sheppard WS (2005) Phylogenetic relationships of honey bees (Hymenoptera: Apinae:Apini) inferred from nuclear and mitochondrial DNA sequence data. Mol Phylogenet Evol 37(1):25–35. Epub 2005 Apr 19. Erratum in: 2006 Jul;40(1):315.

Aronstein KA, Murray KD (2010) Chalkbrood disease in honey bees. J Invertebr Pathol 103(1):S20–S29. https://doi.org/10.1016/j.jip.2009.06.018

Audisio MC, Benítez-Ahrendts MR (2011a) Lactobacillus johnsonii CRL1647, isolated from Apis mellifera L. bee-gut, exhibited a beneficial effect on honeybee colonies. Benef Microbes 2(1):29–34. https://doi.org/10.3920/BM2010.0024

Audisio MC, Torres MJ, Sabaté DC, Ibarguren C, Apella MC (2011b) Properties of different lactic acid bacteria isolated from Apis mellifera L. bee-gut. Microbiol Res 166(1):1–13

Audisio MC, Sabaté DC, Benítez-Ahrendts MR (2015) Effect of CRL1647 on different parameters of honeybee colonies and bacterial populations of the bee gut. Benef Microbes 6(5):687–695

Aydin L, Gulegen E, Cakmak I, Girisgin AO, Wells H (2006) Relation between Nosema and Chalkbrood diseases, and its implication for an apiary management model. Bull Vet Inst Pul 50(4):471

Baffoni L, Gaggìa F, Alberoni D, Cabbri R, Nanetti A, Biavati B, Di Gioia D (2016) Effect of dietary supplementation of Bifidobacterium and Lactobacillus strains in Apis mellifera L. against Nosema ceranae. Benef Microbes 7:45–51. https://doi.org/10.3920/BM2015.0085

Bailey L (1968) Honey bee pathology. Annu Rev Entomol 13:191–212

Bailey L (1974) An unusual type of *Streptococcus pluton* from the Eastern Hive bee. J Inv Pathol 23:246–247

Bailey L (1983) Melissococcus pluton, the cause of European foulbrood of honeybees (Apis ssp.) J Appl Bacteriol 55:65–69

Bailey L, Ball BV (1991) Honey bee pathology, 2nd edn. Academic Press, London

Benitez LB, Velho RV, da Motta AS, Segalin J, Brandelli A (2012) Antimicrobial factor from Bacillus amyloliquefaciens inhibits Paenibacillus larvae, the causative agent of American foulbrood. Arch Microbiol 194(3):177–185

Ben-Ami E, Yuval B, Jurkevitch E (2010) Manipulation of the microbiota of mass-reared Mediterranean fruit flies *Ceratitis capitata* (Diptera: Tephritidae) improves sterile male sexual performance. ISME J 4:28–37

Breeze TD, Vaissière BE, Bommarco R, Petanidou T, Seraphides N, Kozák L, Scheper J, Biesmeijer JC, Kleijn D, Gyldenkærne S, Moretti M, Holzschuh A, Steffan-Dewenter I, Stout JC, Pärtel M, Zobel M, Potts SG (2014) Agricultural policies exacerbate honeybee pollination service supply-demand mismatches across Europe. PLoS One 9(1):e82996. https://doi.org/10.1371/journal.pone.0082996

Brown M, Loosli R, Schmid-Hempel P (2000) Condition dependent expression of virulence in a trypanosome infecting bumblebees. Oikos 91:421–427

Brummel T, Ching A, Seroude L, Simon AF, Benzer S (2004) Drosophila lifespan enhancement by exogenous bacteria. Proc Natl Acad Sci U S A 101(35):12974–12979

Buchon N, Broderick NA, Chakrabarti S, Lemaitre B (2009) Invasive and indigenous microbiota impact intestinal stem cell activity through multiple pathways in Drosophila. Gen Dev 23(19):2333–2344
Butler È, Alsterfjord M, Olofsson TC, Karlsson C, Malmström J, Vásquez A (2013) Proteins of novel lactic acid bacteria from Apis mellifera mellifera: an insight into the production of known extra-cellular proteins during microbial stress. BMC Microbiol 13:235. https://doi.org/10.1186/1471-2180-13-235
Cai J, Collins MD (1994) Evidence for a close phylogenetic relationship between Melissococcus pluton, the causative agent of European foulbrood disease, and the genus Enterococcus. Int J Syst Bacteriol 44:365–367
Cariveau DP, Powell JE, Koch H, Winfree R, Moran NA (2014) Variation in gut microbial communities and its association with pathogen infection in wild bumble bees (Bombus). ISME J 8(12):2369–2379
Celli G, Maccagnani B (2003) Honey bees as bio-indicators of environmental pollution. Bull Insectol 56(1):137–139
Chen YP, Evans JD, Murphy C, Gutell R, Zuker M, Gundensen-Rindal D, Pettis JS (2009) Morphological, molecular and phylogenetic characterization of *Nosema ceranae*, a microsporidian parasite isolated from the European honey bee *Apis mellifera*. J Eukaryot Microbiol 56:142–147. https://doi.org/10.1111/j.1550-7408.2008.00374.x
Clark TB (1977) *Spiroplasma* sp., a new pathogen in honey bees. J Invertebr Pathol 29:112–113
Clark TB (1982) *Spiroplasmas*: diversity of arthropod reservoirs and host–parasite relationships. Science 217:57–59. https://doi.org/10.1126/science.217.4554.57
COLOSS Workshop (2009) Conclusions. In: Proc. Workshop "Nosema disease: lack of knowledge and work standardization" (COST Action-FA0803) Guadalajara, http://www.coloss.org/ news/nosema-workshop-proceedings-online. Accessed 20 Nov 2009
Copping L (2013) Bees and neonicotinoids: the story continues. Outlooks Pest Manag 24(3):109–119. https://doi.org/10.1564/v24_jun_05
Corby-Harris V, Maes P, Anderson KE (2014) The bacterial communities associated with honey bee (Apis mellifera) foragers. PLoS One 9(4):e95056. https://doi.org/10.1371/journal.pone.0095056
Cox-Foster DL, Conlan S, Holmes EC, Palacios G, Evans JD, Moran NA, Quan P-L, Briese T, Hornig M, Geiser DM, Martinson V, vanEngelsdorp D, Kalkstein AL, Drysdale A, Hui J, Zhai J, Cui L, Hutchison SK, Simons JF, Egholm M, Pettis JS, Lipkin WI (2007) A metagenomic survey of microbes in honey bee colony collapse disorder. Science 318:283–287. https://doi.org/10.1126/science.1146498
Crotti E, Rizzi A, Chouaia B, Ricci I, Favia G, Alma A, Sacchi L, Bourtzis K, Mandrioli M, Cherif A, Bandi C, Daffonchio D (2010) Acetic acid bacteria, newly emerging symbionts of insects. Appl Environ Microbiol 76(21):6963–6970. https://doi.org/10.1128/AEM.01336-10
Dahle B, Sørum H, Weideman JE (2011) How to get rid of EFB in Norway. Mellifera 11:22–25
Davidson EW (1973) Ultrastructure of American foulbrood disease pathogenesis in larvae of the worker honey bee, Apis mellifera. J Inv Pathol 21:53–61
De Miranda JR, Bailey L, Ball BV, Blanchard P, Budge G, Chejanovsky N, Chen YP, Gauthier L, Genersch E, De GFD, Bibière M, Ryabov E, De Smet L, Van Der Steen JJM (2013) Standard methods for virus research in *Apis mellifera*. In V Dietemann; J D Ellis; P Neumann (Eds) The COLOSS BEEBOOK, Volume II: standard methods for *Apis mellifera* pest and pathogen research. J Apicul Res 52(4):1. https://doi.org/10.3896/IBRA.1.52.4.22
Dillon R, Charnley K (2002) Mutualism between the desert locust Schistocerca gregaria and its gut microbiota. Res Microbiol 153:503–509. https://doi.org/10.1016/S0923-2508(02)01361-X
Doublet V, Labarussias M, de Miranda JR, Moritz RF, Paxton RJ (2015) Bees under stress: sublethal doses of a neonicotinoid pesticide and pathogens interact to elevate honey bee mortality across the life cycle. Environ Microbiol 17(4):969–983. https://doi.org/10.1111/1462-2920.12426
EFSA (European Food Safety Authority) (2013) Towards holistic approaches to the risk assessment of multiple stressors in bees. pp 1–76. ISBN: 978-92-9199-573-8. https://doi.org/10.2805/53269

Ellegaard KM, Tamarit D, Javelind E, Olofsson TC, Andersson SG, Vásquez A (2015) Extensive intra-phylotype diversity in lactobacilli and bifidobacteria from the honeybee gut. BMC Genomics 16:284. https://doi.org/10.1186/s12864-015-1476-6

Engel MS (1999) The taxonomy of recent and fossil honey bees (Hymenoptera: Apidae; *Apis*). J Hymen Res 8(2):165–196

Engel P, Martinson VG, Moran NA (2012) Functional diversity within the simple gut microbiota of the honey bee. Proc Natl Acad Sci U S A 109(27):11002–11007. https://doi.org/10.1073/pnas.1202970109

European Food Safety Authority (EFSA), Kass G, Moon J, Robinson T (2017) Horizon 2020: EFSA's priority research topics. EFSA Supporting Publ 14:1–11. https://doi.org/10.2903/sp.efsa.2017.EN-1166

Evans JD, Aronstein K, Chen YP, Hetru C, Imler JL, Jiang H, Kanost M, Thompson GJ, Zou Z, Hultmark D (2006) Immune pathways and defence mechanisms in honey bees Apis mellifera. Insect Mol Bio 15:645–656. https://doi.org/10.1111/j.1365-2583.2006.00682.x

Flemming HC, Wingender J (2010) The biofilm matrix. Nat Rev Microbiol 8:623–633. https://doi.org/10.1038/nrmicro2415

Forsgren E (2010) European foulbrood in honey bees. J Inv Pathol 103:5–9. https://doi.org/10.1016/j.jip.2009.06.016

Forsgren E, Fries I (2010) Comparative virulence of *Nosema ceranae* and *Nosema apis* in individual European honey bees. Vet Parasitol 170(3–4):212–217. https://doi.org/10.1016/j.vetpar.2010.02.010

Gaggìa F, Mattarelli P, Biavati B (2010) Probiotics and prebiotics in animal feeding for safe food production. Int J Food Microbiol 141(1):S15–S28. https://doi.org/10.1016/j.ijfoodmicro.2010.02.031

Gaggìa F, Di Gioia D, Baffoni L, Biavati B (2011) The role of protective and probiotic cultures in food and feed and their impact in food safety. Trends Food Sci Tech 22:58–66. https://doi.org/10.1016/j.tifs.2011.03.003

Gaggia F, Baffoni L, Stenico V, Alberoni D, Buglione E, Lilli A, Di Gioia D, Porrini C (2015) Microbial investigation on honey bee larvae showing atypical symptoms of European foulbrood. Bull Insectol 68(2):321–327. ISSN: 1721-8861

Genersch E (2010) Honey bee pathology: current threats to honey bees and beekeeping. Appl Microbiol Biotechnol 87:87–97. https://doi.org/10.1007/s00253-010-2573-8

Genersch E, Forsgren E, Pentikäinen J, Ashiralieva A, Rauch S, Kilwinski J, Fries I (2006) Reclassification of Paenibacillus larvae subsp. pulvifaciens and Paenibacillus larvae subsp. larvae as Paenibacillus larvae without subspecies differentiation. Int J Syst Evol Microbiol 56:501–511

Giersch T, Berg T, Galea F, Hornitzky M (2009) Nosema ceranae infects honey bees (Apis mellifera) and contaminates honey in Australia. Apidologie 40(2):117–123

Gilliam M, Valentine DK (1976) Bacteria isolated from the intestinal contents of foraging worker honey bees, Apis mellifera: the genus Bacillus. J Invertebr Pathol 28(2):275–276

Girolami V, Marzaro M, Vivan L, Mazzon L, Greatti M, Giorio C, Marton D, Tapparo A (2012) Fatal powdering of bees in flight with particulates of neonicotinoids seed coating and humidity implication. J Appl Entomol 136:17–26. https://doi.org/10.1111/j.1439-0418.2011.01648.x

Gisder S, Genersch E (2015) Identification of candidate agents active against N. ceranae infection in honey bees: establishment of a medium throughput screening assay based on N. ceranae infected cultured cells. PLoS One 10(2):e0117200. https://doi.org/10.1371/journal.pone.0117200

Goblirsch M, Huang ZY, Spivak M (2013) Physiological and behavioral changes in honey bees (Apis mellifera) induced by Nosema ceranae infection. PLoS One 8(3):e58165. https://doi.org/10.1371/journal.pone.005816

Goodwin RM, Taylor MA, Mcbrydie HM, Cox HM (2006) Drift of *Varroa destructor*-infested worker honey bees to neighbouring colonies. J Apicult Res 45(3):155–156

Hamdi C, Daffonchio D (2011) Methods for the prevention and control of pathogenic infections in bees and relative composition. Patent Application WO/2011/138310

Hamdi C, Balloi A, Essanaa J, Crotti E, Gonella E, Raddadi N, Ricci I, Boudabous A, Borin S, Manino A, Bandi C, Alma A, Daffonchio D, Cherif A (2011) Gut microbiome dysbiosis and honeybee health. J Appl Entomol 135:524–533. https://doi.org/10.1111/j.1439-0418.2010.01609.x
Hedtke K, Jensen PM, Jensen AB, Genersch E (2008) Evidence for emerging parasites and pathogens influencing outbreaks of stress-related diseases like chalkbrood. J Invertebr Pathol 108(3):167–173. https://doi.org/10.1016/j.jip.2011.08.006
Hedtke SM, Blitzer EJ, Montgomery GA, Danforth BN (2015) Introduction of non-native pollinators can lead to trans-continental movement of bee-associated fungi. PLoS One 10(6):e0130560. https://doi.org/10.1371/journal.pone.0130560
Hider RC (1988) Honeybee venom: a rich source of pharmacologically active peptides. Endeavour 12:60–65
Higes M, Garcia-Palencia P, Martin-Hernandez R, Meana A (2007) Experimental infection of *Apis mellifera* honeybees with *Nosema ceranae* (Microsporidia). J Invertebr Pathol 94:211–217. https://doi.org/10.1016/j.jip.2006.11.001
Higes M, Martín-Hernández R, Botías C, Bailón EG, González-Porto AV, Barrios L, Del Nozal MJ, Bernal JL, Jiménez JJ, Palencia PG, Meana A (2008) How natural infection by Nosema ceranae causes honeybee colony collapse. Environ Microbiol 10(10):2659–2669. https://doi.org/10.1111/j.1462-2920.2008.01687.x
Higes M, Martin-Hernandez R, Meana A (2010) *Nosema ceranae* in Europe: an emergent type C nosemosis. Apidologie 41:375–392
Hölldobler B, Wilson EO, Nelson MC (2009) Chapter 1: the construction of a superorganism. In: The superorganism: the beauty, elegance, and strangeness of insect societies. W. W. Norton & Company, New York. pp 3–10 (522 pp)
Hrabák J, Martínek K (2007) Screening of secreted proteases of Paenibacillus larvae by using substrate-SDS-polyacrylamide gel electrophoresis. J Apicul Res 46:160–164
Ikediobi CO, Obi VC, Achoba IA (1985) Beekeeping and honey production in Nigeria. Nigeria Field 50:59–70
Koch H, Schmid-Hempel P (2011) Socially transmitted gut microbiota protect bumble bees against an intestinal parasite. Proc Natl Acad Sci U S A 108:19288–19292
Kwong WK, Moran NA (2013) Cultivation and characterization of the gut symbionts of honey bees and bumble bees: description of Snodgrassella alvi gen. nov., sp. nov., a member of the family Neisseriaceae of the Betaproteobacteria, and Gilliamella apicola gen. nov., sp. nov., a member of Orbaceae fam. nov., Orbales ord. nov., a sister taxon to the order 'Enterobacteriales' of the Gammaproteobacteria. Int J Syst Evol Microbiol 63(6):2008–2018. https://doi.org/10.1099/ijs.0.044875-0
Lee H, Churey JJ, Worobo RW (2009) Isolation and characterization of a protective bacterial culture isolated from honey active against American Foulbrood disease. FEMS Microbiol Lett 296(1):39–44. https://doi.org/10.1111/j.1574-6968.2009.01615.x
Leven LV, Boot WB, Mutsaers M, Segeren P, Velthuis H (1997) Beekeeping in the tropics. In: Leen van't Leven LV, Segeren P (eds) Agrodok 32 publication. The Netherlands. ISBN Agromisa: 90-8573-043-0
Li J, Qin H, Wu J, Sadd BM, Wang X, Evans JD, Peng W, Chen Y (2012) The prevalence of parasites and pathogens in Asian honeybees Apis cerana in China. PLoS One 7(11):e47955. https://doi.org/10.1371/journal.pone.0047955
Lipa JJ, Triggiani O (1996) *Apicystis* gen nov and *Apicystis bombi* (Liu, Macfarlane & Pengelly) comb nov (Protozoa: Neogregarinida), a cosmopolitan parasite of *Bombus* and *Apis* (Hymenoptera: Apidae). Apidologie 27:29–34
Lopez MI, Pettis JS, Smith IB, Chu PS (2008) Multiclass determination and confirmation of antibiotic residues in honey using LC–MS/MS. J Agric Food Chem 56:1553–1559. https://doi.org/10.1021/jf073236w
Maggi M, Negri P, Plischuk S, Szawarski N, De Piano F, De Feudis L, Eguaras M, Audisio C (2013) Effects of the organic acids produced by a lactic acid bacterium in Apis mellifera colony development, Nosema ceranae control and fumagillin efficiency. Vet Microbiol 167:474–483. https://doi.org/10.1016/j.vetmic.2013.07.030

Martín-Hernández R, Botías C, Barrios L, Martínez-Salvador A, Meana A, Mayack C, Higes M (2011) Comparison of the energetic stress associated with experimental *Nosema ceranae* and *Nosema apis* infection of honeybees (*Apis mellifera*). Parasitol Res 109(3):605–612. https://doi.org/10.1007/s00436011-2292-9

Martinson VG, Moy J, Moran NA (2012) Establishment of characteristic gut bacteria during development of the honeybee worker. Appl Environ Microbiol 78(8):2830–2840. https://doi.org/10.1128/AEM.07810-11

Marzaro M, Vivan L, Mazzon L, Mori N, Greatti M, Petrucco E, Bernardo A, Giorio C, Marton D, Tapparo A, Girolami V (2011) Lethal aerial powdering of honey bees with neonicotinoids from fragments of maize seed coat. Bull Insectol 64(1):119–126. ISSN: 1721-8861

Menzel R (1993) Associative learning in honey bees. Apidologie 24(3):157–168

Mohr KI, Tebbe CC (2006) Diversity and phylotype consistency of bacteria in the guts of three bee species (Apoidea) at an oilseed rape field. Environ Microbiol 8(2):258–272

Moore L, Kosut M (2013) Buzz: urban beekeeping and the power of the bee. New York University Press, New York. http://www.jstor.org/stable/j.ctt9qfrd1. ISBN: 9780814763063

Moran NA (2015) Genomics of the honey bee microbiome. Curr Opin Insect Sci 10:22–28. https://doi.org/10.1016/j.cois.2015.04.003

Morimoto T, Kojima Y, Yoshiyama M, Kimura K, Yang B, Peng G, Kadowaki T (2013) Molecular detection of protozoan parasites infecting Apis mellifera colonies in Japan. Environ Microbiol Rep 5(1):74–77. https://doi.org/10.1111/j.1758-2229.2012.00385.x

Moritz RFA, Southwick EE (1992) Bees as superorganisms: an evolutionary reality. Springer, Berlin

Morse RA (1989) History of subsection; apiculture and social insects. Bull Entomol Soc Am 35:116–118. https://doi.org/10.1093/besa/35.3.115

Mouches C, Bové JM, Albisetti J, Clark TB, Tully JG (1982) A *spiroplasma* of serogroup IV causes a May-disease-like disorder of honeybees in Southwestern France. Microb Ecol 8:387–399. https://doi.org/10.1007/BF02010677

Neumann P, Carreck NL (2010) Honey bee colony losses. J Apicul Res 49(1):1–6. https://doi.org/10.3896/IBRA.1.49.1.01

Oldroyd BP (2007) What's killing American honey bees? PLoS Biol 5(6):e168

Olofsson TC, Vásquez A (2008) Detection and identification of a novel lactic acid bacterial flora within the honey stomach of the honeybee Apis mellifera. Curr Microbiol 57:356–363. https://doi.org/10.1007/s00284-008-9202-0

Olofsson TC, Alsterfjord M, Nilson B, Butler È, Vásquez A (2014) Lactobacillus apinorum sp. nov., Lactobacillus mellifer sp. nov., Lactobacillus mellis sp. nov., Lactobacillus melliventris sp. nov., Lactobacillus kimbladii sp. nov., Lactobacillus helsingborgensis sp. nov. and Lactobacillus kullabergensis sp. nov., isolated from the honey stomach of the honeybee Apis mellifera. Int J Syst Evol Microbiol 64(9):3109–3119. https://doi.org/10.1099/ijs.0.059600-0

Page RE Jr, Peng CY (2001) Aging and development in social insects with emphasis on the honey bee, Apis mellifera L. Exp Gerontol 36(4–6):695–711

Pătruică S, Dumitrescu G, Stancu A, Bura M, Bănătean Dunea I (2012) The effect of prebiotic and probiotic feed supplementation on the wax glands of worker bees (Apis mellifera). J Anim Sci Biotech 45:268–271

Paxton RJ, Klee J, Korpela S, Fries I (2007) Nosema ceranae has infected Apis mellifera in Europe since at least 1998 and may be more virulent than Nosema apis. Apidologie 38(6):558–565

Peng YCS, Marston JM (1986) Filtering mechanism of the honey bee proventriculus. Physiol Entomol 11(4):433–439

Plischuk S, Meeus I, Smagghe G, Lange CE (2011) Apicystis bombi (Apicomplexa: Neogregarinorida) parasitizing Apis mellifera and Bombus terrestris (Hymenoptera: Apidae) in Argentina. Env Microbiol Rep 3:565–568. https://doi.org/10.1111/j.1758-2229.2011.00261.x

Porrini MP, Audisio MC, Sabaté DC, Ibarguren C, Medici SK, Sarlo EG, Garrido PM, Eguaras MJ (2010) Effect of bacterial metabolites on microsporidian Nosema ceranae and on its host Apis mellifera. Parasitol Res 107:381–388. https://doi.org/10.1007/s00436-010-1875-1

Potts SG, Biesmeijer JC, Kremen C, Neumann P, Schweiger O, Kunin WE (2010) Global pollinator declines: trends, impacts and drivers. Trends Ecol Evol 25(6):345–353. https://doi.org/10.1016/j.tree.2010.01.007

Prell H (1926) The amoeba-disease of adult bees: a little-noticed spring-time disease. Bee World 8:10–13

Ptaszyńska AA, Borsuk G, Mułenko W, Olszewski K (2013) Impact of ethanol on Nosema spp. infected bees. Med Weter 69:736–741

Ptaszyńska AA, Borsuk G, Zdybicka-Barabas A, Cytryńska M, Małek W (2016) Are commercial probiotics and prebiotics effective in the treatment and prevention of honeybee nosemosis C? Parasitol Res 115:397–406. https://doi.org/10.1007/s00436-015-4761-z

Raffiudin R, Crozier RH (2007) Phylogenetic analysis of honey bee behavioral evolution. Mol Phylogenet Evol 43(2):543–552

Ravoet J, Maharramov J, Meeus I, De Smet L, Wenseleers T, Smagghe G, de Graaf DC (2013) Comprehensive bee pathogen screening in Belgium reveals Crithidia mellificae as a new contributory factor to winter mortality. PLoS One 8(8):e72443. https://doi.org/10.1371/journal.pone.0072443

Regassa LB, Gasparich GE (2006) Spiroplasmas: evolutionary relationships and biodiversity. Front Biosci 11:2983–3002

Roetschi A, Berthoud H, Kuhn R, Imdorf A (2008) Infection rate based on quantitative real-time PCR of Melissococcus plutonius, the causal agent of European foulbrood, in honeybee colonies before and after apiary sanitation. Apidologie 39:362–371. https://doi.org/10.1051/apido:200819

Rosenkranz P, Aumeier P, Ziegelmann B (2010) Biology and control of *Varroa destructor*. J Inv Pathol 103:96–119. https://doi.org/10.1016/j.jip.2009.07.016

Runckel C, Flenniken ML, Engel JC, Ruby JG, Ganem D, Andino R, DeRisi JL (2011) Temporal analysis of the honey bee microbiome reveals four novel viruses and seasonal prevalence of known viruses, Nosema, and Crithidia. PLoS One 6(6):e20656. https://doi.org/10.1371/journal.pone.0020656

Ruottinen L, Berg P, Kantanen J, Kristensen TN, Praebel A, Groeneveld L (2014) Status and conservation of the nordic brown bee: final report. Nordic Genetic Resource Center (NordGen). http://vbn.aau.dk/ws/files/207960984/Ruottinen_et_al_2014.pdf. Accessed 31 May 2016

Ryu JH, Kim SH, Lee HY, Bai JY, Nam YD, Bae JW, Lee DG, Shin SC, Ha EM, Lee WJ (2008) Innate immune homeostasis by the homeobox gene caudal and commensal-gut mutualism in *Drosophila*. Science 319(5864):777–782. https://doi.org/10.1126/science.1149357

Sabaté DC, Carrillo L, Audisio MC (2009) Inhibition of Paenibacillus larvae and Ascosphaera apis by Bacillus subtilis isolated from honeybee gut and honey samples. Res Microbiol 160(3):193–199. https://doi.org/10.1016/j.resmic.2009.03.002

Sabaté DC, Cruz MS, Benítez-Ahrendts MR, Audisio MC (2012) Beneficial effects of Bacillus subtilis subsp. subtilis Mori2, a honey-associated strain, on honeybee colony performance. Probiotics Antimicrob Proteins 4:39–46. https://doi.org/10.1007/s12602-011-9089-0011-9089-0

Saraiva MA, Zemolin APP, Franco JL, Boldo JT, Stefenon VM, Triplett EW, De Oliveira FA, Roesch LFW (2015) Relationship between honeybee nutrition and their microbial communities. Antonie Van Leeuwenhoek 107:921–933. https://doi.org/10.1007/s10482-015-0384-8

Scardovi V, Trovatelli LD (1969) New species of bifidobacteria from Apis mellifica L. and Apis indica F. A contribution to the taxonomy and biochemistry of the genus Bifidobacterium. Zentralbl Bakteriol Parasitenkd Infektionskr Hyg Abt II 123:64–88

Schwarz RS, Evans JD (2013) Infections with single and mixed-species of trypanosome and microsporidia elicit distinct, ephemeral cellular and humoral immune responses in honey bees. Dev Comp Immunol 40:300–310. https://doi.org/10.1016/j.dci.2013.03.010

Sharon G, Segal D, Ringo JM, Hefetz A, Zilber-Rosenberg I, Rosenberg E (2010) Commensal bacteria play a role in mating preference of Drosophila melanogaster. Proc Natl Acad Sci U S A 107(46):20051–20056

Shin SC, Kim SH, You H, Kim B, Kim AC, Lee KA, Yoon JH, Ryu JH, Lee WJ (2011) *Drosophila* microbiome modulates host developmental and metabolic homeostasis via insulin signaling. Science 334:670–674. https://doi.org/10.1126/science.1212782

Storelli G, Defaye A, Erkosar B, Hols P, Royet J, Leulier F (2011) Lactobacillus plantarum promotes drosophila systemic growth by modulating hormonal signals through TOR-dependent nutrient sensing. Cell Metabol 14(3):403–414

Tontou R, Gaggìa F, Baffoni L, Devescovi G, Venturi V, Giovanardi G, Stefani E (2015) Molecular characterisation of an endophyte showing a strong antagonistic activity against Pseudomonas syringae pv. actinidiae. Plant Soil 405:97. https://doi.org/10.1007/s11104-015-2624-0

Vanengelsdorp D, Meixner MD (2010) A historical review of managed honey bee populations in Europe and the United States and the factors that may affect them. J Invertebr Pathol 103:S80–S95. https://doi.org/10.1016/j.jip.2009.06.011

Vásquez A, Olofsson TC (2009) The lactic acid bacteria involved in the production of bee pollen and bee bread. J Apicul Res 48(3):189–195

Vásquez A, Forsgren E, Fries I, Paxton RJ, Flaberg E, Szekely L, Olofsson TC (2012) Symbionts as major modulators of insect health: lactic acid bacteria and honeybees. PLoS One 7(3):e33188. https://doi.org/10.1371/journal.pone.0033188

Vidau C, Diogon M, Aufauvre J, Fontbonne R, Viguès B, Brunet JL, Texier C, Biron DG, Blot N, El Alaoui H, Belzunces LP, Delbac F (2011) Exposure to sublethal doses of fipronil and thiacloprid highly increases mortality of honeybees previously infected by *Nosema ceranae*. PLoS One 6:e21550. https://doi.org/10.1371/journal.pone.0021550

Wallberg A, Han F, Wellhagen G, Dahle B, Kawata M, Haddad N, Simões ZL, Allsopp MH, Kandemir I, De la Rúa P, Pirk CW, Webster MT (2014) A worldwide survey of genome sequence variation provides insight into the evolutionary history of the honeybee *Apis mellifera*. Nat Genet 46(10):1081–1088. https://doi.org/10.1038/ng.3077

Wilkins S, Brown M, Andrew A, Cuthbertson GS (2007) The incidence of honey bee pests and diseases in England and Wales. Pest Manag Sci 63:1062–1068. https://doi.org/10.1002/ps.1461

Williams GR, Sampson MA, Shutler D, Rogers R (2008) Does fumagillin control the recently detected invasive parasite *Nosema ceranae* in western honeybees (*Apis mellifera*)? J Invertebr Pathol 99:342–344. https://doi.org/10.1016/j.jip.2008.04.005

Winston ML (1987) The biology of the honey bee. Harvard Univ. Press, Cambridge, MA

Wu M, Sugimura Y, Takaya N, Takamatsu D, Kobayashi M, Taylor D, Yoshiyama M (2013) Characterization of bifidobacteria in the digestive tract of the Japanese honeybee, Apis cerana japonica. J Invertebr Pathol 112:88–93. https://doi.org/10.1016/j.jip.2012.09.005

Yoshiyama M, Kimura K (2009) Bacteria in the gut of Japanese honeybee, Apis cerana japonica, and their antagonistic effect against Paenibacillus larvae, the causal agent of American foulbrood. J Invertebr Pathol 102:91–96. https://doi.org/10.1016/j.jip.2009.07.005

# 10 Paraprobiotics as Potential Agents for Improving Animal Health

Caroline Nunes de Almada, Carine Nunes de Almada, and Anderson de Souza Sant'Ana

## 10.1 Introduction

The most common types of microorganisms used as probiotic include species of the genus *Lactobacillus*, *Bifidobacterium*, *Lactococcus*, *Streptococcus*, *Enterococcus*, *Saccharomyces*, *Bacillus*, *Brevibacillus*, and *Sporolactobacillus*, among others (Borchers et al. 2009). Probiotic microorganisms are characterized according to their source of origin and physiology and should exert health benefits without any toxic property or virulence factor/pathogenicity (Nogueira and Gonçalves 2011; Sanders et al. 2007). However, the beneficial effects of probiotics are specific to a particular strain. Thus, these effects cannot be extrapolated to other strains, species, or genera (Almada et al. 2015).

The use of probiotics is associated with a variety of benefits in animal production, such as increased resistance to disease (Da Silva Almeida et al. 2015), an increase in reproductive performance (Abdelrahman et al. 2014), growth improvement (Dehaghani et al. 2015; Hu et al. 2015), and reduction in perinatal mortality (Kritas et al. 2006). Other documented benefits include the improvement of immune function (Altmeyer et al. 2014), morphology of the gastrointestinal tract (Di Giancamillo et al. 2008), and better product quality (Mappley et al. 2013). In addition, there is evidence to support the clinical applications of probiotics in the prevention and treatment of diseases such as gastrointestinal diseases (Amit-Romach et al. 2010) and to combat pathogen infections (Castillo et al. 2012).

Throughout the last decades, probiotics have been defined as live microorganisms which when administered in adequate amounts confer health benefits to the host (Hill et al. 2014). However, in recent years, studies have reported that not only

C.N. de Almada • C.N. de Almada • A.S. de Sant'Ana (✉)
Department of Food Science, Faculty of Food Engineering (FEA), University of Campinas (UNICAMP), Campinas, SP, Brazil
e-mail: carol_ald@hotmail.com; carinealmada86@gmail.com; and@unicamp.br



D. Di Gioia, B. Biavati (eds.), *Probiotics and Prebiotics in Animal Health and Food Safety*, https://doi.org/10.1007/978-3-319-71950-4_10

live probiotic cells but also "dead" or "inactivated" probiotic cells have the ability to provide health benefits (Nakamura et al. 2012; Tanzer et al. 2010).

Therefore, the aim of this chapter is to discuss the use of live and dead probiotics in animals. First, a general description of the dead probiotics will be presented, followed by description of the main probiotics and their use in animal production. Finally, studies that correlate live and dead probiotics with the animal health will be addressed.

## 10.2 Dead Probiotics

Dead probiotics are nonviable microbial cells, which when administered in appropriate amounts and frequency are able to confer benefits on human or animal health, also known as paraprobiotics, ghost probiotics, or postbiotics (Patel and Denning 2013; Raz and Rachmilewitz 2005; Taverniti and Guglielmetti 2011).

The inactivation of probiotic bacteria can be achieved by heat, high pressure, ultraviolet, irradiation, sonication, drying, acid, and formalin, among other methods (Ananta and Knorr 2009; Kamiya et al. 2006; Newaj-Fyzul et al. 2007; Shin et al. 2010; Taverniti and Guglielmetti 2011). Although several inactivation treatments have been used, the most appropriate method will depend on the microorganism and the expected benefit. In addition, it is fundamental that the method retains the health benefits of probiotcs (Raz and Rachmilewitz 2005). Thus, the choice of the inactivation method should be carefully evaluated, once the type of treatment influences the cell structure and the probiotic properties (Ananta and Knorr 2009).

The use of dead probiotic bacteria offers advantages when compared with live microorganisms. Live microorganisms lose viability at elevated temperatures. In contrast, dead microorganisms may remain stable during storage over a wide temperature range. Additionaly, dead probiotics are more easily stored, transported, and handled, besides the possibility of application to animals through feed and water. Moreover, dead probiotics present less or no interaction with other feed components (Chuang et al. 2007; Ishikawa et al. 2010), which does not impact the durability of feeds and supplements used in animal husbandry, for example. Dead and live probiotics can also provide benefits in animal health such as modulation of intestinal microbiota (Yang et al. 2014), prevention of bacterial infections (Grzeskowiak et al. 2014), and modulation of the immune system (Biswas et al. 2013) among others.

## 10.3 Mechanism of Action (Alive Versus Dead Probiotic Bacteria)

Probiotics may provide benefits through several mechanisms. Live probiotics act on the improvement of the intestinal epithelial barrier and increase in adhesion to the intestinal mucosa, thus inhibiting adhesion of pathogens, competitive exclusion of pathogen (competition for nutrients and adhesion sites), and production of antimicrobial substances (organic acids and other bacteriocins). Other important benefit

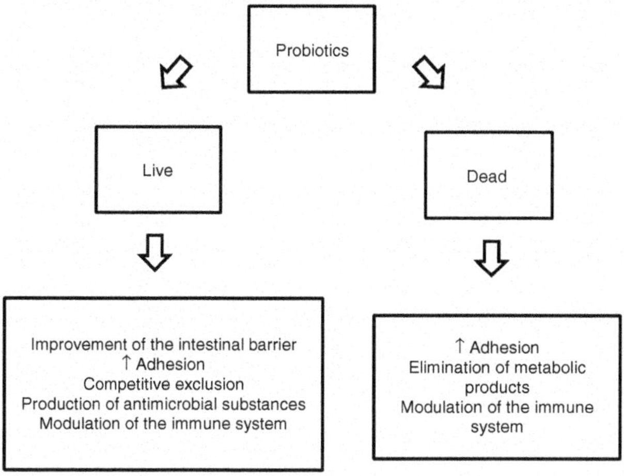


**Fig. 10.1** Mechanism of action of live and dead probiotics

includes the modulation of the immune system (interaction between dendritic and intestinal epithelial cells, macrophages, and lymphocytes) (Bermudez-Brito et al. 2012). In relation to dead probiotics, the mechanisms of action involve an increase in their adhesion and concomitant inhibition of pathogens, secretion of metabolic products, and modulation of the immune system (Fig. 10.1) (Grzeskowiak et al. 2014; Kim et al. 2013a, b; Shin et al. 2010).

In the literature, there are several reports on the mechanisms of action of probiotics. Youn et al. (2012) have shown that live and dead lactobacilli species were effective against influenza virus infection in mice. This effect was due to a direct modulation of the immune system of the respiratory tract (increased level of IgA antibodies and reduction of pro-inflammatory cytokines in the lung).

Dead and live *Lactobacillus acidophilus* was able to prevent *Salmonella* infection in mice once it facilitates the excretion of *Salmonella* due to property coaggregation and adherence to this pathogen. However, dead *Lactobacillus acidophilus* cells have percentages of coaggregation and larger adhesion than live cells. On the other hand, live *Lactobacillus acidophilus* produces lactic acid, which changes the intestinal pH and helps to inhibit *Salmonella* growth (Kim et al. 2013a, b).

Live and heat-killed *Lactobacillus casei* prevented the invasion of *Staphylococcus aureus* (aureus-induced mastitis) in bovine mammary epithelial cells. The live cells of *Lactobacillus casei* probiotics were able to reduce the adhesion and internalization of *Staphylococcus* to bovine mammary epithelial cells, while dead probiotic cells only inhibited the adhesion. This reduction may be due to a modulation of integrity or physiology of mammalian cells and direct effect of *Lactobacillus* casei on *Staphylococcus* such as coaggregation or competition for binding sites used in internalization. Inhibition of adhesion is achieved through competitive exclusion, with saturation of the binding sites in bovine mammary epithelial cells by live and dead probiotics, preventing *Staphylococcus* to bind to these cells (Bouchard et al. 2013).

Dead and live *Lactobacillus gasseri* TMC0356 protected mice against infection by influenza virus. *Lactobacilli* can exert such protection due to cell wall components such as peptide glycan that act to improve the intestinal and respiratory immune response at a local and systemic manner (Kawase et al. 2010, 2012).

Sonication-killed *Bifidobacterium longum* has hypocholesterolemic effect in rats, probably due to the elimination of metabolic products from dead cells, inhibiting the enzyme responsible for cholesterol synthesis or cholesterol absorption into the body or facilitating the cholesterol elimination (Shin et al. 2010).

However, the mechanisms of action of dead probiotics have not been fully elucidated. Thus, further studies are needed to understand how the dead probiotics behave and interact with the gastrointestinal system in order to result in health benefits to animals.

## 10.4 Probiotics in Animal Production

A diverse group of microorganisms has been studied as probiotics for both human and animal use. Among these, *Lactobacillus*, *Bifidobacterium*, *Enterecoccus*, *Bacillus*, and *Saccharomyces* stand out (Sanders et al. 2003).

*Lactobacillus* are widely distributed in nature and occur in a variety of environments such as soil, water, plants, fruits, vegetables, milk, dairy products, meat, cereals, and fermented beverages. In addition, some members of this genus naturally reside in the gastrointestinal, vaginal, and oral tract of humans and animals (Herbel et al. 2013). These microorganisms are Gram-positive, nonspore-formers, nonpathogenic, catalase negative, anaerobic or aero-tolerant, and acid tolerant, have bacillary or cocobacilar form, and possess fermentative metabolism (Salvetti et al. 2012). *Lactobacillus* species used as probiotics include *Lactobacillus acidophilus*, *Lactobacillus reuteri*, *Lactobacillus rhamnosus*, *Lactobacillus casei*, *Lactobacillus paracasei*, *Lactobacillus plantarum*, and *Lactobacillus helveticus* (Gaggìa et al. 2010).

*Bifidobacterium* genus includes more than 40 species. These microorganisms were first isolated from the feces of newborns and belong to the microbiota dominant (>1%) of adults (Andriantsoanirina et al. 2013). Bifidobacteria are of great importance in the gastrointestinal tract of humans and other mammals and are dependent upon the age and diet (Morais and Jacob 2006). They are Gram-positive microorganisms, non-sporulating, strictly anaerobic, and catalase negative (Holzapfel et al. 2001; Khan et al. 2013). The main species are *Bifidobacterium bifidum*, *B. breve*, *B. longum*, *B. animalis*, *B. infantis*, and *B. adolescentis* (Andriantsoanirina et al. 2013). Microorganisms of the genera *Lactobacillus* and *Bifidobacterium* are most commonly used as probiotics (Sanders et al. 2003).

*Enterococcus* belongs to the group of lactic acid bacteria. These microorganisms are found naturally in foods and are commensal gut bacteriain humans and animals. In some cases, they have been associated with infections (Gaggìa et al. 2010).

*Bacillus* is a genus of Gram-positive rod-shaped bacteria, catalase positive, aerobic or facultative anaerobic, mobile, and ubiquitous (Nakano and Zuber 1998; Whitman and William 2009). Although the soil has been most often considered as a primary source of *Bacillus*, these microorganisms have also been isolated from the gastrointestinal tract of animals and humans (Fakhry et al. 2008; Hong et al. 2005, 2009). In recent years, various *Bacillus* species have been reported as probiotics, including *Bacillus clausii*, *Bacillus cereus*, *Bacillus coagulans*, *Bacillus mojavensis* KJS-3, *Bacillus flexus* Hk1, *Bacillus licheniformis* Me1, *Bacillus subtilis* Bn1, *Bacillus subtilis* natto, *Bacillus subtilis* KD1, *Bacillus subtilis* PY79, and *Bacillus indicus* HU36 (Cutting 2011; Hong et al. 2008; Kim et al. 2011; Nithya and Halami 2013; Riazi et al. 2009; Ripamonti et al. 2009; Sanders et al. 2003; Sun et al. 2013; Wu et al. 2011; Zhou et al. 2010).

*Saccharomyces* is a genus of yeasts and can be found in plants, fruits, and soil. Some species have been used in animal feed, including *Saccharomyces cerevisiae* in the diet of broiler chickens (Iraqi and Fayed 2012) and *Saccharomyces boulardii* in feed for pigs and ruminants (Gaggia et al. 2010).

The use of probiotics in animal breeding emerged as an alternative to antibiotics. According to the FDA (US Food and Drug Administration), the use of antibiotics is allowed in food-producing animals to control specific diseases. However, to promote growth, increase in reproductive performance, and improve feed efficiency, its use will no longer be allowed (FDA 2012). The restrictions on antibiotic usage are due to the development of antimicrobial resistance and transfer of antibiotic resistance genes from animals to human microbiota (Mathur and Singh 2005; Salyers et al. 2004). Thus, there is an increased need for alternative methods to control and prevent the colonization of gut by pathogenic bacteria as well as to growth promotion and performance without the side effects of antibiotics. Given these aspects, several authors have been investigated the applications of probiotics in animal production (Table 10.1).

A mixture of live probiotics (*Bifidobacterium animalis* DSM 16284, *Lactobacillus salivarius* DSM 16351, and *Enterococcus faecium* DSM 21913) when administered as supplements in male broilers for 42 days showed a similar effect on feed conversion and decrease rate in occistos count and intestinal lesions (intestinal health indicators) when compared with the salinomycin-treated group. The mechanism of action can be the competitive exclusion, once the probiotic cells will compete with the *Eimeria* species (intracellular parasite that causes the enteric disease coccidiosis) for adhesion sites on intestinal epithelial cells. Thus, the probiotics may be a possible alternative to control coccidiosis in broilers (Abdelrahman et al. 2014).

Food additives containing live probiotics (*Bacillus toyonenses*) improved growth performance and protect against enteric pathogens when administered in postweaning piglets. Thus, it was observed an increase in the dietary intake, reduction in enteric pathogens, and an increase in the number of lactic acid bacteria, with a beneficial effect on the balance of the intestinal microbiota (Kantas et al. 2015). In neonate broiler chickens, diet supplemented with live probiotics (FloraMax-B11) together with 5% glycerol decreased the incidence of *Salmonella* in those animals, whose effect was not observed when the probiotics or glycerol was administered alone, thus evidencing a synergistic effect between FloraMax-B11 and glycerol (Delgado et al. 2014).

**Table 10.1** Probiotics for use in animal production

| Effects | Live probiotics | Dead probiotics | References |
|---|---|---|---|
| Positive performance in coccidiosis control in broiler | Mixture of probiotics (*Bifidobacterium animalis* DSM 16284, *Lactobacillus salivarius* DSM 16351, and *Enterococcus faecium* DSM 21913) | – | Abdelrahman et al. (2014) |
| Increase in egg production performance in layer hens | *Bacillus subtilis* | *Lactobacillus salivarius* (heat killed at 80 °C/30 min) | Zhang et al. (2012) |
| Increased diet intake in pigs and weight performance in the period of pregnancy and lactation | *Enterococcus faecium* DSM 7134 | – | Böhmer et al. (2006) |
| Increase in growth performance, improving weight gain, feed conversion efficiency, and stimulation of serum IgG secretion and Th1 cytokine levels, including IFN-γ in weaning calves | *Bacillus subtilis* natto | – | Sun et al. (2010) |
| Weight gain performance in piglets | *Lactobacillus* ssp. and *Lactobacillus plantarum* | *Lactobacillus* ssp. and *Lactobacillus plantarum* (heat killed at 121 °C/15 min) | Busanello et al. (2015) |
| Inhibition of *Staphylococcus aureus* invasion in bovine mammary epithelial cells | *Lactobacillus casei* CIRM-BIA 667, *Lactobacillus casei* BL23, and *Lactobacillus casei* CIRM-BIA 1542 | *Lactobacillus casei* CIRM-BIA 667 (heat killed at 95 °C/15 min) | Bouchard et al. (2013) |
| Improvement in growth capacity and immune response against *Eimeria* and *Clostridium perfringens* in broilers | *Bacillus subtilis* | – | Lee et al. (2015) |
| Improvement in carcass characteristics and quality of broiler meat | *Enterococcus faecium* CGMCC 2516 | – | Zheng et al. (2014) |

**Table 10.1** (continued)

| Effects | Live probiotics | Dead probiotics | References |
|---|---|---|---|
| Improvement in production performance in broilers | *Saccharomyces cerevisiae* | *Saccharomyces cerevisiae* (Thepax®) | Iraqi and Fayed (2012) |
| Improvement in reproductive performance in terms of greater GSI, fecundity, survival, and morphometric characteristics in fish | *Bacillus subtilis* | – | Ghosh et al. (2007) |
| Improvement in growth performance and protection against enteric pathogens in postweaning piglets | *Bacillus toyonensis* | – | Kantas et al. (2015) |
| Improvement in performance and relief from the negative effects of *Eimeria* infection in broilers | Mixtureof probiotics (*Bifidobacterim animalis* DSM 16284, *Lactobacillus salivarius* DSM 16351, and *Enterococcus faecium* DSM 21913) | – | Ritzi et al. (2012) |
| Growth promotion in post-weaned mice | *Lactobacillus rhamnosus* MA27/6B and *Lactobacillus acidophilus* MA27/6R | *Lactobacillus rhamnosus* MA27/6B and *Lactobacillus acidophilus* MA27/6R (heat killed at 120 °C/15 min) | Bernardeau et al. (2002) |
| Reduction in the incidence of *Salmonella* in neonate broilers | FloraMax-B11 (*Lactobacillus salivarius* and *Pediococcus parvulus*) + glycerol | – | Delgado et al. 2014 |
| Reduction of *Salmonella* infection in pigs | *Lactobacillus zeae* and *Lactobacillus casei* | – | Yin et al. (2014) |

– Not used

Sun et al. (2010) have reported that live *Bacillus subtilis* natto in milk to weaning calves increased the growth performance, improving weight gain and feed efficiency and stimulating serum IgG secretion and Th1 cytokines levels, including IFN-γ. Thus, there was also an activation of the immune system and immune enhancement. Increased dietary intake and weight gain was also observed in pigs fed diet supplemented with live probiotics (*Enterococcus faecium* DSM 7134) during pregnancy and lactation (Böhmer et al. 2006). Live *Bacillus subtilis* administered to fish as dietary supplement improved reproductive performance in terms of greater GSI, fecundity, survival, and morphometric characteristics (Ghosh et al. 2007).

The ingestion of live and dead *Lactobacillus* strains was able to increase pigs' weight. This performance may be due to the action of probiotics to improve the absorption of nutrients in combination with a balance of intestinal microflora (Busanello et al. 2015). Live and dead *Lactobacillus rhamnosus* MA27/6B and *Lactobacillus acidophilus* MA27/6R cells promoted the growth of weanling mice, which showed an increase in weight gain after ingesting these probiotics (Bernardeau et al. 2002).

Broiler fed diet supplemented with live probiotics (*Enterococcus faecium* CGMCC 2516) showed an improvement in the characteristics of carcass and meat quality. The probiotic improved meat color, water-holding capacity, pH of the pectoralis major muscle, and lower abdominal fat. These meat quality changes were related to the altered abundance of proteins in the pectoral muscle that regulate meat quality (Zheng et al. 2014). Broilers supplemented with live yeast cells (*Saccharomyces cerevisiae*) and dead cells (*Saccharomyces cerevisiae* var. *ellipsoides* (Thepax®)) had an improvement in behavior and productive performance (Iraqi and Fayed 2012).

Although many genera and species share several health benefit properties, these effects are characteristic of each strain and should not be attributed to other strains (Almada et al. 2015).

## 10.5 Probiotics and Health Effects in Animals

The consumption of probiotics is associated with a variety of benefits to the animal health (Table 10.2). These bacteria act in maintaining the balance of intestinal microbiota (Sarkar 2013), prevention and relief of gastrointestinal (Im et al. 2009) and respiratory diseases (Hougee et al. 2010), as well as protection against pathogen infections (Naqid et al. 2015). Other important effects are prevention of diabetes (Tabuchi et al. 2003), reduction in cholesterol levels (Huang et al. 2010), and modulation of the immune system (Panigrahi et al. 2005). Several benefits are reported below according to the kind of animal, trying to focus on the effects of dead microorganisms of on the comparison beetwen live and dead.

### 10.5.1 Mice

Live *Lactobacillus reuteri* CRL 1324 decreased vaginal colonization by *Enterococcus* in mice. The mechanism of action can involve the formation of biofilms on the vaginal epithelium leading to a competition for nutrients and with pathogens for binding sites. Thus, this probiotic bacterium was capable of reducing infections caused by *Enterococcus* (De Gregorio et al. 2015). Fooladi et al. (2015) found that live *Lactobacillus acidophilus* ATCC4356 could stimulate immune cell proliferation (increased IFN-γ and IL-4) in mice with breast cancer, showing that

**Table 10.2** Probiotics for use in animal health

| Effects | Live probiotics | Dead probiotics | References |
|---|---|---|---|
| Relief of alcoholic liver disease in mice | – | *Lactobacillus brevis* SBC8803 (heat killed at 121 °C/20 min) | Segawa et al. (2008) |
| Gastric pain relief in rats | *Lactobacillus reuteri* | – | Duncker et al. (2011) |
| Modification in fecal bacterial community (increase in the number of *Lactobacillus*) and lower incidence of diarrhea in weanling pigs | *Enterococcus faecalis* LAB31 | – | Hu et al. (2015) |
| Attenuation in necrotizing enterocolitis in newborn mice and premature piglets | *Lactobacillus rhamnosus* HN001 | *Lactobacillus rhamnosus* HN001(ultraviolet light radiated) | Good et al. (2014) |
| Attenuation in airway inflammation in mice | – | *Lactobacillus plantarum* KTCT3104 and *Lactobacillus curvatus* KTCT3767 (heat killed at (100 °C/30 min) | Hong et al. (2010) |
| Immunomodulatory activity in fish | – | *Lactobacillus paracasei* spp. *paracasei* 06TCa22 and *Lactobacillus plantarum* 06CC2 (heat killed in boiling in water for 30 min) | Biswas et al. (2013) |
| | *Pseudomonas* sp. (GP21) and *Psychrobacter* sp. (GP12) | *Pseudomonas* sp. (GP21) and *Psychrobacter* sp. (GP12) (heat killed at 60 °C/60 min) | Lazado and Caipang (2013) |
| | *Lactobacillus rhamnosus* JCM1136 | *Lactobacillus rhamnosus* JCM1136 (heat killed at 75 °C/60 min) | Panigrahi et al. (2005) |
| | – | *Lactobacillus delbrüeckii* ssp. *lactis* CECT 287 and *Bacillus subtilis* CECT 35(heat killed at 60 °C/60 min) | Salinas et al. (2008) |

(continued)

**Table 10.2** (continued)

| Effects | Live probiotics | Dead probiotics | References |
|---|---|---|---|
| Increased levels of certain bacteria in the intestinal microbiota (*Enterococcus faecium*) in rabbits (modulation of intestinal microbiota) | *Enterococcus faecium* NCIMB 30183 | – | Benato et al. (2014) |
| Control of furunculosis in fish | – | Gram-positive Coccus A1–6 not identified, *Vibrio fluvialis* A3-47S, *Aeromonas Hydrophila* A3–51, and *Carnobacterium* BA211 (formalin killed) | Irianto and Austin (2003) |
| Control of *Aeromonas* infection in fish | *Bacillus subtilis* AB1 | *Bacillus subtilis* AB1 (formalin inactivated sonicated) | Newaj-Fyzul et al. (2007) |
| Antidiabetic effect in rats | *Lactobacillus* CG | – | Tabuchi et al. (2003) |
| Antidiabetic effect, lipid metabolism modulation, and protection of renal function in rats | Probiotic fermented milk (*Lactobacillus plantarum*, *Lactobacillus helveticus*, *Lactococcus lactis*, *Lactobacillus harbinensis*, *Lactobacillus hilgardii*, *Lactobacillus rhamnosus*, *Lactobacillus mucosae*, *Lactobacillus par*, *Lactobacillus paracasei tolerans*, *Lactobacillus pentosus*, *Kluyveromyces marxians*, *Pichia membranifaciens*, *Candida ethanolica*, and *Issatchenkia orientalis*) | – | Manaer et al. (2015) |
| Antiaging effect in mice (reduction of hair loss and suppression of the incidence of skin ulcers) | *Lactococcus lactis* subsp. *cremoris* H61 | *Lactococcus lactis* subsp. *cremoris* H61 (heat killed at 100 °C/30 min) | Kimoto-Nira et al. (2007) |
| Anti-inflammatory effect in mice with mucositis | *Lactobacillus acidophilus* | – | Justino et al. (2015) |
| Anti-inflammatory effect in rats | *Lactobacillus rhamnosus* GG | *Lactobacillus rhamnosus* GG (heat killed at 80 °C/20 min) | Li et al. (2009) |

**Table 10.2** (continued)

| Effects | Live probiotics | Dead probiotics | References |
|---|---|---|---|
| Anti-inflammatory effect in rats | *Bioflora* (*Lactobacillus casei*, *Lactobacillus plantarum*, *Streptococci faecalis*, and *Bifidobacterium brevis*) | *Bioflora* (*Lactobacillus casei*, *Lactobacillus plantarum*, *Streptococci faecalis*, and *Bifidobacterium brevis*) (heat killed at 60 °C/30 min) | Laudanno et al. (2006) |
| Hypoglycemic effect in rats (reduction in blood glucose levels) | *Lactobacillus acidophilus*, *Bifidobacterium lactis*, and *Lactobacillus rhamnosus* | – | Al-Salami et al. (2008) |
| Immunomodulatory effect and improvement of disease resistance (*Vibrio anguillarum* and *Aeromonas* | *Clostridium butyrium* CB2 | *Clostridium butyrium* CB2 (heat killed at 150 °C/15 min) | Pan et al. (2008) |
| Protective effect against cadmium toxicity in mice | *Lactobacillus plantarum* CCFM 8610 | *Lactobacillus plantarum* CCFM 8610 heat killed at 100 °C/60 min) | Zhai et al. (2013) |
| Protective effect against lead toxicity in mice | *Lactobacillus plantarum* CCFM8661 | *Lactobacillus plantarum* CCFM8661 heat killed at 115 °C/20 min) | Tian et al. (2012) |
| Protective effect in mice with ulcerative colitis | *Lactobacillus plantarum* 21 | – | Kumar et al. (2015) |
| Beneficial effects of rotavirus infection in neonate mice | *Lactobacillus rhamnosus* GG | *Lactobacillus rhamnosus* GG | Ventola et al. 2012 |
| Enteropathogens exclusion (*Enterococcus canis*, *Salmonella enterica* serovar *Typhimurium* and *Clostridium perfringens*) in dogs | *Lactobacillus fermentum* VET9A, *Lactobacillus plantarum* VET14A, *Lactobacillus rhamnosus* VET16A, and their blends | *Lactobacillus fermentum* VET9A, *Lactobacillus plantarum* VET14A, *Lactobacillus rhamnosus* VET16A, and their mixtures (heat killed at 80 °C/30 min) | Grzeskowiak et al. (2014) |
| Inhibition of visceral pain in rats | *Lactobacillus reuteri* | *Lactobacillus reuteri* (gamma irradiated and heat killed at 80 °C/20 min) | Kamiya et al. (2006) |

(continued)

**Table 10.2** (continued)

| Effects | Live probiotics | Dead probiotics | References |
|---|---|---|---|
| Inhibition of caries in rats | – | *Lactobacillus paracasei* DSMZ16671 (prolonged pasteurization at 80 °C) | Tanzer et al. (2010) |
| Inhibition of respiratory allergic responses in mice | *Lactobacillus reuteri* | – | Forsythe et al. (2007) |
| Inhibition of symptoms of atopic dermatitis (inflammation and lesions of the skin) in mice | *Lactobacillus sakei* probio 65 | *Lactobacillus sakei* probio 65 (heat killed at 121 °C/15 min) | Kim et al. (2013a, b) |
| Maintenance of gut integrity and modulation of immune system in mice | *Saccharomyces boulardii* | *Saccharomyces boulardii* (heat killed at 121 °C/15 min) | Generoso et al. (2011) |
| Improved immune response to infection by *Salmonella typhimurium* in pigs | *Lactobacillusplantarum* B2984 | – | Naqid et al. (2015) |
| Improvement in inflammation of the colon (mechanism for suppressing apoptosis and proliferation/migration of epithelial cells) in mice | *Bacillus polyfermenticus* | – | Im et al. (2009) |
| Improvement in the growth performance and immunomodulatory activity in piglets | – | *Enterococcus faecium* NHRD IHARA (heat killed at 85 °C/30 min) | Sukegawa et al. (2014) |
| | *Enterococcus faecium* NHRD IHARA | – | Ihara et al. (2013) |
| Improvement of the intestinal tract in chickens | *Bacillus coagulans* ATTCC7050 | – | Hung et al. (2012) |
| Improvement of symptoms of colitis in mice | – | *Lactobacillus brevis* SBC8803 (heat killed at 121 °C/20 min) | Ueno et al. (2011) |
| Modulation of intestinal microbiota in piglets | – | *Enterococcus faecium* NHRD IHARA (heat killed at 85 °C/30 min) | Sukegawa et al. (2014) |
| Modulation of intestinal microbiota in fish | *Bacillus pumilus* SE5 | *Bacillus pumilus* SE5 (heat killed at 95 °C/60 min) | Yang et al. (2014) |
| Modulation of intestinal microbiota in fish | *Bacillus pumilus* SE5 | – | Sun et al. (2011) |

**Table 10.2** (continued)

| Effects | Live probiotics | Dead probiotics | References |
|---|---|---|---|
| Modulation of immune response (increase in IFN-γ and IL-4) in mice with breast cancer | *Lactobacillus acidophilus* ATCC4356 | – | Fooladi et al. (2015) |
| Modulation of physiological functions (lipid profile and antioxidant) in hypercholesterolemic rats | *Bacillus polyfermenticus* SDC | – | Paik et al. (2005) |
| Prevention of vaginal colonization by *Streptococcus* and reduction of the risk of infections caused by *Streptococcus* in mice | *Lactobacillus reuteri* CRL1324 | – | De Gregorio et al. (2015) |
| Prevention of *Salmonella* infection in mice | *Lactobacillus acidophilus* 11869BP | *Lactobacillus acidophilus* 11869BP (heat killed at 121 °C/15 min) | Kim et al. (2013a, b) |
| Promotion protective response against infection by nematodes (*Trichinella spiralis*) in mice | *Lactobacillus casei* | *Lactobacillus casei* (heat killed by boiling in water for 30 min) | Bautista-Garfias et al. (2001) |
| Protection against influenza virus infection in mice | *Lactobacillus rhamnosus* | *Lactobacillus rhamnosus* (formalin killed) | Youn et al. (2012) |
| Protection against influenza virus infection in mice | – | *Lactobacillus gasseri* TMC0356 (heat killed at 70 °C/30 min or 90 °C/5 min) | Kawase et al. (2012) |
| | *Lactobacillus gasseri* TMC0356 | – | Kawase et al. (2010) |
| Protection against *Salmonella enterica* serovar Typhimurium in mice | – | *Lactobacillus plantarum* b240 (heat killed at 121 °C/15 min) | Ishikawa et al. (2010) |
| Reduction of inflammation and dysfunction of the gastrointestinal tract in mice with mucositis | *Saccharomyces boulardii* | – | Justino et al. (2014) |
| Reduction of respiratory allergy symptoms in mice | *Bifidobacterium breve* M-16V | – | Hougee et al. (2010) |

(continued)

**Table 10.2** (continued)

| Effects | Live probiotics | Dead probiotics | References |
|---|---|---|---|
| Reduction of cholesterol in rats | – | *Bifidobacterium longum* SPM1207 (sonicated) | Shin et al. (2010) |
| | *Bifidobacterium longum* SPM1207 | – | Lee et al. (2009) |
| Reduction of cholesterol in rats | *Lactobacillus acidophilus* 4356 | – | Huang et al. (2010) |
| Suppression of inflammatory diseases (colitis) in rats | Probiotic IRT5 (*Lactobacillus casei*, *Lactobacillus acidophilus*, *Lactobacillus reuteri*, *Bifidobacterium bifidum*, and *Streptococcus thermophilus*) | – | Jeong et al. (2015) |
| Suppression of *Listeria monocytogenes* infection in mice | *Enterococcus faecium* JWS 833 | – | Choi et al. (2012) |

– Not used

probiotics were able to modulate the immune response. Live *Lactobacillus plantarum* 21 when administered to rats with ulcerative colitis recovered the damaged tissue together with an improvement in the production of mediators involved in the inflammatory response of the intestine (free radicals, nitric oxide, and cytokines). Therefore, *Lactobacillus plantarum* 21 was effective in improving the inflammatory response of the intestine due to the immunomodulating and antioxidant properties (Kumar et al. 2015).

Live or dead *Lactobacillus plantarum* CCFM 8610, when administered in mice intoxicated with cadmium, was able to reduce the cadmium absorption in the intestine and the cadmium accumulation in tissues. In addition, the administration of live or dead cells of this microorganism resulted in decrease of oxidative stress in the liver and kidneys and alleviated histological changes in the liver. Despite this, live *Lactobacillus* was most effective than dead *Lactobacillus*. The effect of live and dead *Lactobacillus* on cadmium toxicity is because *Lactobacillus* can bind cadmium ion before absorption by the intestine, favoring excretion via feces. However, it was also found that the administration of live *Lactobacilli* also stimulated the intestinal peristalsis and the uptake of divalent essential elements (Ca, Mg, and Fe) (Zhai et al. 2013). In addition, live and dead *Lactobacillus plantarum* CCFM8661 decrease lead toxicity in mice, due to the lead binding ability of these bacteria (Tian et al. 2012). Thus, both live and dead probiotics can act on animal protection against metal toxicity.

Heat-killed *Lactobacillus plantarum* KTCT3104 and *Lactobacillus curvatus* KTCT3767 attenuated airway inflammation in mice. These probiotics modulated the intestinal immunity, exerting a protective effect against inflammation (Hong et al. 2010).

Heat-killed *Lactobacillus plantarum* B240 has protected against Salmonella typhimurium infection in mice, once *Lactobacillus* B240 can inhibit the binding and invasion of *Salmonella typhimurium* into the cells. Thus, dead probiotic bacteria are able to act on the protection against pathogen infection (Ishikawa et al. 2010).

Dead and live *Saccharomyces boulardii* cells were able to maintain intestinal integrity and modulate the immune system in mice. These benefits are not dependent on the yeast viability, since both live and dead bacteria showed beneficial effects. Possibly some structural components of the yeast cell wall are responsible for reducing intestinal damage and modulation of the immune system (Generoso et al. 2011).

### 10.5.2 Rats

Al-Salami et al. (2008) showed hypoglycemic effect of live probiotics (*Lactobacillus acidophilus*, *Bifidobacterium lactis*, and *Lactobacillus rhamnosus*) when administered in diabetic rats, once probiotics reduced the blood glucose levels. Live *Bacillus polyfermenticus* SDC was used as dietary supplement in rats, with significant health benefits by modulating physiological functions, including antioxidant and lipid profile in hypercholesterolemic rats (Paik et al. 2005). Live and dead *Lactobacillus rhamnosus* GG presented beneficial effects against rotavirus infection in rats. Live *Lactobacillus* decreased rotavirus levels in the colon. However, dead *Lactobacillus* only alleviated the infection by reducing the swelling of tissues (Ventola et al. 2012).

Treatment with live, heat-killed, or gamma-irradiated *Lactobacillus reuteri* was able of inhibiting visceral pain in rats, due to the beneficial effect of probiotics on direct components of the nervous system (Kamiya et al. 2006). Heat-killed *Lactobacillus paracasei* DSMZ16671 inhibited the formation of cavities in rats induced by *Streptococcus mutans*. This inhibition was due to an increased specificity of *Streptococcus* with the probiotic rather than the teeth, allowing *Streptococcus* join the probiotic bacteria, thereby preventing its colonization in teeth (Tanzer et al. 2010).

### 10.5.3 Piglets and Chickens

Live *Enterococcus faecalis* LAB31 supplemented in diet of weanling foals reduced the incidence of diarrhea, probably due to the changes in the microbiota, including an increase in the number of *Lactobacillus* (Hu et al. 2015). Live and ultraviolet light-irradiated *Lactobacillus rhamnosus* reduced the severity of necrotizing enterocolitis in preterm piglets. This benefit was observed by reducing the damage to the intestinal mucosa observed by histology. In addition, expression of proinflammatory molecules was attenuated in the mucosa, improving intestinal morphology (Good et al. 2014).

Live and heat-killed *Enterococcus faecium* improved pig growth and immunomodulatory activity of piglets. The consumption of this probiotics led to increase in weight of the piglets and production of IgA in feces and serum. Moreover, dead *Enterococcus faecium* modulated the intestinal microbiota of piglets increasing the numbers of *Lactobacillus* ssp., *Enterococcus* ssp., *Clostridium cluster* XIV, and *Enterobacteriaceae* (Ihara et al. 2013; Sukegawa et al. 2014). Hung et al. (2012) have reported that live *Bacillus coagulans* ATCC 7050 supplemented in chicken diet improved intestinal microflora balance.

### 10.5.4 Fish

*Bacillus pumilus* SE5, as a supplement in diet of fish, was able to modulate the gut microbiota, due to a competitive effect between the probiotics and pathogenic bacteria present in the gastrointestinal tract (Sun et al. 2011). Live and dead *Bacillus pumilus* SE5 modulated the intestinal microbiota of fish, due to a competition for nutritional substances or secretion of inhibitory compounds. However, the mechanism of action of dead *Bacillus* may be related to activation of mucosal immunity (Yang et al. 2014).

Heat-killed *Lactobacillus delbrüeckii* ssp. *lactis* CECT 287 and *Bacillus subtilis* CECT 35 modulated the immune system of fish. These bacteria can induce the effects on innate cellular and humoral immune system, thus affecting intestinal immune cells, with local and systemic immunostimulatory effects on the immune system (Salinas et al. 2008). Live and dead *Bacillus subtilis* AB1 protected fish against the virulence of *Aeromonas*. This protection has been achieved due to a stimulation of the humoral and cellular immune system. The intake of *Bacillus* increased white blood cell counts, stimulated phagocytic activity, and raised the lysozyme level in the gut and serum (Newaj-Fyzul et al. 2007).

Live and dead *Clostridium butyrium* CB2 supplemented in fish diet had immunomodulatory effect on fish challenged with *Vibrio anguillarum* and *Aeromonas hydrophila*. The probiotic bacteria administration increased the phagocytic activity of macrophages, the lysozyme activity in the gut and serum, and the immunoglobulin levels (Ig). Moreover, the intake of *Clostridium butyrium* CB2 decreased mortality of fish challenged by *Vibrio anguillarum* or *Aeromonas hydrophila* (Pan et al. 2008).

### 10.5.5 Others

Different species of live and heat-killed *Lactobacillus* have positive effect against enteropathogenic bacteria (*Enterococcus canis*, *Salmonella enterica* serovar Typhimurium, and *Clostridium perfringens*) in dogs. However, the heat inactivation of probiotics increased exclusion of pathogen when compared to live probiotics, possibly due to the action of dead probiotics on the modulation of host immunity (Grzeskowiak et al. 2014). Live *Enterococcus faecium* NCIMB 30183 as a dietary supplement increased fecal levels of certain intestinal bacteria (*Enterococcus faecium*) in rabbits, thus modulating the gut microbiota (Benato et al. 2014).

## Conclusion

Both live and dead probiotic bacteria can be a strategy for prevention and control of diseases and infections in various animals. In addition, they can promote the growth and performance in animal production. These bacteria can also be used as an alternative to antibiotics, without negative effects, thus providing economic benefits to farmers. However, when choosing a probiotic bacteria (live and dead), it is fundamental to take into account the dosage, frequency of consumption, mechanism of action, method, and intensity of the inactivation process (dead probiotics), in addition to considering that the beneficial effect is strain dependent.

The mechanisms of action of live and dead probiotics have not been completely elucidated. Therefore, further studies are needed to better understand the action of these bacteria and select the most suitable probiotics for obtaining certain benefits in animal husbandry.

**Acknowledgments** The authors acknowledge the financial support of "Fundação de Amparo a Pesquisa do Estado de São Paulo" (FAPESP), "Conselho Nacional de Desenvolvimento Científico e Tecnológico" (CNPq) (Grant #302763/2014-7), and Coordenação de Aperfeiçoamento de Pessoal de Nível Superior (CAPES).

## References

Abdelrahman W, Mohnl M, Teichmann K et al (2014) Comparative evaluation of probiotic and salinomycin effects on performance and coccidiosis control in broiler chickens. Poult Sci 93:3002–3008

Almada CN, Almada CN, Martinez RCR et al (2015) Characterization of the intestinal microbiota and its interaction with probiotics and health impacts. Appl Microbiol Biotechnol 99:4175–4199

Al-Salami H, Butt G, Fawcett JP et al (2008) Probiotic treatment reduces blood glucose levels and increases systemic absorption of gliclazide in diabetic rats. Eur J Drug Metab Pharmacokinet 33:101–106

Altmeyer S, Kröger S, Vahjen W et al (2014) Impact of a probiotic Bacillus Cereus strain on the jejunal epithelial barrier and on the NKG2D expressing immune cells during the weaning phase of piglets. Vet Immunol Immunopathol 161:57–65

Amit-Romach E, Uni Z, Reifen R (2010) Multistep mechanism of probiotic bacterium, the effect on innate immune system. Mol Nutr Food Res 54:277–284

Ananta E, Knorr D (2009) Comparison of inactivation pathways of thermal or high pressure inactivated *Lactobacillus rhamnosus* ATCC 53103 by flow cytometry analysis. Food Microbiol 26:542–546

Andriantsoanirina V, Allano S, Butel MJ et al (2013) Tolerance of *Bifidobacterium* human isolates to bile, acid and oxygen. Anaerobe 21:39–42

Bautista-Garfias CR, Ixta-Rodríguez O, Martínez-Gómez F et al (2001) Effect of viable or dead *Lactobacillus casei* organisms administered orally to mice on resistance against *Trichinella spiralis* infection. Parasite 8:S226–S228

Benato L, Hastie P, O'Shaughnessy P et al (2014) Effects of probiotic *Enterococcus faecium* and *Saccharomyces cerevisiae* on the faecal microflora of pet rabbits. J Small Anim Pract 55:442–446

Bermudez-Brito M, Plaza-Díaz J, Muñoz-Quezada S et al (2012) Probiotic mechanisms of action. Ann Nutr Metab 61:160–174

Bernardeau M, Paul J, Gueguen M (2002) Safety and efficacy of probiotic lactobacilli in promoting growth in post-weaning Swiss mice. Int J Food Microbiol 77:19–27

Biswas G, Korenaga H, Nagamine RT et al (2013) Cytokine responses in the Japanese pufferfish (*Takifugu rubripes*) head kidney cells induced with heat-killed probiotics isolated from the Mongolian dairy products. Fish Shellfish Immunol 34:1170–1177

Böhmer BM, Kramer W, Roth-Maier DA (2006) Dietary probiotic supplementation and resulting effects on performance, health status, and microbial characteristics of primiparous sows. J Anim Physiol Anim Nutr 90:309–315

Borchers AT, Selmi C, Meyers FJ et al (2009) Probiotics and immunity. J Gastroenterol 44:26–46

Bouchard DS, Rault L, Berkova N et al (2013) Inhibition of *Staphylococcus aureus* invasion into bovine mammary epithelial cells by contact with live *Lactobacillus casei*. Appl Environ Microbiol 79:877–885

Busanello M, Pozza MSS, Pozza PC et al (2015) Probiotics: viable and inactivated cells on the performance, microflora and blood parameters of piglets. Rev Bras Saude Prod Anim 16:387–396

Castillo NA, LeBlanc AM, Galdeano CM et al (2012) Probiotics: an alternative strategy for combating salmonellosis: immune mechanisms involved. Food Res Int 45:831–841

Center for Veterinary Medicine, Food and Drug Administration (2012) The judicious use of medically important antimicrobial drugs in food-producing animals. Guidance for Industry #209, 1–26, http://www.fda.gov/downloads/AnimalVeterinary/GuidanceComplianceEnforcement/GuidanceforIndustry/UCM216936.pdf

Choi HJ, Shin MS, Lee SM et al (2012) Immunomodulatory properties of *Enterococcus faecium* JWS 833 isolated from duck intestinal tract and suppression of *Listeria monocytogenes* infection. Microbiol Immunol 56:613–620

Chuang L, Wu K-G, Pai C et al (2007) Heat-killed cells of lactobacilli skew the immune response toward T helper 1 polarization in mouse splenocytes and dendritic cell-treated T cells. J Agr Food Chem 55:11080–11086

Cutting SM (2011) *Bacillus* probiotics. Food Microbiol 28:214–220

Da Silva Almeida AP, Avi CM, Barbisan LF et al (2015) Yacon (*Smallanthus Sonchifolius*) and *Lactobacillus acidophilus* CRL 1014 reduce the early phases of colon carcinogenesis in male Wistar rats. Food Res Int 74:48–54

De Gregorio PR, JuárezTomás MS, LecceseTerraf MC et al (2015) Preventive effect of *Lactobacillus reuteri* CRL1324 on Group B *Streptococcus* vaginal colonization in an experimental mouse model. J Appl Microbiol 118:1034–1047

Dehaghani PG, Baboli MJ, Moghadam AT et al (2015) Effect of synbiotic dietary supplementation on survival, growth performance, and digestive enzyme activities of common carp (Cyprinuscarpio) fingerlings. Czech J Anim Sci 60:224–232

Delgado R, Latorre JD, Vicuña E et al (2014) Glycerol supplementation enhances the protective effect of dietary FloraMax-B11 against *Salmonella Enteritidis* colonization in neonate broiler chickens. Poult Sci 93:2363–2369

Di Giancamillo A, Vitari F, Savoini G et al (2008) Effects of orally administered probiotic *Pediococcus acidilactici* on the small and large intestine of weaning piglets. A qualitative and quantitative micro-anatomical study. Histol Histopathol 23:651–664

Duncker SC, Kamiya T, Wang L et al (2011) Probiotic *Lactobacillus reuteri* alleviates the response to gastric distension in rats. J Nutr 141:1813–1818

Fakhry S, Sorrentini I, Ricca E et al (2008) Characterization of spore forming *Bacilli* isolated from the human gastrointestinal tract. J Appl Microbiol 105:2178–2186

Fooladi AAI, Yazdi MH, Pourmand MR et al (2015) Th1 cytokine production induced by *Lactobacillus acidophilus* in BALB/c mice bearing transplanted breast tumor. Jundishapur J Microbiol 8:1–5

Forsythe P, Inman MD, Bienenstock J (2007) Oral treatment with live *Lactobacillus reuteri* inhibits the allergic airway response in mice. Am J Respir Crit Care Med 175:561–569

Gaggìa F, Mattarelli P, Biavati B (2010) Probiotics and prebiotics in animal feeding for safe food production. Int J Food Microbiol 141(Suppl 1):S15–S28. https://doi.org/10.1016/j.ijfoodmicro.2010.02.031

Generoso SV, Viana ML, Santos RG et al (2011) Protection against increased intestinal permeability and bacterial translocation induced by intestinal obstruction in mice treated with viable and heat-killed *Saccharomyces boulardii*. Eur J Nutr 50:261–269

Ghosh S, Sinha A, Sahu C (2007) Effect of probiotic on reproductive performance in female livebearing ornamental fish. Aquac Res 38:518–526

Good M, Sodhi CP, Ozolek JA et al (2014) *Lactobacillus rhamnosus* HN001 decreases the severity of necrotizing enterocolitis in neonatal mice and preterm piglets: evidence in mice for a role of TLR9. Am J Physiol Gastrointest Liver Physiol 306:G1021–G1032

Grzeskowiak K, Collado MC, Beasley S et al (2014) Pathogen exclusion properties of canine probiotics are influenced by the growth media and physical treatments simulating industrial processes. J Appl Microbiol 116:1308–1314

Herbel SR, Vahjen W, Wiele LH et al (2013) Timely approaches to identify probiotic speciesof the genus *Lactobacillus*. Gut Path 5:1–13

Hill C, Guarner F, Reid G et al (2014) Expert consensus document: The International Scientific Association for Probiotics and Prebiotics consensus statement on the scope and appropriate use of the term probiotic. Nat Rev Gastroenterol Hepathol 11:506–514

Holzapfel WH, Haberer P, Geisen R et al (2001) Taxonomy and important features of probiotic microorganisms in food and nutrition. Am J Clin Nutr 73:365S–373S

Hong HA, Duc LH, Cutting SM (2005) The use of bacterial spore formers as probiotics. FEMS Microbiol Rev 29:813–835

Hong HA, Huang JM, Khaneja R et al (2008) The safety of *Bacillus subtilis* and Bacillus indicus as food probiotics. J Appl Microbiol 105:510–520

Hong HA, Khaneja R, Tama NMK et al (2009) *Bacillus subtilis* isolated from the human gastrointestinal tract. Res Microbiol 160:134–143

Hong H-J, Kim E, Cho D et al (2010) Differential suppression of heat-killed lactobacilli isolated from kimchi, a Korean traditional food, on airway hyper-responsiveness in mice. J Clin Immunol 30:449–458

Hougee S, Vriesema AJ, Wijering SC et al (2010) Oral treatment with probiotics reduces allergic symptoms in ovalbumin-sensitized mice: a bacterial strain comparative study. Int Arch Allergy Immunol 151:107–117

Hu Y, Dun Y, Li S et al (2015) Dietary *Enterococcus faecalis* LAB31 improves growth performance, reduces diarrhea, and increases fecal *Lactobacillus* number of weaned piglets. PLoS One 10:e0116635

Huang Y, Wang J, Cheng Y et al (2010) The hypocholesterolaemic effects of *Lactobacillus acidophilus* American type culture collection 4356 in rats are mediated by the down-regulation of Niemann-Pick C1-like 1. Br J Nutr 104:807–812

Hung AT, Lin SY, Yang TY et al (2012) Effects of *Bacillus coagulans* ATCC 7050 on growth performance, intestinal morphology, and microflora composition in broiler chickens. Anim Prod Sci 52:874–879

Ihara Y, Hyodo H, Sukegawa S et al (2013) Isolation, characterization, and effect of administration in vivo, a novel probiotic strain from pig feces. Anim Sci J 84:434–441

Im E, Choi YJ, Pothoulakis C et al (2009) *Bacillus polyfermenticus* ameliorates colonic inflammation by promoting cytoprotective effects in colitic mice. J Nutr 139:1848–1854

Iraqi EKG, Fayed RH (2012) Effect of yeast as feed supplement on behavioural and productive performance of broiler chickens. Life Sci J 9:4026–4031

Irianto A, Austin B (2003) Use of dead probiotic cells to control furunculosis in rainbow trout, *Oncorhynchus mykiss* (Walbaum). J Fish Dis 26:59–62

Ishikawa H, Kutsukake E, Fukui T et al (2010) Oral administration of heat-killed *Lactobacillus plantarum* strain b240 protected mice against *Salmonella enterica* serovar Typhimurium. Biosci Biotechnol Biochem 74:1338–1342

Jeong J-J, Woo J-Y, Ahn Y-T et al (2015) The probiotic mixture IRT5 ameliorates age-dependent colitis in rats. Int Immunopharmacol 26:416–422

Justino PFC, Melo LFM, Nogueira AF et al (2014) Treatment with *Saccharomyces boulardii* reduces the inflammation and dysfunction of the gastrointestinal tract in 5-fluorouracil-induced intestinal mucositis in mice. Br J Nutr 111(9):1611–1621

Justino PFC, Melo LFM, Nogueira AF et al (2015) Regulatory role of *Lactobacillus acidophilus* on inflammation and gastric dysmotility in intestinal mucositis induced by 5-fluorouracil in mice. Cancer Chemother Pharmacol 75:559–567

Kamiya T, Wang L, Forsythe P et al (2006) Inhibitory effects of *Lactobacillus reuteri* on visceral pain induced by colorectal distension in Sprague-Dawley rats. Gut 55:191–196

Kantas D, Papatsiros VG, Tassis PD et al (2015) A feed additive containing *Bacillus toyonensis* (Toyocerin®) protects against enteric pathogens in postweaning piglets. J Appl Microbiol 118:727–738

Kawase M, He F, Kubota A et al (2010) Oral administration of lactobacilli from human intestinal tract protects mice against influenza virus infection. Lett Appl Microbiol 51:6–10

Kawase M, He F, Kubota A et al (2012) Heat-killed *Lactobacillus gasseri* TMC0356 protects mice against influenza virus infection by stimulating gut and respiratory immune responses. FEMS Immunol Med Microbiol 64:280–288

Khan AA, Khurshid M, Khan S et al (2013) Gut microbiota and probiotics: current status and their role in cancer therapeutics. Drug Dev Res 74:365–375

Kim KM, Jung TS, Ok S et al (2011) *In vitro* characterization study of *Bacillus mojavensis* KJS-3 for a potential probiotic. Food Sci Biotechnol 20:1155–1159

Kim J-Y, Park B-K, Park H-J et al (2013a) Atopic dermatitis-mitigating effects of new *Lactobacillus* strain, *Lactobacillus sakei* probio 65 isolated from Kimchi. J Appl Microbiol 115:517–526

Kim M, Yoon Y, Seo J et al (2013b) A study on the prevention of *Salmonella* infection by using the aggregation characteristics of lactic acid bacteria. Toxicol Res 29:129–135

Kimoto-Nira H, Suzuki C, Kobayashi M et al (2007) Anti-ageing effect of a lactococcal strain: analysis using senescence-accelerated mice. Br J Nutr 98:1178–1186

Kritas SK, Govaris A, Christodoulopoulos G et al (2006) Effect of Bacillus licheniformis and Bacillus subtilis supplementation of ewe's feed on sheep milk production and young lamb mortality. J Vet Med Ser A Physiol Pathol Clin Med 53:170–173

Kumar CSVS, Reddy KK, Reddy AG et al (2015) Protective effect of *Lactobacillus plantarum* 21, a probiotic on trinitrobenzenesulfonic acid-induced ulcerative colitis in rats. Int Immunopharmacol 25:504–510

Laudanno O, Vasconcelos L, Catalana J et al (2006) Anti-inflammatory effect of bioflora probiotic administered orally or subcutaneously with live or dead bacteria. Dig Dis Sci 51:2180–2183

Lazado CC, Caipang CMA (2013) Bacterial viability differentially influences the immunomodulatory capabilities of potential host-derived probiotics in the intestinal epithelial cells of Atlantic cod *Gadus morhua*. J Appl Microbiol 116:990–998

Lee DK, Jang S, Baek EH et al (2009) Lactic acid bacteria affect serum cholesterol levels, harmful fecal enzyme fecal activity, and fecal water content. Lipids Health Dis 8:21

Lee K-W, Kim DK, Lillehoj HS et al (2015) Immune modulation by *Bacillus subtilis*-based direct-fed microbials in commercial broiler chickens. Anim Feed Sci Technol 200:76–85

Li N, Russell WM, Douglas-Escobar M et al (2009) Live and heat-killed *Lactobacillus rhamnosus* GG: effects on proinflammatory and anti-inflammatory cytokines/chemokines in gastrostomy-fed infant rats. Pediatr Res 66:203–207

Manaer T, Yu L, Zhang Y et al (2015) Anti-diabetic effects of shubat in type 2 diabetic rats induced by combination of high-glucose-fat diet and low-dose streptozotocin. J Ethnopharmacol 169:269–274

Mappley LJ, Tchórzewska MA, Nunez A et al (2013) Oral treatment of chickens with *Lactobacillus reuteri* LM1 reduces Brachyspirapilosicoli-induced pathology. J Med Microbiol 62:287–296

Mathur S, Singh R (2005) Antibiotic resistance in food lactic acid bacteria—a review. Int J Food Microbiol 105:281–295

Morais MB, Jacob CMA (2006) The role of probiotics and prebiotics in pediatric practice. J Pediatr 82:S189–S197

Nakamura S, Kuda T, An C et al (2012) Inhibitory effects of *Leuconostoc mesenteroides* 1RM3 isolated from narezushi, a fermented fish with rice, on *Listeria monocytogenes* infection to CaCo-2 cells and A/J mice. Anaerobe 18:19–24

Nakano MM, Zuber P (1998) Anaerobic growth of a "strict aerobe" (*Bacillus subtilis*). Annu Rev Microbiol 52:165–190

Naqid IA, Owen JP, Maddison BC et al (2015) Prebiotic and probiotic agents enhance antibody-based immune responses to *Salmonella* Typhimurium infection in pigs. Anim Feed Sci Technol 201:57–65
Newaj-Fyzul A, Adesiyun AA, Mutani A et al (2007) *Bacillus subtilis* AB1 controls Aeromonas infection in rainbow trout (*Oncorhynchus mykiss*, Walbaum). J Appl Microbiol 103:1699–1706
Nithya V, Halami PM (2013) Evaluation of the probiotic characteristics of *Bacillus* species isolated from different food sources. Ann Microbiol 63:129–137
Nogueira JCR, Gonçalves MCR (2011) Probióticos—Revisão da Literatura. Rev Bras Ciên Saúde 15:487–492
Paik H-D, Park J-S, Park E (2005) Effects of *Bacillus polyfermenticus* SCD on lipid and antioxidant metabolisms in rats fed a high-fat and high-cholesterol diet. Biol Pharm Bull 28:1270–1274
Pan X, Wu T, Song Z et al (2008) Immune responses and enhanced disease resistance in Chinese drum, *Miichthys miiuy* (Basilewsky), after oral administration of live or dead cells of *Clostridium butyrium* CB2. J Fish Dis 31:679–686
Panigrahi A, Kiron V, Puangkaew J et al (2005) The viability of probiotic bacteria as a factor influencing the immune response in rainbow trout *Oncorhynchus mykiss*. Aquaculture 243: 241–254
Patel RM, Denning PW (2013) Therapeutic use of prebiotics, probiotics, and postbiotics to prevent necrotizing enterocolitis. What is the current evidence? Clin Perinatol 40:11–25
Raz E, Rachmilewitz D (2005) Inactivated probiotic bacteria and methods of use thereof. Patent n. US20050180962A1, 18 ago
Riazi S, Wirawan RE, Badmaev V et al (2009) Characterization of lactosporin, a novel antimicrobial protein produced by *Bacillus coagulans* ATCC 7050. J Appl Microbiol 106:1370–1377
Ripamonti B, Agazzi A, Baldi A et al (2009) Response of *Lactobacillus casei* BL23 to phenolic compounds. J Appl Microbiol 111:1473–1481
Ritzi MM, Abdelrahman W, Mohnl M et al (2012) Effects of probiotics and application methods on performance and response of broiler chickens to an *Eimeria* challenge. Poult Sci 93(11):2772–2778
Salinas I, Abelli L, Bertoni F et al (2008) Monospecies and multispecies probiotic formulations produce different systemic and local immunostimulatory effects in the gilthead seabream (*Sparu saurata* L.) Fish Shellfish Immunol 25:114–123
Salvetti E, Torriani S, Felis GE (2012) The genus *Lactobacillus*: a taxonomic update. Probiotics Antimicrob Proteins 4:217–226
Salyers AA, Gupta A, Wang Y (2004) Human intestinal bacteria as reservoirs for antibiotic resistance genes. Trends Microbiol 12:412–416
Sanders ME, Morelli L, Tompkins TA (2003) Sporeformers as human probiotics: *Bacillus, Sporolactobacillus,* and *Brevibacillus*. Compr Rev Food Sci Food Saf 2:101–110
Sanders ME, Gibson G, Gill HS, Guarner F (2007) Probiotics: their potential to impact human health. CAST 36
Sarkar S (2013) Probiotics as functional foods: gut colonization and safety concerns. Nutr Food Sci 43:496–504
Segawa S, Wakita Y, Hirata H et al (2008) Oral administration of heat-killed *Lactobacillus brevis* SBC8803 ameliorates alcoholic liver disease in ethanol-containing diet-fed C57BL/6N mice. Int J Food Microbiol 128:371–377
Shin HS, Park SY, Lee DK et al (2010) Hypocholesterolemic effect of sonication-killed *Bifidobacterium longum* isolated from healthy adult Koreans in high cholesterol fed rats. Arch Pharm Res 33:1425–1431
Sukegawa S, Ihara Y, Yuge K et al (2014) Effects of oral administration of heat-killed *Enterococcus faecium* strain NHRD IHARA in post-weaning piglets. Anim Sci J 85:454–460
Sun P, Wang JQ, Zhang HT (2010) Effects of *Bacillus subtilis* natto on performance and immune function of preweaning calves. J Dairy Sci 93:5851–5855
Sun YZ, Yang HL, Ma RL et al (2011) Molecular analysis of autochthonous microbiota along the digestive tract of juvenile grouper *Epinephelus coioides* following probiotic *Bacillus pumilus* administration. J Appl Microbiol 110:1093–1103

Sun P, Wang JQ, Deng LF (2013) Effects of *Bacillus subtilis* natto on milk production, rúmen fermentation and ruminal microbiome of dairy cows. Animal 7:216–222

Tabuchi M, Ozaki M, Tamura A et al (2003) Antidiabetic effect of *Lactobacillus* GG in streptozotocin-induced diabetic rats. Biosci Biotechnol Biochem 67:1421–1424

Tanzer JM, Thompson A, Lang C (2010) Caries inhibition by and safety of *Lactobacillus paracasei* DSMZ16671. J Dent Res 89:921–926

Taverniti V, Guglielmetti S (2011) The immunomodulatory properties of probiotic microorganisms beyond their viability (ghost probiotics: proposal of paraprobiotic concept). Genes Nutr 6:261–274

Tian F, Zhai Q, Zhao J (2012) *Lactobacillus plantarum* CCFM8661 alleviates lead toxicity in mice. Biol Trace Elem Res 150:264–271

Ueno N, Fujiya M, Segawa S (2011) Heat-killed body of *Lactobacillus brevis* SBC8803 ameliorates intestinal injury in a murine model of colitis by enhancing the intestinal barrier function. Inflamm Bowel Dis 17:2235–2250

Ventola H, Lehtoranta L, Madetoja M (2012) Effects of the viability of *Lactobacillus rhamnosus* GG on rotavirus infection in neonatal rats. World J Gastroenterol 18:5925–5931

Wu BQ, Zhang T, Guo LQ et al (2011) Effects of *Bacillus subtilis* KD1 on broiler intestinal flora. Poult Sci 90:2493–2499

Yang H-L, Xia H-Q, Ye Y-D et al (2014) Probiotic *Bacillus pumilus* SE5 shapes the intestinal microbiota and mucosal immunity in grouper *Epinephelus coioides*. Dis Aquat Org 111:119–127

Yin F, Farzan A, Wang QC et al (2014) Reduction of *Salmonella enterica* serovar Typhimurium DT104 infection in experimentally challenged weaned pigs fed a *Lactobacillus*-fermented feed. Foodborne Pathog Dis 11:628–634

Youn H-N, Lee D-H, Lee Y-N et al (2012) Intranasal administration of live *Lactobacillus* species facilitates protection against influenza virus infection in mice. Antivir Res 93:138–143

Zhai Q, Wang G, Zhao J et al (2013) Protective effects of *Lactobacillus plantarum* CCFM8610 against acute cadmium toxicity in mice. Appl Environ Microbiol 79:1508–1515

Zhang JL, Xie QM, Ji J et al (2012) Different combinations of probiotics improve the production performance, egg quality, and immune response of layer hens. Poultry Sci 91:2755–2760

Zheng A, Luo J, Meng K et al (2014) Proteome changes underpin improved meat quality and yield of chickens (*Gallus gallus*) fed the probiotic *Enterococcus faecium*. BMC Genomics 15:1167

Zhou X, Wang Y, Gu Q, Li W (2010) Effect of dietary probiotic, Bacillus coagulans, on growth performance, chemical composition, and meat quality of Guangxi Yellow chicken. Poult Sci 89(3):588–593. https://doi.org/10.3382/ps.2009-00319

# Probiotics and Prebiotics in Animal Health and Food Safety: Conclusive Remarks and Future Perspectives

# 11

Diana Di Gioia and Bruno Biavati

It is undeniable that food safety is of fundamental importance to the consumer, food industry and economy. Increasing consumer awareness and desire for natural products and processes, coupled to the EU legislation that bans the use of chemotherapeutic agents at subtherapeutic levels as growth promoters in animals, has given strength to the use of alternatives to "traditional" techniques to ensure animal health. On the other hand, the increase in world population and the reduction of agricultural production areas will require the intensification of production systems. Therefore, it will be essential to have tools as substitutes to antibiotics to help animal and zoonotic pathogens control and thus improve animal and public health. Beneficial microorganisms and protective cultures belong to this approach. This book has been designed to increase awareness of the idea that probiotics and prebiotics can be used as effective intervention measures in primary production, as substitutes of antibiotics. This approach is becoming more and more popular for the health of all animals, even if not linked to the production chain. Moreover, an additional added value is that probiotics and prebiotics are becoming crucial elements to reduce the spread of antibiotic-resistant bacteria.

Chapter 1 of the present book has underlined the positive aspects of the use of probiotics and prebiotics as novel additives in animal feeding to maintain animal health and welfare. However, several bottlenecks are also present, not only at the science, technology and application level but also at the regulatory level. The regulatory frame and its limitations are described.

D. Di Gioia (✉)
Department of Agricultural Sciences, University of Bologna, Bologna, Italy
e-mail: diana.digioia@unibo.it

B. Biavati
Division of Rural Sciences and Food Systems, Institute of Earth Systems, University of Malta, Msida, Malta
e-mail: bbiav01@um.edu.mt



D. Di Gioia, B. Biavati (eds.), *Probiotics and Prebiotics in Animal Health and Food Safety*, https://doi.org/10.1007/978-3-319-71950-4_11

The key role of a healthy gut microbiota in the overall health of the host is the factor behind the development of the probiotic and prebiotic strategy, and this has been the main focus of Chap. 2. The balance among the so-called "major" phyla of the gut microbiota and other "minor" phyla is crucial for the host health, and it is influenced by several factors including the diet, the use of antibiotics, the stressing rearing conditions and the environment. The chapter also stresses the key role of new available technologies, which can be summarised as "the metagenomic approach", to study the gut microbiota and to understand its decisive role in host-microbe interactions. Improved understanding of the mechanisms supporting a "normal" gut microbiota opens new ways to the modulation of the gut microbiota, as a potential therapeutic option, for the improvement of animal performance and the general well-being.

The increasing number of studies in the field of protective cultures confirms that the microorganism-based technology is gaining more and more consensus. This book, particularly in Chap. 3, has highlighted the potential use of protective cultures both for prolonging the shelf life of cooked products, such as meat, and for ready-to-eat foods. Fish and seafood products and dairy products have also been the target of this technology, with good results. However, their implementation should support, but not substitute, good manufacturing practices, and protective cultures can be used as part of the concept of hurdle technology. Research in the future years will have to focus in a better understanding of the mechanism of action of the inhibiting activity of the used strains, to improve the knowledge about possible resistance by target bacteria and to define the most suitable technologies to preserve this activity when the microorganisms are produced on the large scale by industries. In addition, new perspectives of using such cultures are rising in the area of fungal control in certain foods.

The efficacy of probiotics to reduce pathogen load at the primary production level has been demonstrated for several farm animals, such as pigs, poultry and ruminants; therefore, probiotics represent a real intervention strategy alternative to the use of antibiotics to prevent zoonotic disease from occurring. Chapter 4 of this book focuses on probiotic application on different productive stage of pigs, including sows, suckling piglets, nursery and growing-finishing pigs. Each stage is characterised by a diverse microbiota and suffers peculiar stressors and diseases; therefore, different strains have to be used. Moreover, the farm context in which the organism is used is critical: the use of probiotics is more likely to result in measurable economic gains in animals living in suboptimal conditions rather than in those reared in the highest welfare and husbandry conditions. On the whole, positive results have been obtained in all stages of pig's life.

The literature on poultry, as outlined in Chap. 5, is very rich, and contrasting results are also present. On the whole, it is undeniable that probiotic and prebiotic, also in formulations that contain both of them (synbiotics), can increase the production performance of broiler chickens. The post-hatching period is undoubtedly a critical one because of several stressors such as feed changes or imbalances, transportation, processing at the hatchery and high stocking densities. These factors make young chicks very susceptible to diseases. The positive effects of probiotic and prebiotics have been ascribed to the suppression of pathogenic bacteria, reduction of pH value in the intestine, creating a harsh environment for pathogens, as well as an increase of short-chain fatty acids in the intestine. These beneficial mechanisms lead to an increased intestinal absorption surface and villus height thus improving animal health status and overall performance.

Regarding cattle, Chap. 6 has shown that the probiotic/prebiotic approach has been extensively investigated in beef and calves compared to other ruminants such as sheep and goats where only a few data on the control of pathogens are available. Some of the major achievements in the field suggest to combine the administration of prebiotics with probiotics. The evolution of the intestinal microbiota is a crucial point to analyse the effect of the administration on animal health as well as other intestinal health indicators, e.g. parameters related to the immune system and local response of the intestine. On this point, useful information has been derived from experimental models that induce nutritional diarrhoea. A well-designed trial, with the monitoring of all the important parameters, can reduce the number of animals required and be very useful to evaluate the effect of probiotic/prebiotic administration on growth performance and health in young calves.

This book has also focused on a less explored field, i.e. the administration of probiotics/prebiotics/synbiotics to companion animals (Chap. 7). The use of microorganism and prebiotic-based formulations in the animal field has always been linked to productivity and profit for farmers, whereas their use in companion animals, such as dogs and cats, is aimed at reducing the use of medicines if health troubles occur or to prevent infections. In humans, the link between gut and nervous system has posed the basis to understand the effects of the gut microbiota on several brain-related functions. Studies are trying to elucidate which are the microbial species that are critical for the development of a healthy phenotype and those that may have negative impacts on behaviour, mood and emotion. This may also apply for companion animals. Although not so many data exist on the effects of probiotics and prebiotics supplementation in companion animals, studies performed mainly on cats and dogs underline that this strategy is promising on healthy animals as a preventive approach. Studies concerning the effects on pathogen loads are also sparse. With respect to the studies on animals involved in production processes, administration to companion animals has the main drawback that the knowledge about gut microbiota is not as detailed as it is for other animals. In addition, and probably as a consequence of that, commercial probiotic formulations often are not rigorous in the declaration of the microorganism contained in the products. As the gut microbial composition is elucidated, it will become more certain which strains are better to use. This will help companies in the design and production of new targeted formulations that, hopefully, will be based on the synbiotic approach, considering the good results obtained in the few experimental trials existing up to now.

Chapter 8 has underlined the troubles concerning the application of probiotics in fish rearing. Despite the number of research works that demonstrates possible benefits, several studies show a lack of effect, either positive or negative, as well as difficulties in obtaining reproducible results. It is becoming clearer that biogeography, life history, seasonality and diet influence the composition or activity of the microbiota of fishes, and, therefore, variations under different trial conditions are likely to influence the efficacy of probiotics and thus hamper reproducibility. It is difficult, therefore, to find probiotic strains that have the versatility to work across multiple fish species, rearing conditions and life stages. At present, only one strain is authorised for use in aquaculture in the EU. It is undeniable that we have to increase our understanding of the composition and functionality of the gut microbiota of fishes to improve the efficacy of probiotic intervention.

The health status of honeybees, as outlined in Chap. 9, is an increasing economic problem as it can compromise the pollination service, with damages for agriculture, and the hive productivity, with damages for all the bee-related product, the most popular of which is honey. We are now assisting to the transfer of the probiotic concept into the bee science. The increasing knowledge on the composition and functions of the bee gut microbiota and the link between a balanced gut microbiota and health status have encouraged researchers on the use of gut microorganisms to improve bee health. Most of the used strains are isolated from honeybee crop or gut, but some applications involve environmental strains or formulation for animal and human consumption. The overall results show the favourable effect of applied microbial strains on bee health and productivity, in particular if strains of bee origin are used. However, the attention should be posed on beekeepers and on their real needs: what the market requires are high-quality, cost-effective and easy-to-use products.

Finally, the use of paraprobiotics has been discussed (Chap. 10). This is an interesting point as it allows to overcome many drawbacks concerning the use of live cells, in particular in the production and storage of cells that require to be lively when administered to the target animals. Paraprobiotics are finding increasing applications in animal production mainly because they stimulate the gut functions, gut health and for their immunomodulatory activity, as shown by the studies performed in mice and rats. However, several studies have also been performed in animal husbandry, including pigs, chickens and fish. Further studies are necessary to apply to full scale breeding this technology.

As a general conclusion, we must say that a lot of new achievements in the research on the use of "beneficial microorganisms" to animals have been made in the past years, and most of these studies are supported by in vivo trials that pave the way to a widespread use of these microbial agents. As already pointed out, we still need a lot of research on this topic: new investigation techniques and advanced molecular tools will allow scientists to acquire more information on the strains with particular regard to their interaction with pathogens and the host. In this way, the selection of new probiotic strains, targeting selected food-borne pathogens and suitable to the host, will be more effective.

Nevertheless, research efforts are needed on the technological side, in particular on the way of administrating them to animals, in order to maintain both viability and functionality. One of the major challenges in the commercialization of protective and probiotic cultures is the technology used to produce them at the large scale, as well as the development of suitable methods to guarantee both storage for long time and maintenance of the viability and efficacy of the initial population. Encapsulated strains to enhance microorganism storage time in the feed and for controlled release in the gut after the gastric barrier represent a promising way of administering probiotics to animals. The extreme acidic environment of the stomach can seriously decrease the number of living cells reaching the intestine. Microencapsulation is based on the use of sealed capsules that can release their contents at controlled rates under specific conditions (Anal and Singh 2007). Spray drying is one of the most used techniques to perform microencapsulation (Calo-Mata et al. 2008). This technology allows the incorporation of cryo- and osmo-protective components into the matrix containing the microbial cells, enhancing their survival during processing and storage. Once the microcapsules have been dried, a further surface coating can be applied to improve the sensory properties of the product and also provide an extra level of protection. Last but not least, the coating layer can have desirable dissolution properties, thus permitting release of the cells only in particular condition, such as, for example, after gastric transit when pH gets higher again (Martin et al. 2015). Microencapsulation can protect microorganisms not only during its production process but also during its incorporation into the feed matrix, also with protective effects during storage. In conclusion, microencapsulation is of great interest since it could allow a wider application of probiotics in the feed market.

## References

Anal AK, Singh H (2007) Recent advances in microencapsulation of probiotics for industrial applications and targeted delivery. Trends Food Sci Technol 18:240–251

Calo-Mata P, Arlindo S, Boehme K, de Miguel T, Pascoal A, Barros-Velazquez J (2008) Current applications and future trends of lactic acid bacteria and their bacteriocins for the biopreservation of aquatic food products. Food Bioprocess Technol 1:43–63

Martin MJ, Lara-Villoslada F, Ruiz MA, Morales ME (2015) Microencapsulation of bacteria: a review of different technologies and their impact on the probiotic effects. Innovative Food Sci Emerg Technol 27:15–25

The manufacturer's authorised representative in the EU is Springer Nature Customer Service Centre GmbH, Europaplatz 3, 69115 Heidelberg, Germany. If you have any concerns regarding our products, please contact ProductSafety@springernature.com

Printed and bound by CPI Group (UK) Ltd, Croydon, CR0 4YY
05/10/2026
02228134-0003